Cambridge Studies in Biological and Evolutionary Anthropology 32

Primate Dentition: An Introduction to the Teeth of Non-human Primates

Primate dentitions vary widely both between genera and between species within a genus. This book is a comparative dental anatomy of the teeth of living non-human primates that brings together information from many disciplines to present the most useful and comprehensive database possible in one consolidated text. The core of the book consists of comparative morphological and metrical descriptions with analyses, reference tables and illustrations of the permanent dentitions of 85 living primate species to establish a baseline for future investigations. The book also includes information on dental microstructure and its importance in understanding taxonomic relationships between species, data on deciduous dentitions, prenatal dental development and ontogenetic processes, and material to aid age estimation and life history studies. *Primate Dentition* will be an important reference work for researchers in primatology, dental and physical anthropology, comparative anatomy and dentistry as well as vertebrate paleontology and veterinary science.

DARIS R. SWINDLER is Professor Emeritus of Anthropology at the University of Washington, Seattle. His main research interests are in primate anatomy and dental studies of early primate dental development, comparative dental morphology and odontometrics of living and fossil primates, and Pacific dental anthropology. He has written or edited seven previous books including *An Atlas of Primate Gross Anatomy: Baboon, Chimpanzee, and Man* (1973) with C. D. Wood, *The Dentition of Living Primates* (1976), *Systematics, Evolution and Anatomy: Comparative Primate Biology* Vol. 1 (1986) with J. Erwin, *Paleontologia Umana, Evoluzione, Addamento, Cultura* (1996) with A. Drusini, and *Introduction to the Primates* (1998).

Cambridge Studies in Biological and Evolutionary Anthropology

Series Editors

HUMAN ECOLOGY
C. G. Nicholas Mascie-Taylor, University of Cambridge
Michael A. Little, State University of New York, Binghamton
GENETICS
Kenneth M. Weiss, Pennsylvania State University
HUMAN EVOLUTION
Robert A. Foley, University of Cambridge
Nina G. Jablonski, California Academy of Science
PRIMATOLOGY
Karen B. Strier, University of Wisconsin, Madison

Consulting Editors
Emeritus Professor Derek F. Roberts
Emeritus Professor Gabriel W. Lasker

Primate Dentition

An Introduction to the Teeth of Non-human Primates

DARIS R. SWINDLER

Professor Emeritus
Department of Anthropology
University of Washington, USA

ILLUSTRATED BY

ROBERT M. GEORGE
Department of Biology
Florida International University, USA

CAMBRIDGE
UNIVERSITY PRESS

CAMBRIDGE UNIVERSITY PRESS
Cambridge, New York, Melbourne, Madrid, Cape Town, Singapore, São Paulo

Cambridge University Press
The Edinburgh Building, Cambridge CB2 2RU, UK

Published in the United States of America by Cambridge University Press, New York

www.cambridge.org
Information on this title: www.cambridge.org/9780521652896

First published 2002
This digitally printed first paperback version 2005

A catalogue record for this publication is available from the British Library

Library of Congress Cataloguing in Publication data

Swindler, Daris Ray
Primate dentition: an introduction to the teeth of nonhuman primates /
Daris R. Swindler; illustrated by Robert M. George.
    p.   cm. – (Cambridge studies in biological and evolutionary
anthropology; 32)
Includes bibliographical references (p.   ).
ISBN 0 521 65289 8 (hb)
1. Primates – Anatomy.   2. Dentition.   3. Teeth.   I. Title.   II. Series.
QL737.P9 S823 2002
573.3´56198 – dc21   2001037356

ISBN-13 978-0-521-65289-6 hardback
ISBN-10 0-521-65289-8 hardback

ISBN-13 978-0-521-01864-7 paperback
ISBN-10 0-521-01864-1 paperback

Primate Dentition

An Introduction to the Teeth of Non-human Primates

DARIS R. SWINDLER

Professor Emeritus
Department of Anthropology
University of Washington, USA

ILLUSTRATED BY

ROBERT M. GEORGE
Department of Biology
Florida International University, USA

CAMBRIDGE
UNIVERSITY PRESS

CAMBRIDGE UNIVERSITY PRESS
Cambridge, New York, Melbourne, Madrid, Cape Town, Singapore, São Paulo

Cambridge University Press
The Edinburgh Building, Cambridge CB2 2RU, UK

Published in the United States of America by Cambridge University Press, New York

www.cambridge.org
Information on this title: www.cambridge.org/9780521652896

First published 2002
This digitally printed first paperback version 2005

A catalogue record for this publication is available from the British Library

Library of Congress Cataloguing in Publication data

Swindler, Daris Ray
Primate dentition: an introduction to the teeth of nonhuman primates /
Daris R. Swindler; illustrated by Robert M. George.
  p.   cm. – (Cambridge studies in biological and evolutionary
anthropology; 32)
Includes bibliographical references (p.   ).
ISBN 0 521 65289 8 (hb)
1. Primates – Anatomy.   2. Dentition.   3. Teeth.   I. Title.   II. Series.
QL737.P9 S823 2002
573.3´56198 – dc21   2001037356

ISBN-13 978-0-521-65289-6 hardback
ISBN-10 0-521-65289-8 hardback

ISBN-13 978-0-521-01864-7 paperback
ISBN-10 0-521-01864-1 paperback

This book is dedicated to the memory of Linda E. Curtis

Contents

Contents xi

Preface

Since 1976, when I published the *Dentition of Nonhuman Primates*, much has happened in the field of dental anthropology, creating a tremendous array of new information available to student and field researcher alike. This volume combines basic material available to me then with knowledge gleaned from more recent research in an attempt to gather the most useful and comprehensive data in one consolidated text.

The organization of the book is taxonomic, beginning with the prosimians and ending with the great apes. There is no temporal aspect to the data presented. It is solely heuristic, not evolutionary, nor does the taxonomic organization of the book intend to suggest in any way that the dentition of one group gave rise to that of another group.

Chapter 1 introduces the primates studied, organized in the Linnean system, i.e. a hierarchy of levels that group organisms into larger and larger units (see Table 1.1). This chapter continues with a discussion of dental anatomy and terminology as well as a section reporting recent information in the field of dental genetics. Several of the taxa presented have genera and species names, or even hierarchical positions, different from those that they did several years ago. I have attempted to follow the latest information regarding their rank and scientific names. However, primate classifications continue to change, particularly at the family and genus levels; for example, *Cercocebus albigena*, the gray-cheeked mangabey, is now considered to be *Lophocebus albigena* (Disotell, 1994, 1996). This chapter also details the extensive dental cast collection used for much of the study and illustrations. In addition, the origin and usefulness of mammalian dental terminology is discussed here.

Chapters 2, 3 and 4 present material not addressed in my earlier book. Chapter 2 contains recent information on the microstructure (histology) of non-human primate teeth and the importance of this information for better understanding taxonomic relationships among the taxa. Chapter 3 presents data on the prenatal development of non-human primate teeth in several taxa, from initial calcification to crown completion, and considers some of the ontogenetic processes that play such important roles in the growth and development of teeth as well as material for age estimations

xiii

and life history investigations. Chapter 4 offers morphologic and metric descriptions of the deciduous teeth of several taxa with comparative discussions.

Morphological data were collected and measurements were made of the dental casts of each species. This material forms the core of the remaining chapters, which, in turn, offer expanded analyses and tables and hand-drawn illustrations of upper and lower dentitions of the various genera and species. One illustrator was employed, using actual specimens and the dental casts, so as to provide consistency of detail and style useful for comparative studies. In Chapters 7 and 9, illustrations and tables show the crown variables that occur on the lingual surface of the upper incisors as well as extra cusps on the premolars and molars, and these will enhance the reader's ability to see and appreciate the magnitude of dental variability among living non-human primates.

Odontometrics of both the deciduous and permanent teeth are presented in the odontometric appendix (Appendix 1). There is also a tooth eruption appendix (Appendix 2) to facilitate finding the eruption sequences of many non-human taxa. There is also a large amount of useful statistical data throughout this book on the incidence of various dental traits.

Many people have assiduously proofread the data several times, but mistakes are inevitable where so many morphological details and statistical tables must be read and reread. I take full responsibility for any mistakes and hopefully, no major error has found its way into the final publication.

Daris R. Swindler
Seattle, Washington
April, 2001

Acknowledgements

I thank the following museums and staff for the generous use of collections and facilities: American Museum of Natural History, New York; Field Museum of Natural History, Chicago; Smithsonian Institution, Washington, D.C.; and the Cleveland Museum of Natural History, Cleveland. Casts were made of the collections of Dr Neil C. Tappen and Dr Henry C. McGill.

In addition to the individuals that assisted me in so many ways in the publication of the earlier book, I have had the good fortune to have had the help and advice of M. Christopher Dean and David G. Gantt, who read earlier chapters of the manuscript and made many helpful comments and suggestions. I am particularly indebted to Mark Terry, who read drafts that resulted in many fruitful discussions. Mark's support has been invaluable to me, especially during those last important, and I might add hectic, weeks of the project.

Special acknowledgement goes to the illustrator, Robert M. George, for the quality of the original illustrations of the dental casts. Bob has been a colleague and friend for many years and without his time, effort, and support this project would probably never have been completed. Thanks also to Erik McArthur for scanning and print assistance.

My wife, Kathy Swindler, has assisted me greatly with the writing of this book. It was her fortitude and patience that helped me in so many different ways to get on with the job. Thank you.

I thank my editor at Cambridge University Press, Dr Tracey Sanderson, for her patience, understanding and guidance throughout the preparation of this book. I also thank my subeditor, Lynn Davy.

Finally, I acknowledge the original support of the National Institutes of Health from 1962 to 1968 for grant DE-02955, which supported the collecting of the dental casts, and NIH grant RR-00166, administered through the Washington Regional Primate Research Center, which sponsored other portions of the project.

1   *Introduction*

Order Primates

Primates are a diverse group of mammals that have evolved from a group of insectivorous mammals some 60 million years ago. Indeed, it is difficult to define primates since they lack a single feature that separates them from other mammalian groups. At the same time, primates have remained plesiomorphic, retaining many ancestral features, rather than becoming highly apomorphic as did many groups of mammals, for example, the horse with a single digit in each foot.

Today, there are nearly 300 primate species grouped into about 80 genera (depending on the source), most of which live in tropical or subtropical regions of the world. The majority of living primate taxa are monkeys, and are present in both the New and Old Worlds, while prosimians are found in Madagascar, Africa, and Asia, the great apes inhabit Africa, Borneo, and Sumatra, and the lesser apes live in many regions of Southeast Asia. The remaining primate species, *Homo sapiens*, is the only living hominid and is found in most regions of the world. The primate classification presented here is often referred to as the traditional one since it is based on the level or grade of organization of the different primate groups. Table 1.1 presents a classification of living primates. This list includes only the primates examined in this book, and therefore does not represent a complete list of all extant genera. Classifications and scientific names often change through time; I have therefore attempted to include the changes that have occurred since the original version of this book appeared in 1976.

Dental cast collection

The basic data presented in this book were taken from plaster casts made from alginate impressions. The impressions and casts were made of the permanent and deciduous teeth of primate skulls housed in the following museums: American Museum of Natural History, National Museum of

1

Table 1.1. *Classification of living primates studied in this book*

===

ORDER PRIMATES

Suborder: Prosimii
  Infraorder: Lemuriformes
    Superfamily: Lemuroidea
      Family: Lemuridae
        Lemur catta
        Eulemur macaco
        E. rubiventer
        E. mongoz
        Varecia variegata
        Hapalemur griseus
      Family: Lepilemuridae
        Lepilemur mustelinus
      Family: Cheirogaleidae
        Microcebus murinus
        Cheirogaleus major
        Phaner furcifer
      Family: Indriidae
        Indri indri
        Propithecus verreauxi
        Avahi laniger
      Family: Daubentoniidae
        Daubentonia madagascariensis
    Superfamily: Lorisoidea
      Family: Lorisidae
        Loris tardigradus
        Nycticebus coucang
        Perodicticus potto
        Arctocebus calabarensis
      Family: Galagidae
        Otolemur crassicaudatus
        Galago senegalensis
  Infraorder: Tarsiiformes
    Superfamily: Tarsioidea
      Family: Tarsiidae
        Tarsius spectrum
        T. bancanus
        T. syrichta
Suborder: Anthropoidea
  Infraorder: Platyrrhini
    Superfamily: Ceboidea
      Family: Cebidae
        Subfamily: Callitrichinae
          Saguinus geoffroyi
          Leontopithecus rosalia
          Callithrix penicillata
          Cebuella pygmaea
          Callimico goeldii

        Subfamily: Cebinae
          Cebus apella
          Saimiri sciureus
          S. oerstedii
        Subfamily: Aotinae
          Aotus trivirgatus
      Family: Atelidae
        Subfamily: Callicebinae
          Callicebus moloch
        Subfamily: Atelinae
          Ateles geoffroyi
          A. belzebuth
          A. paniscus
          A. fusciceps
          Lagothrix lagotricha
          Alouatta palliata
          A. seniculus
          A. belzebul
          Brachyteles arachnoides
        Subfamily: Pitheciinae
          Cacajo calvus
          Chiropotes satanas
          Pithecia pithecia
  Infraorder: Catarrhini
    Superfamily: Cercopithecoidea
      Family: Cercopithecidae
        Subfamily: Cercopithecinae
          Macaca nemestrina
          M. mulatta
          M. fascicularis
          M. nigra
          Lophocebus albigena
          L. aterrimus
          Cercocebus torquatus
          C. galeritus
          Papio cynocephalus
          Theropithecus gelada
          Mandrillus sphinx
          Cercopithecus nictitans
          C. cephus
          C. mona
          C. mitis
          C. lhoesti
          C. neglectus
          C. ascanius
          Chlorocebus aethiops
          Erythrocebus patas
          Miopithecus talapoin

Table 1.1 (*cont.*)

ORDER PRIMATES (*cont.*)

Subfamily: Colobinae	Superfamily: Hominoidea
Piliocolobus badius	Family: Hylobatidae
Colobus polykomos	*Hylobates klossi*
Presbytis comata	*H. moloch*
Trachypithecus pileatus	*H. lar*
T. cristata	*H. syndactylus*
T. phyrei	Family: Pongidae
Pygathrix nemaeus	*Pongo pygmaeus*
Simias concolor	*Gorilla gorilla*
Nasalis larvatus	*Pan troglodytes*
Rhinopithecus roxellanae	*P. paniscus*
Kasi johnii	

Sources: Martin (1990), Swindler (1998), Fleagle (1999).

Natural History (Smithsonian Institution), Chicago Field Museum, and The Cleveland Museum of Natural History. Casts were also made of specimens in the collections of Dr Neil C. Tappen and Henry C. McGill.

The casting technique is relatively simple and provides permanent material for detailed study in the laboratory. All casts were poured within five to ten minutes after the impressions were made; this minimizes the possibility of dimensional change (Skinner, 1954). In addition, a study has shown that measurements taken on dental casts are directly comparable to measurements of the original teeth (Swindler, Gavan and Turner, 1963). The observed differences are more likely due to instrumentation than to dimensional change resulting from the dental materials. All casts were made by my assistants and myself.

The original specimens were collected for the respective museums and come from many different geographic areas of the world. In the majority of cases, species are represented from a wide range within their normal geographic range, although in certain groups, e.g. *Papio cynocephalus,* the animals were collected from a more limited area and may well approximate an interbreeding population. *Macaca mulatta* specimens from Cayo Santiago, Puerto Rico were used as well as *M. nemestrina* from the Regional Primate Research Center at the University of Washington. Unfortunately, several species were represented by only a few specimens, or in one or two cases, by a single specimen. This usually meant that these species were rare in museum collections and because of constraints of money and time, it was impossible to increase the sample. Also, the manner in which specimens were collected in the field influenced the randomness of a sample and anyone who has used museum collections is quick to realize this fact. There are obviously other biases in such a collection of specimens (ca. 2000) as studied in this book. However, since the principal objective of this work is to describe the normal dentition and present a statement of the range and magnitude of dental variability within the major genera and species of extant primates, the influences of these unavoidable biases should be mitigated.

The sex of the animals was determined in the field at the time of collection and any specimen of doubtful sex was excluded from the study. In the analytical descriptions in each section of the book the sexes are pooled unless otherwise stated.

The illustrations of the upper and lower teeth of all of the species in the book, unless otherwise stated, were drawn by Dr Robert M. George, Department of Biology, Florida International University, Miami, Florida. The number in millimeters (mm) that appears in the caption of each illustration represents the length of the maxillary arch of the original specimen measured from the mesial surface of the upper central incisors to a line perpendicular to the distal surfaces of the maxillary third molars or, in the callitrichids, the second molars.

Odontometry

All tooth measurements were taken with a Helios caliper. The arms were ground to fine points for greater accuracy. Mesiodistal and buccolingual dimensions of maxillary and mandibular teeth were taken. In all odontometric calculations, the sample size (n) refers to the number of animals measured. This procedure is more realistic than presenting the number of teeth measured since it is well known that there are very few significant differences between the dimensions of right and left teeth. The right side is presented here; however, if a tooth was badly worn or absent its antimere was used. Also, teeth exhibiting noticeable wear were excluded. It should also be mentioned that in many cases the n presented in Appendix 1 differs from the number of animals studied in the morphological section for a given species. This is due to the fact that in many cases the teeth could be examined for a particular morphological trait, yet were too worn to measure, or vice versa.

Repeated measurements taken on the same teeth revealed an average difference between measurements of 0.2 mm. The teeth were measured by

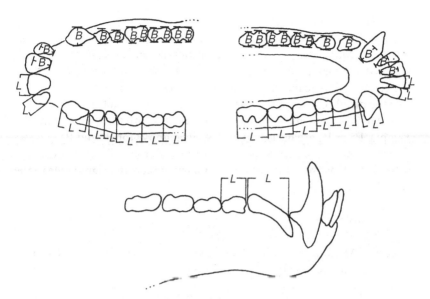

Fig. 1.1. Odontometric landmarks. B, breadth; L, length. Reprinted from Swindler, D. R. (1976) *Dentition of Living Primates*, with the permission of Academic Press Inc. (London) Ltd.

the author and research assistants, each of whom was trained by the author. The measurements are shown in Fig. 1.1 and are defined as follows.

Incisors

LENGTH. Mesiodistal diameter taken at the incisal edge of the upper and lower incisors.

BREADTH. Buccolingual diameter taken at the cementoenamel junction at a right angle to the mesiodistal diameter.

Canines

LENGTH. Upper canine: diameter from the mesial surface to the distolingual border. Lower canine: mesiodistal diameter measured at the level of the mesial alveolar margin.

BREADTH. Buccolingual diameter taken at the cementoenamel junction at a right angle to the mesiodistal diameter.

Premolars

LENGTH. Maximum mesiodistal diameter taken between the contact points. If the mesial contact is lacking on P^3 or P_3 owing to a diastema between it and the canine, the maximum horizontal distance is measured from the distal contact point to the most mesial point on the surface of the premolar. The same method is used on primates with three premolars, for example, P^2 or P_2 are measured as described above for P^3 and P_3.

BREADTH. Maximum buccolingual diameter taken at a right angle to the mesiodistal diameter.

Molars

LENGTH. Maximum mesiodistal diameter taken on the occlusal surface between the mesial and distal contact points.

BREADTH. Maximum buccolingual diameter measured at a right angle to the mesiodistal dimension. The breadths of both the trigon (trigonid) and the talon (talonid) were taken in this manner.

Statistical calculations for means and standard deviations (s.d.) were performed for the dental measurements of each species by sex. Hypotheses of equality of means between sexes of each species, where the samples were large enough, were tested by using the appropriate small sample t-test statistic (Sokal and Rohlf, 1969). The results of the t-tests for sexual dimorphism are presented for each species in the odontometric tables in Appendix 1.

Dental terminology

The incisors and canines are known as the anterior teeth; premolars and molars are the posterior teeth (Fig. 1.2). The tooth surfaces facing toward the cheek are called the buccal surfaces (odontologists often distinguish between the buccal and labial surfaces, labial being limited to the incisor and canine surfaces facing the lips, i.e. labia). All surfaces facing the tongue are referred to as lingual. The mesial (anterior) surface of a tooth faces toward the front of the oral cavity; those more distant are called the distal (posterior) surfaces. A cusp is defined as having structural or functional occlusal areal components delimited by developmental grooves and having independent apexes. The principal cusps, conules and styles, as well

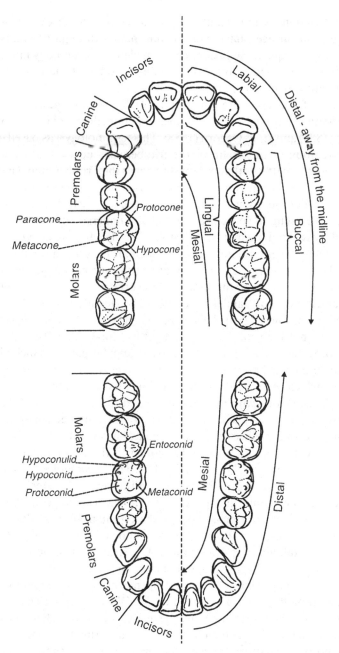

Fig. 1.2. The permanent upper and lower dental arches of the gorilla, with cusp terminology and terms of position within the oral cavity. Drawn by Linda E. Curtis.

as the functionally important crests connecting them in various ways on the occlusal surfaces, have had different names through the years. Since there is still frequent confusion regarding dental terminology in the literature, the terms used in this book as well as some of the more common synonymies are presented in Table 1.2.

Many of the terms presented in Table 1.2 were suggested in the late nineteenth century for the cusp names of mammalian molar teeth by E. D. Cope (1888) and H. F. Osborn (1888). This terminology was based on their interpretation of the origin of the tritubercular upper and lower mammalian molar patterns that became known as the Tritubercular Theory. As mentioned, other terms have been proposed for the major cusps through the years, but the original names are so well entrenched, and their weaknesses and strengths so well recognized and understood by odontologists, that I have used them in this book. This is often referred to as the Cope–Osborn nomenclature for the principal cusps of mammalian molars (Fig. 1.2). For other views on the fascinating subject of the evolution of mammalian molars and the naming of the principal cusps, one should read the contributions of Vandebroek (1961) and Hershkovitz (1971).

Today it is known that the reptilian single cusp of the upper jaw is the paracone, not the protocone as thought by Cope. After the evolution of the protocone the molars evolved into a triangular pattern which Simpson (1936) termed tribosphenic (from the Greek *tribein,* to rub; *sphen,* a wedge), which better describes the grinding functions of the protocone and talonid basin along with the alternating and shearing action of the trigon and trigonid (Fig. 1.3). The upper molar is a three-cusped triangle, formed by the paracone, metacone, and protocone, known as the trigon (Fig. 1.3). The hypocone appears later and is added to the distolingual surface of upper molars forming the talon (Fig. 1.3). The lower molar has a trigonid (*-id* is the suffix added to the terms of all lower teeth) consisting of the paraconid, metaconid and protoconid (note, protoconid is the correct designation for the original cusp of the reptilian lower molars). In the majority of extant primates, with the exception of *Tarsius,* the paraconid is absent. The talonid develops on the distal aspect of the trigonid and often bears three cusps, hypoconid, entoconid, and hypoconulid (Fig. 1.3). A cingulum (cingulid) girdled the tribosphenic molar and portions of it may be present on the teeth of extant primates. When present, these structures are known as styles (stylids) and are named for their related cusp, e.g., paracone (parastyle). The molars of all extant primates are derived from the tribosphenic pattern; indeed, all living mammals, with the exception of monotremes, are descended from Cretaceous ancestors with tribosphenic molars (Butler, 1990). This has been the accepted theory of mammalian

Table 1.2. *Tooth nomenclature*

This book	Synonymy
Upper teeth	
Paracone (O)	Eocone (Vb)
Protocone (O)	Epicone (Vb)
Metacone (O)	Distocone (Vb)
Hypocone (O)	Endocone (Vb)
Metaconule (O)	Plagioconule (Vb)
Protoconule (O)	Paraconule (Vb)
Distoconulus (R)	Postentoconule (H)
Parastyle (O)	Mesiostyle (Vb)
Mesostyle (O)	Ectostyle-1 (H)
Metastyle (O)	Distostyle (Vb)
Distostyle (K)	—
Carabelli cusp	Protostyle (O)
Postprotostyle (K)	Interconule (R)
Preprotocrista (VV)	Protoloph (O)
Crista obliqua (R)	Postprotocrista (VV)
Entocrista (H)	—
Premetacrista (S)	—
Postmetacrista (S)	—
Trigon basin (S)	Protofossa (VV)
Lower teeth	
Paraconid (O)	Mesioconid (Vb)
Protoconid (O)	Eoconid (Vb)
Metaconid (O)	Epiconid (Vb)
Entoconid (O)	Endoconid (Vb)
Hypoconid (O)	Teloconid (Vb)
Hypoconulid (O)	Distostylid (Vb)
Mesiostylid (Vb)	—
Ectostylid (K)	—
Protostylid (K)	Postmetaconulid (H)
Tuberculum intermedium (R)	Postentoconulid (H)
Tuberculum sextum (R)	Protolophid (VV)
Protocristid (S)	Premetacristid (H)
Cristid obliqua (S)	—
Postentocristid (H)	Paralophid (VV)
Paracristid (S)	—
Postmetacristid (S)	—
Trigonid basin (S)	Prefossid (VV)
Talonid basin (S)	Postfossid (VV)

Sources: H, Hershkovitz (1971); K, Kinzey (1973); O, Osborn (1907); R, Remane (1960); S, Szalay (1969); Vb, Vandebroek (1961); VV, Van Valen (1966). Reprinted from Swindler, D. R. (1976) *Dentition of Living Primates*, with permission of Acadmic Press Inc. (London) Ltd.

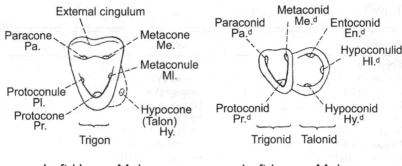

Fig. 1.3. The generalized upper and lower mammalian molar pattern. Reprinted from Simpson, G. G. (1937). The beginning of the age of mammals, *Biological Reviews of the Cambridge Philosophical Society* **12**, 1–47, Fig. 11, p. 28. Reprinted with the permission of Cambridge University Press.

molar evolution until recently. The discovery of new pre-Cretaceous fossils with fully developed tribosphenic molars challenges the current idea concerning the timing of divergence of the main extant mammalian groups. It now appears that there may have been a dual origin of the tribosphenic molar (Luo, Cifelli and Kielan-Jawarowska, 2001). According to their hypothesis, a lineage with a tribosphenic molar radiated in southern Gondwanaland, giving rise to monotremes. The other lineage with a tribosphenic molar, evolved in the northern landmass of Laurasia, into the marsupials and placental mammals of today. This new paradigm will certainly engender controversy until it is either accepted or rejected. The evolutionary significance of the tribosphenic molar was clearly stated by Simpson (1936, p. 810) when he wrote 'This is the most important and potent type of molar structure that has ever been evolved.'

One of the most comprehensive studies of the evolution of primate teeth still remains *The Origin and Evolution of the Human Dentition* (1922) by William King Gregory. Of course, this book is dated; however, it still contains much of interest to all students of primate dental evolution.

The original permanent mammalian dental formula was:

$$I^3\text{-}C^1\text{-}P^4\text{-}M^3 \,/\, I_3\text{-}C_1\text{-}P_4\text{-}M_3 \times 2 = 44$$

The majority of early primates had lost one incisor, and by the Eocene, premolar reduction had begun. The first (central) incisor is usually considered the missing member of the group and premolar reduction occurs from mesial to distal as explained below.

Primates have two sets of teeth, deciduous (also known as milk or primary) and permanent. The deciduous teeth, incisors, canines, and premolars emerge into the oral cavity before the permanent incisors, canines, and premolars that replace them, while the permanent molars emerge distally to the deciduous teeth (Chapter 3). The number and class of teeth in each quadrant of the jaw can be written for both deciduous and permanent teeth as a dental formula:

$$di^2\text{-}dc^1\text{-}dp^2 / di_2\text{-}dc_1\text{-}dp_2 \times 2 = 20$$

$$I^2\text{-}C^1\text{-}P^2\text{-}M^3 / I_2\text{-}C_1\text{-}P_2\text{-}M_3 \times 2 = 32$$

In this example, the formulas represent the deciduous and permanent dentitions of all catarrhine primates. A question arises regarding the designation of the deciduous cheek teeth. Since these teeth are actually the deciduous third and fourth premolars of other mammals and are replaced by the permanent third and fourth premolars, they should be called deciduous premolars (Delson, 1973; Hillson, 1996). However, the terms of human dentistry have not included mammalian dental evolutionary theory and have generally prevailed through the years. A similar issue exists regarding the permanent premolars: as noted above, there were originally four premolars in each quadrant of ancient mammalian jaws, but in all living primates, at least the first premolar has been lost. Hence, most prosimians and all platyrrhines have three premolars. Some genera have also lost more than one. A second premolar has been lost in all catarrhines. The remaining permanent premolars then are properly identified as P2, P3 and P4 or P3 and P4; however, traditional dentistry refers to them as P1 and P2. The zoological terminology is used in this book.

2   *Dental anatomy*

Anatomy

A typical mammalian tooth (Fig. 2.1) consists of a crown formed of enamel that covers the exposed, oral portion of the tooth. The principal mass of a tooth is composed of dentine, which is covered by the protective enamel crown; cementum surrounds the dentine of the tooth root. The central portion of the tooth is the pulp made up of soft tissues containing blood vessels and nerves which enter the tooth through the apical foramen. The mammalian tooth has a crown, cervix (neck), and root; this structure results from the way the tooth attaches to the jaw (Peyer, 1968). In most mammals, and all primates, the tooth root is anchored in a bony alveolus by a suspensory ligament, the periodontal ligament, which forms a fibrous joint known as a gomphosis. The bony alveolus covers the root up to the region of the cementoenamel junction, that is, the neck or cervical portion of the tooth. According to Peyer (1968), the neck is present only if its diameter is smaller than the crown. This definition eliminates the teeth of non-mammals and some mammals, but includes the teeth of all primates.

We have seen that primate teeth can be separated on the basis of form, position, and function into incisors, canines, premolars, and molars. This is a heterodont dentition and can be contrasted with a homodont dentition where the teeth consist of a single cusp that is similar in shape from incisors to molars, as for example, those found in many sea mammals. In primates, the upper incisors are generally chisel-shaped cutting and nibbling teeth; the lower incisors in most prosimians form a dental comb consisting of procumbent (extending forward from the lower jaw) incisors and canines, if present. Primate upper and lower canines are cone-shaped piercing teeth except in those prosimian genera where the lower canines join the incisors to form the dental comb. Premolars are transitional teeth situated between the anterior cutters and piercers and the posterior grinders and may have one to multiple cusps. In some species, there is a tendency for the most anterior premolar to become caniniform, and the most posterior premolar may be molariform. The molars possess anywhere from three to five major cusps. The premolars and molars, often called cheek teeth, form the main

12

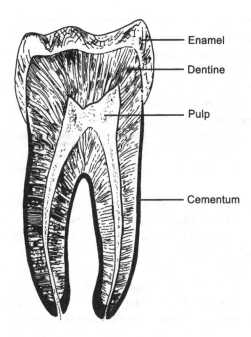

Fig. 2.1. The components of a mammalian tooth. Reprinted from Swindler, D. R. (1998) *Introduction to the Primates*, with permission of the University of Washington Press.

crushing and grinding teeth. There is recent evidence, however, suggesting that these terms do not describe what teeth do and that 'Such terms are members of a large family of words that simply denote fracture and, by masquerading as explanation, have stunted the understanding of how teeth work' (Lucas and Teaford, 1994, p. 183; see below and Chapter 4).

Mammalian molars display a variety of shapes and forms adapted for mastication. Although primate molars never express some of the extreme conditions seen in other mammals, the following terms are useful in describing their morphology. Molar crowns that tend to be wide and low are brachydont, whereas a hypsodont molar possesses a relatively high crown. When molars have separate, low, and moderately rounded cusps they are termed bunodont. If the cusps are aligned in linear ridges either transverse or oblique to the long axis of the occlusal surface the tooth is said to be lophodont. A final cusp arrangement, not found in primates, results when the cusps expand into crescents forming a selenodont molar, as found in deer, goats, and sheep.

A useful method enabling odontologists, particularly paleoanthropologists, to identify isolated teeth is to proceed in the following order. First, establish whether the tooth is deciduous or permanent (*set* trait); second, decide whether it is an incisor, canine, premolar, or molar (*class* trait); third, establish whether it is a maxillary or mandibular tooth (*arch* trait); and fourth, determine its position, e.g. first, second or third permanent molar (*type* trait). Because this book is concerned primarily with the teeth of non-human primates, there are also differences among families, genera, and species that will be discussed in later chapters.

Enamel

The tooth crown is covered with a normally smooth layer of enamel that is semi-translucent, varying in color from a light yellow to a grayish white. Enamel is the hardest biological structure in an animal's body, varying from 5 to 8 on the Mohs scale of hardness of minerals, where talc is 1 and diamond is 10. Enamel attains its full thickness before the teeth emerge into the oral cavity. Enamel is thicker over the cusps of unworn permanent premolars and molars, thinner around the cervical region. Deciduous teeth have thinner enamel than permanent teeth and may be slightly whiter.

Mature enamel is mostly inorganic calcium phosphate, about 96%, and belongs to the hydroxyapatite mineral group, which is found only in mammalian tissues (Hillson, 1996). Enamel formation is initiated along the dentinoenamel junction (Fig. 2.2) between the ameloblasts (enamel-forming cells) and odontoblasts (dentine-forming cells) when the latter commence to secrete predentine, which, in turn, almost immediately stimulates the ameloblasts to secrete the enamel matrix. It is interesting to note that enamel will not form in the absence of odontoblasts. Transplanted ameloblasts fail to form enamel unless accompanied by odontoblasts. The process begins at the incisal or cusp tips and proceeds down along the sides of the tooth. Soon after enamel matrix formation begins, inorganic calcified crystals appear, indicating the beginning of calcification. Thus, enamel formation involves matrix secretion followed by maturation. Enamel forms incrementally, reflecting the speeding up and slowing down of enamel secretion, and the closer the enamel layer is to the surface of the crown the more mineralized (denser) it becomes.

The basic histological structure of mammalian enamel is calcified rods or prisms that extend from the dentinoenamel junction to the external surface of the tooth. Enamel prisms have received a great deal of attention through the years because they appear in three patterns that may differ among primates and thus have proved useful for taxonomic allocations

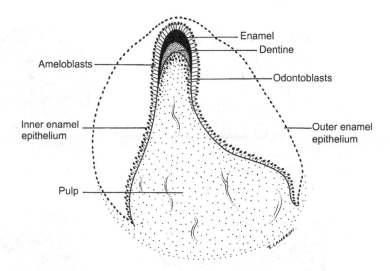

Fig. 2.2. Tooth germ showing formation of enamel and dentine. Reprinted from Aiello, L. and Dean, C. (1990) *An Introduction to Human Evolutionary Anatomy.* Reprinted with the permission of Academic Press.

(Boyde, 1976; Boyde and Martin, 1984; Gantt, 1982, 1986; Martin, 1985; Hillson, 1996). The three major shapes or patterns of prism cross-sections have been identified by Boyde (1969, 1976). These are: pattern 1, round closed circular enamel prisms formed by medium-sized ameloblasts; pattern 2, prisms arranged in alternate rows with an open surface and formed by small ameloblasts; pattern 3, prisms keyhole-shaped and formed by the largest ameloblasts (Fig. 2.3). The three patterns have been found in all primates; however, patterns 1 and 3 are more common in hominoids, and pattern 2 seems to be more frequent in Old World monkeys (Aiello and Dean, 1990).

Other important structures in enamel are the striae of Retzius and incremental lines. The striae of Retzius are lines of light brown to near black when seen in transmitted light microscopy, which cross the prism boundaries in arc-like layers beginning at the dentinoenamel junction and terminating at the enamel surface. The regular striae of Retzius are formed incrementally and are associated with surface perikymata. They vary in periodicity among primates during odontogenesis from 3 days in *Victoriapithecus* to 9 or 10 days in humans and great apes (personal communication, C. Dean). There are also other 'accentuated markings' that look like striae of Retzius but these result from various disturbances that occur during odontogenesis, however, they are 'crucial to histological studies of

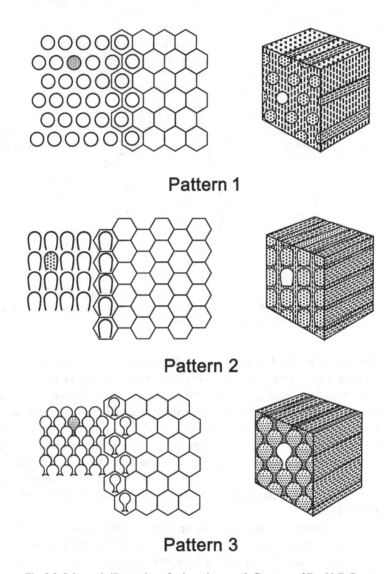

Pattern 1

Pattern 2

Pattern 3

Fig. 2.3. Schematic illustration of prismatic enamel. Courtesy of David G.Gantt.

tooth growth since they mark both the internal structure and the surfaces of enamel (*perikymata*) and dentine forming at one time period' (Dean, 2000, p. 120, italics mine). Other lines, the incremental lines (known as short period lines or daily cross-striations) are visible between the long-period striae of Retzius when viewed with polarized light and other forms of microscopy (Dean, 1989, 2000). Cross-striations or varicosities represent the circadian rhythmic nature of the secretory cell cycle during enamel formation and appear as dark lines crossing the prisms. Although there is some debate regarding the nature of the cellular mechanisms controlling these circadian cycles 'these short-period incremental lines allow estimates of the linear daily secretion rate of enamel' (Schwartz and Dean, 2000, p. 214). Indeed, Dean (2000, p. 129) is hopeful that incremental lines, also found in dentine and cementum, along with 'New discoveries in developmental biology and evolutionary biology will eventually mean that we can ask more focused questions about the nature of the relationship between ontogeny and phylogeny.'

As we shall see in later chapters, enamel thickness varies a great deal among extant and extinct primates. Much evolutionary and taxonomic importance has been attached to the plethora of studies that have appeared during the past several decades (Martin, 1985; Shellis *et al.*, 1998). Because teeth are the primary organs responsible for reducing a wide variety of foods with different physical properties to digestible particles during mastication, it is understandable that there have been so many studies of primate enamel thickness. There is no doubt that primate taxa feeding on hard objects have relatively thicker enamel than taxa feeding on softer substances, but what is important to remember taxonomically is that these taxa may none the less be closely related (Dumont, 1995). What is becoming ever more clear to students studying the causal relation between tooth form and function is clearly stated by Strait (1997, p. 199): 'The physical properties of the foods that teeth encounter during mastication may be the primary factor affecting changes in dental morphology.' Indeed, this was appreciated by Lucas and Teaford (1994, p. 183) when they wrote 'The first step toward understanding dental-dietary adaptations is to consider the fracture properties of foods, because it is to these that teeth are ultimately adapted.' In addition, as we study primate teeth it will be well to remember that the endurance of a tooth can be extended by having thicker enamel or increasing the size of the tooth (Lucas, Corlett and Luke, 1986a). Variations in enamel thickness among primates will be considered further in the chapters on the different primate taxa.

There are several types of enamel defect present in primate teeth (Brook, 1998). One important defect is enamel hypoplasia, a deficiency of enamel thickness, which occurs during the secretory phase of amelogenesis (Goodman and Rose, 1990; this is also an excellent review of enamel hypoplasias in humans, as is the recent contribution of Guatelli-Steinberg and Lukacs (1999) for non-human primates). Enamel hypoplasias may be expressed in several ways, e.g. pits, single sharp horizontal lines, single grooves or furrows in the crown surface (Hillson, 1996; Hillson and Bond, 1997; Guatelli-Steinberg, 2000). Reid and Dean (2000) have also presented new methods for estimating the timing of linear enamel hypoplasia (LEH) in the anterior teeth of humans. Although used for years as an indicator of non-specific systemic stresses during crown development in humans, studies of enamel hypoplasia have a rather checkered history among non-human primates (Guatelli-Steinberg and Lukacs, 1999; Guatelli-Steinberg, 2000). The term linear enamel hypoplasia (LEH) is generally used to designate faint or deeper lines or grooves on the surface of a tooth crown (Goodman and Rose, 1990; Guatelli-Steinberg, 2000) and this is the definition accepted here. The defect forms due to a physiological stress, e.g. disease or poor nutrition, that disturbs enamel matrix formation, resulting in a deficiency of enamel thickness. In contrast to bone, enamel does not remodel, and once formed, the defect is permanently implanted in the enamel.

The most comprehensive studies and reviews of the prevalence and incidence of LEH in non-human primates are those of Guatelli-Steinberg (1998; Guatelli-Steinberg and Lukacs, 1999; Guatelli-Steinberg and Skinner, 2000). More details relating to the occurrence and frequency of LEH among different primate taxa will be considered in the general dental information sections of later chapters.

Dentine

Dentine (ivory) is not as mineralized as enamel, therefore it is softer than enamel but harder than bone. Dentine makes up the bulk of a tooth and its root. It is composed of collagen and hydroxyapatite and is more compressible and elastic than enamel. It is covered by the enamel crown. Cementum surrounds the root (Fig. 2.1). As with enamel, dentine formation occurs in two stages, organic matrix secretion and mineralization. Dentine tubules containing odontoblasts transverse the dentine from the dentinoenamel junction and dentinocemental junction to the pulp chamber.

It is also well known that during dentine formation the various systemic rhythms that affect the development of enamel result in incremental lines in dentine (Dean, 2000). In contrast to ameloblasts, odontoblasts remain capable of being productive throughout their lives. After the completion of tooth growth they produce secondary dentine. As enamel wears down during attrition, the production of secondary dentine increases, recognizable as small areas of dark tissue on the occlusal surface. Eventually, it may form the entire occlusal surface of the tooth. The presence of secondary dentine protects the pulp chamber from exposure as enamel is worn away.

Another physiological product of the odontoblasts is a hyper-mineralized form of dentine containing few, if any, collagen fibers, known as peritubular dentine. More than fifty years ago Gustafson (1950) and later Miles (1963) noted that, in ground sections of teeth, the roots tend to become more translucent in older teeth, which can be useful in estimating the approximate age of an individual. This transparency or sclerosis is due to the peritubular dentine matrix that continues to form as teeth get older. It is interesting to note that Drusini, Calliari and Volpe (1991), using an image analyzer, did not find any difference between this method and that of measuring the transparent area directly on tooth sections.

We have mentioned a few of the many investigations of the microstructure of enamel in the last several decades that have aided in sorting out some of the knotty problems in primate systematics of both extinct and extant primates, and considered their usefulness in establishing criteria for better understanding their growth, development, and maturation. For one reason or another, investigation of the microanatomy of dentine has not kept pace with that of enamel, although the importance of dentine for daily incremental studies has been known since Schour and Hoffman (1939) (reported in Dean (1993)) as has the probable taxonomic significance of dentine organization (Hildebolt *et al.*, 1986). In the study by Dean (1993 p. 199) he found that the daily rates of dentine formation on macaque tooth roots were consistently between 3 and 4 μm per day, suggesting 'a consistent rate of dentine formation in permanent macaque teeth.' Hildebolt *et al.* (1986, p. 45) studied dentine tubule density and patterning among *Canis, Papio* and *Homo*; allowing for the small sample sizes, the authors concluded that 'dentine does have important taxon-specific structural characteristics that are of use in phylogenetic and taxonomic studies.'

Cementum

Cementum develops over the root dentine and generally extends around the enamel at the cervix. Cementum is produced by cementoblasts that form two types of cementum, cellular and acellular. Cellular cementum is less hard than enamel or dentine, being more similar to bone except that it does not resorb and reform, rather, it grows by apposition, layer by layer. Remodeling can occur, however. When cementum is destroyed by odontoclasts, cementoblasts can repair the damaged areas (Hillson, 1996). Since cementum surrounds the tooth root, fibers of the periodontal ligament pass from the alveolar bone to attach to the cementum in helping to support the tooth in its alveolus. Cementum adds to the size and strength of a tooth as well as protecting the underlying dentine.

Cementum lines or annuli have proven useful as a means of estimating chronological age in primates (Wada, Ohtaishi and Hachiya, 1978; Yoneda, 1982; Kay *et al.*, 1984). As noted by Kay *et al.* (1984) there are still several questions regarding the interpretation of cementum annuli for age estimations. These and related problems are considered more thoroughly by Hillson (1996) and Dean (2000).

Tooth roots

Primate teeth usually have one, two, or three roots, although there may be extra or reduced root formation in most classes of teeth. In general, root reduction is more common than root increase, although both variations of root numbers have been reported for most classes of primate teeth (Bennejeant, 1936; Remane, 1960; James, 1960; Alexandersen, 1963; Miles and Grigson, 1990). An unusually high incidence (40%) of two-rooted maxillary canines in female *Macaca fuscata* was described by Yoshikawa and Deguchi (1992). All males had single roots. Two-rooted maxillary and mandibular canines have been reported in primates (De Terra, 1905; Alexandersen, 1963), but evidently it is quite rare (Swindler, 1995).

Roots begin to develop after enamel and dentine have reached the future cementoenamel junction. At this time the epithelial root sheath is formed, which, in turn, initiates root formation. The root consists of the pulp cavity surrounded by dentine and cementum. On occasion, cells of the epithelial root sheath may form ameloblasts and produce droplets of enamel between or on the surface of the roots. These are known as 'enamel pearls'.

Dental pulp

Dental pulp occupies the pulp cavity and is surrounded by dentine (see Fig. 2.1). Pulp consists of a variety of tissues among which are arteries and veins, nerves, and lymphatic vessels that become more fibrous and less vascular with age. These structures enter and leave the tooth through the apical foramen situated at the tip of the root. The pulp projects toward the cusps of multicusped teeth and is known as pulp horns. When pulp is exposed, whatever the cause, the result is severe pain.

3　*Dental development*

According to Butler (1967), Bateson in 1894 first noticed that teeth resembled vertebrae, digits, and other serial structures. He called this phenomenon 'merism' and described the properties of a meristic series. The term is still used today to describe a heterodont dentition as a metameric system, i.e. different teeth have different morphologies and functions along the tooth row (Weiss, 1990).

The teeth of placental mammals have evolved from teeth possessing a single cusp (homodont). Moreover, morphological differences among the different classes of teeth, e.g. between canines and premolars, are more abrupt in some primate taxa than in others. Generally, the differences among classes of teeth are more distinct in the human dentition (Scott and Turner, 1997) than in the dentitions of many non-human primate taxa, particularly in the posterior teeth. In studying heterodont teeth through the years various explanations have been offered regarding the control of their complex development. Butler (1939) proposed the morphogenetic field theory in which there are three developmental fields: incisor, canine, and molar. Within each field there is a polar tooth affecting tooth development; for example, the upper first incisor is more stable than the second incisor and in both the upper and lower jaws the first molar is more stable than the second molar, which in turn is more stable than the third molar. Thus in each field the influence or gradient becomes less as the distance becomes greater from the polar tooth. In other words, there is the assumption that extrinsic field substances mediate the gradients along the tooth row.

Another theory that offered an explanation for the development of heterodont tooth morphology and gradients is known as the clone model (Osborn, 1978). Osborn suggested that originally there were three primordia associated with the different tooth classes from which additional primordia were added by cloning, which produces tooth-specific morphology along the tooth row. A major difference between the clone model and the field theory is that the former posits that development of the tooth is induced from within the tooth germ, i.e. intrinsic, whereas the latter holds that control of development is extrinsic, being entirely within the field substance or morphogenes.

It is recognized that both theories have certain strengths as well as weaknesses. Indeed, Greenfield (1992, p. 118), in explaining the development of incisification of the deciduous mandibular canines in various anthropoid taxa, proposed a model in which field effects as well as those of cloning occur, 'but at generally different stages of dental development.' After careful analyses of both models and of much experimental dental research at that time, Scott and Turner (1997, p. 83) concluded that 'Both models offer useful insights into the nature of gradients in the dentition, but some pieces of the intriguing puzzle continue to elude researchers.'

Genetics

Within the past couple of decades there has been a veritable explosion of research in the fields of developmental biology and genetics that has had an enormous influence on studies of tooth formation (see Hall, 1992; Schwartz, 1999, for general studies; Weiss, 1990, 1993; Sharpe, 2000; Jernvall and Thesleff, 2000; Zhao, Weiss and Stock, 2000 for tooth formation studies). A great deal of this research has been stimulated by the discovery of a class of regulatory genes known as homeobox genes. These genes control an animal's early development, and although much of this material is beyond the scope of this book, a better understanding of the mechanisms involved in the early development of teeth is being achieved as a result of this research (Tucker, Matthews and Sharpe, 1998; Sharpe, 2000). For example, regarding mammalian dental development, Weiss (1990, p. 19) stated 'that genes like those discussed above (*Hox genes*) [italics mine] are responsible for the critical events in neural crest specialization that define the commitment to dental development, setting in motion the cascade of gene regulation that leads to morphological specialization of the teeth.' This is an exciting time in the field of basic dental research, and perhaps before too long explanations will be forthcoming regarding how both continuous (metrical) and quasi-continuous (non-metrical) traits are controlled and to what extent each is due to 'nature or nurture.'

The majority of genetic investigations of primate teeth have been conducted on one primate, *Homo sapiens*, and have mostly been concerned with developmental defects, dental anomalies, and epidemiological surveys. For example, Scott and Turner (1997) note that in McKusick's (1990) book on human inheritance, none of the commonly occurring dental traits has well-established modes of inheritance. However, it has long been accepted that tooth development, morphology, and size are genetically determined (Lasker, 1950; Kraus, 1957; Moorrees, 1957). This information is based on population studies (Campbell, 1925; Pedersen, 1949; Selmer-Olsen, 1949; Moorrees, 1957), as well as family and twin studies (Osborne, Horowitz and De George, 1958; Saheki, 1958; Biggerstaff, 1970; Sofaer *et al.*, 1972; Townsend *et al.*, 1988, 1992).

There is disagreement regarding the inheritance of human tooth size as to which sex chromosome the gene(s) is (are) on. Some maintain it is on the X chromosome (Garn, Lewis and Kerewsky, 1964, 1965, 1967; Aas, 1983); other investigators conclude that genes on both the X and Y chromosomes influence tooth size (Alvesalo, 1971; Alvesalo and Tigerstedt, 1974). And, of course, some researchers hypothesize that tooth size is governed by genes on the autosomal chromosomes and that neither the X or Y chromosomes expresses any influence on tooth size (Niswander and Chung, 1968; Goose, 1971). There are, as far as I know, two studies on the mediation of tooth size in non-human primates. Sirianni and Swindler (1972, 1975) studied the inheritance of deciduous tooth dimensions in *Macaca nemestrina* and concluded that there was little evidence to support the X chromosome hypothesis, although Y chromosome involvement was considered to be a possibility in these monkeys. To date, there is little if any information relating to the mode of inheritance of dental traits in non-human primates. It is assumed that tooth size and non-metric traits in non-human primates are under genetic controls similar to those operating in humans and other animals. This seems a reasonable assumption in view of our knowledge of genetics today. For this reason, I have not felt it necessary to go into detailed discussions of human dental genetics. This topic is clearly presented in both detail and breadth by Hillson (1996) and Scott and Turner (1997). Indeed, Hillson (1996, p. 100) offered a plausible interpretation of the role of genetics on the dentition when he wrote

> It does seem clear that there is a strong genetic component in the distribution of at least some non-metrical features but, if they really are inherited in a quasi-continuous fashion, then all of the factors that control inheritance of continuous variants must similarly apply.

Odontogenesis

Teeth are considered skin derivatives, as are hair, feathers, and scales. But are they? This traditional scenario is currently being challenged. For example, Smith and Coates (2000 p. 147) maintain that much evidence today demonstrates that many of the processes involved in the development of teeth 'occurred at loci early in vertebrate phylogeny, preceding the origin of jaws and challenging accepted assumptions about dermal denticles and armour preceding initial stages of dental evolution.' In other words, teeth and jaws evolved as relatively independent events.

Teeth form only at the junction between ectoderm and mesoderm (mesenchyme). Enamel is ectodermal in origin whereas dentine, pulp, and cementum originate from the oral mesenchyme that originates from neural crest cells near the midbrain at a very early stage of development. From here, the neural crest cells migrate down along the side of the head to come to lie in the maxillary and mandibular processes that already contain epithelial cells. During their journey the mesenchymal cells, now known as ectomesenchyme, separate into bone primordia and tooth primordia. Butler (1967) suggested that this early sorting process into tooth primordia may represent the fundamental differentiation of the dentition. In any case, the interactions between the epithelium and ectomesenchyme direct odontogenesis, beginning with the formation of the dental lamina, which is a thickening of the ectodermal part of the oral epithelium. Soon after the appearance of the dental lamina, enlargements form along it in both jaws, corresponding to the future locations of the deciduous teeth. These are the tooth buds. Sometime later, successional laminae appear as lingual extensions from each of the deciduous tooth buds representing the permanent incisors, canines, and premolars. Because the permanent molars do not have deciduous predecessors the dental lamina extends distally, allowing for the development of the permanent molars.

The life cycle of a tooth is shown in Fig. 3.1 and consists of the following stages.

1. Growth: (a) initiation; (b) proliferation; (c) histodifferentiation; (d) morphodifferentiation; (e) apposition
2. Calcification (mineralization)
3. Emergence
4. Attrition

These sequential events are not completely separated; rather, there is considerable overlap and many are continuous over several stages as depicted in Fig. 3.1. Only a few comments will be made here regarding these stages since there are many textbooks on dental embryology and oral histology available for more complete discussions of these important issues regarding the life cycles of teeth (Ten Cate, 1994).

As mentioned, tooth buds form along the dental lamina representing the deciduous and permanent teeth. Lack of initiation of a tooth bud,

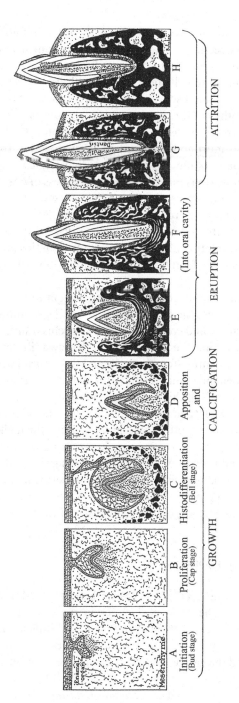

Fig. 3.1. The life cycle of a tooth. Reprinted from Schour, I. and Massler, M. (1940) Studies in tooth development: the growth patterns of human teeth, *Journal of the American Dental Association* **27**, 1778–93, Fig. 1. Copyright © 1940 American Dental Association. Reprinted by Permission of ADA Publishing, a Division of ADA Business Enterprises, Inc.

however, results in absence of a tooth, whereas abnormal initiation may result in supernumerary teeth, e.g. the appearance of fourth molars in orangutans. As cell proliferation continues the bud stage forms the cap stage, which develops into the bell stage as the tooth germ continues to increase in size and mature. It is also well known that explants of the dental lamina (even before tooth-bud formation) continue to develop into their respective teeth when grown in tissue cultures (Glasstone, 1963, 1966). Histodifferentiation and morphodifferentiation occur together from the bud stage to the final phases of the bell stage. Perturbations in morphodifferentiation can result in various malformed teeth, e.g. loss of cusps, extra cusps, and peg-shaped teeth. Appositional growth of enamel and dentine is layer-like and known as additive growth. These periods of activity and rest alternate at intervals that are useful in aging teeth as discussed in Chapter 2. Enamel and dentine formation always begins at the cusp tips and along the incisal borders of incisors. Insufficient enamel apposition can result in enamel hypoplasia.

As the cells of the inner enamel epithelium (IEE) begin to differentiate and divide (mitosis) they cause the IEE to fold and buckle, resulting in the formation of grooves, cusps, and crests on the future occlusal surface (Fig. 3.2) which ultimately determine the form of the tooth crown (Butler, 1956). The junction between the dentine-forming cells and those forming enamel is known as the dentinoenamel junction (DEJ). Dentine and enamel continue to form along either side of this basement membrane, and do so until the crown is formed, at which time the ameloblasts die. The odontoblasts, however, continue to produce dentine (secondary dentine) throughout the life of the tooth as discussed in the previous chapter. Enamel apposition begins at the cusp tips and spreads downward over the surface and sides of the crown to the cementoenamel junction. During this period of mineralization the cusps continue to grow apart and enamel thickens over the tooth crown until the cusps coalesce, signaling the final size of the tooth crown.

Ontogeny of crown patterns

Almost fifty years ago Butler (1956, p. 45) declared that odontologists 'have been interested for the most part either in the earliest stages of development or in histological problems, and have paid little attention to morphological features.' A few years later, the late Bertram S. Kraus (1959a,b, 1963) accepted the gantlet with a series of papers concerned with the morphological development of the deciduous teeth of humans. Then Kraus and Jordan (1965) published *The Human Dentition Before Birth*, a

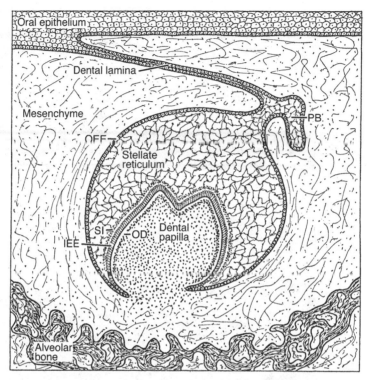

Fig. 3.2. Histodifferentiation (bell stage of development). Reprinted from
Schour, I. and Massler, M. (1940) Studies in tooth development: the growth of
human teeth, *Journal of the American Dental Association* **27**, 1778–93, Fig. 2.
Copyright © 1940 American Dental Association. Reprinted by Permission of
ADA Publishing, a Division of ADA Business Enterprises, Inc.

seminal study of the morphogenesis of the human deciduous dentition. It
was during this time that the author began his studies of the different
mineralization patterns of the deciduous and permanent teeth of non-
human primates (Swindler, 1961).

In studies of crown odontogenesis, the tooth germ is removed by dissec-
tion from its crypt and placed in an aqueous solution of alizarin red S.
After remaining in the solution for variable periods of time, the follicle is
removed by microdissection. All mineralized areas are stained red and the
uncalcified regions are clear (Fig. 3.3). The occlusal surfaces are then ready
for detailed studies of dental morphogenesis, photography, illustrations
and measurements (Swindler and McCoy, 1964; Kraus and Jordan, 1965;
Swails, 1993).

Radiographs are used to study tooth formation in longitudinal or

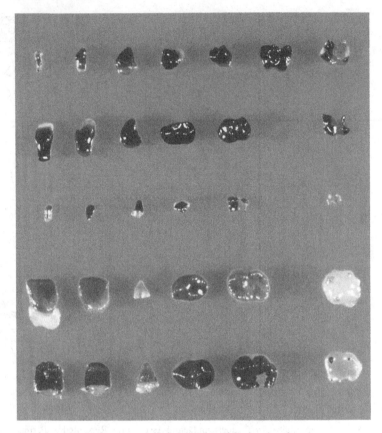

Fig. 3.3. Stained primate tooth buds (alizarin red S) showing the morphodifferentiation of the lower deciduous teeth at birth plus M_1. From top to bottom: *Alouatta caraya, Macaca nemestrina, Hylobates klossi, Pan troglodytes* and *Pongo pygmaeus*.

cross-sectional investigations of living animals and simply represent an extension of the studies of prenatal crown formation mentioned above (Dean and Wood, 1981; Sirianni and Swindler, 1985; Anemone, Watts and Swindler, 1991; Swindler and Meekins, 1991; Winkler, Schwartz and Swindler, 1996). It is well known that the various stages of tooth development are not comparable between radiographs and alizarin red S studies, because mineralization is demonstrable histologically before becoming radio-opaque. This means assessments based on radiographs will tend to underestimate the age at which a tooth reaches a particular stage of development (Fig. 3.4).

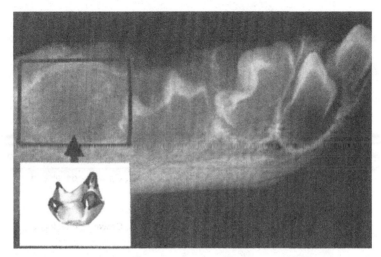

Fig. 3.4. Initial calcification of the mandibular first molar. Reprinted from
Sirianni and Swindler (1985). Reprinted with permission of CRC Press, Inc.

Dental development is a continuous process from initial crypt formation
to the apical closure that signals the termination of root growth. This
developmental scenario is not only a measure of the biological maturity of
the dentition, but also a comparative yardstick that is useful for measuring
the timing and sequence of other developmental systems (Swindler, 1985;
Smith, 1989; Smith, Crummett and Brandt, 1994). It is convenient, however, to separate the continuum into stages of crown and root formation as
shown in Fig. 3.5. All primate teeth pass through these stages, although if a
tooth begins to form at the same time in two species it may take a longer
proportion of the growth period to develop in one species than in the other.
For example, the permanent canines of apes take longer to develop than
the permanent canines of humans, whereas the roots of monkeys and apes
grow more quickly than the roots of humans (Dean and Wood, 1981;
Swindler, 1985).

What is known concerning the order of mineralization of the deciduous
teeth in primates would suggest that it is the same in Old World monkeys
and hominoids, i.e. di1 dp1 di2 dc dp2 (Swindler, 1961; Swindler and
McCoy, 1964; Turner, 1963; Kraus and Jordan, 1965; Oka and Kraus,
1969; Tarrant and Swindler, 1972; Siebert and Swindler, 1991; Swindler
and Beynon, 1993). The sequence of molar cusp mineralization is similar
among these taxa. In the mandible it is protoconid, metaconid, hypoconid,
entoconid, and hypoconulid (hominoids only) with some variation between the metaconid and hypoconid. In the maxilla the sequence is

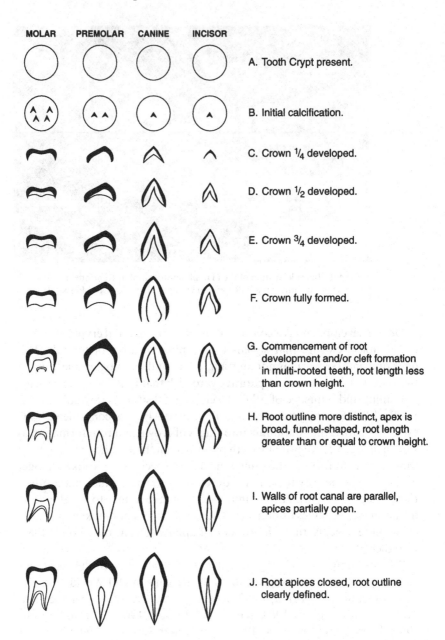

Fig. 3.5. Stages of tooth formation from the tooth crypt to apical closure of the roots. Reprinted from Winkler, L. A. *et al.* (1996) Development of the Orangutan permanent dentition: Assessing patterns and variation in tooth development, *American Journal of Physical Anthropology* **99**, 205–20, Fig. 1, p. 209. Reprinted with the permission of John Wiley & Sons, Inc.

paracone, protocone, metacone, and hypocone while the variable cusps are the protocone and metacone (references as above).

The only study of prenatal dental development in New World monkeys that the author is aware of is that of Tarrant and Swindler (1973), who examined 19 fetuses of black howler monkeys (*Alouatta caraya*). Because all teeth in the specimens showed some degree of mineralization it was impossible to determine the exact order of initial mineralization. It was found that the dis, dcs, and dp2s calcified from a single center whereas the dp3s had two centers, and the dp4s and M1s had four separate centers of calcification. The order of cusp calcification in dp4s was the same as in catarrhine primates.

It seems clear that the order of cusp mineralization is similar among the primate taxa discussed above, although the pattern of cusp coalescence of New World monkeys is different from that of Old World monkeys, which in turn is different from that of the hominoids. It has been suggested that the differences between these coalescence patterns are probably associated with the importance of the various ridges and crests connecting the cusps on the occlusal surfaces of the molars (Tarrant and Swindler, 1973; Swindler, 1985).

The only permanent tooth to begin development prenatally in the non-human primates discussed here is the first molar (see Fig. 3.3), which may have one to three cusps mineralized at birth (Swindler, 1961, *Macaca mulatta*; Oka and Kraus, 1969, *Pan troglodytes*; Tarrant and Swindler, 1973, *Alouatta caraya*; Sirianni and Swindler, 1985, *Macaca nemestrina*; Siebert and Swindler, 1991, *Pan troglodytes*; Swindler and Beynon, 1993, *Theropithecus gelada*). This is also the situation in *Homo sapiens* (Kraus and Jordan, 1965).

Age estimation

'How old is this animal?' is a question frequently asked when one is dealing with living non-human primates or working with their skeletons in museums. A quick way to answer this question is to look at the teeth present in an animal's mouth, a method, as noted by Smith *et al.* (1994), first used by Aristotle, who was aware that teeth erupt at different ages in different animals and could therefore be used to estimate the animal's age. We still use this basic biological information when we compile the tables and graphs employed today to estimate an animal's chronological age based on tooth eruption.

Adolph Schultz (1935) published an extensive paper that became a landmark contribution for studies of the eruption of primate teeth. For decades this paper served as a yardstick by which to measure the maturation of living primates. Through the years, however, many papers have appeared reporting study results regarding the ages of tooth eruption in many primate species. This research has been reviewed and discussed in an extensive compendium that has put together a data base 'that can serve both practical needs and theoretical interests' (Smith *et al.*, 1994 p. 181). This is an extremely important contribution that will be of value to all evolutionary and developmental biologists interested in the maturation and life histories of primates.

The estimation of chronological age of non-human primates of feral origin is often critical in research studies. In addition, it is also important to be able to estimate the developmental or maturational stage of the dentition of animals of known age. Such information is valuable when designing experiments concerned with the causal factors influencing, for example, the histo- and morphodifferentiation of the dentition. In other words, knowledge of the timing of events during the growth process is often crucial to the success of the experiment. A physiological perturbation (stress) during the early stages of tooth calcification (secretory phase of amelogenesis) can have results quite different from those that occur if the same metabolic event happens later in the mineralization process. Linear enamel hypoplasia is a good example of such time-dependent processes.

In addition to tooth eruption, intra-osseous tooth formation has proven useful in estimating the chronological age of primates. Indeed, there is evidence that tooth formation is a more reliable method of assessing dental maturation than tooth emergence (Gleiser and Hunt, 1955; Fanning, 1961; Moorrees, Fanning and Hunt, 1963; Haavikko, 1970). A radiographic longitudinal study of the dental development of male and female pigtailed macaques (*Macaca nemestrina*) of known ages from 6 months to 7 years is presented in Sirianni and Swindler (1985). The dental developmental stages employed in this investigation were similar to those shown in Fig. 3.5. Figures 3.6 and 3.7 summarize the dental development of these laboratory-born and -reared monkeys. The three development stages in the figures are represented by diamonds: open indicates initial calcification; stippled, crown completion; and solid, root completion with apical closure. The width of the diamond represents the mean age; the height, ± 1 S.D, 'E' is approximate age of eruption.

The majority of information regarding primate tooth emergence as well as radiographic studies of tooth formation is from laboratory-born primates. An obvious question is whether there is any difference in the timing

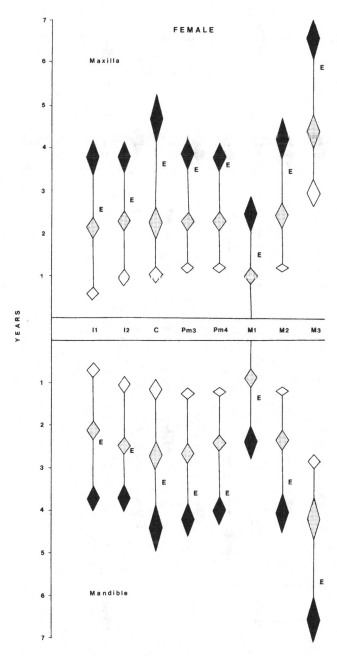

Fig. 3.6. Dental development and the time of tooth emergence of female *Macaca nemestrina*. Reprinted from Sirianni and Swindler (1985), with Permission of CRC Press, Inc.

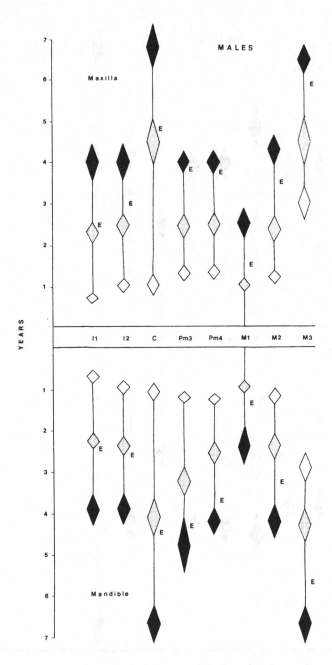

Fig. 3.7. Dental development and the time of tooth emergence of male *Macaca nemestrina*. Reprinted from Sirianni and Swindler (1985), with Permission of CRC Press, Inc.

Table 3.1. *Sequence of permanent mandibular tooth formation in primates*

Stage	Sequence	Taxon
Initial calcification	M1 I1 I2 C P3 [P4 M2] M3	*Papio cynocephalus*[1]
	M1 I1 I2 C P3 [P4 M2] M3	*Macaca nemestrina*[2]
	M1 I1 I2 C P3 [P4 M2] M3	Pongid[3]
	M1 I1 I2 C P3 [P4 M2] M3	*Homo sapiens*[4]
Crown complete	M1 I1 I2 [P3 P4 M2] C M3	*Papio cynocephalus*
	M1 I1 I2 [M2 P3 P4] C M3	*Macaca nemestrina*
	M1 I1 I2 P3 [M2 P4] C M3	Pongid
	M1 I1 I2 C P3 [P4 M2] C M3	*Homo sapiens*
Root formation	M1 I1 I2 M2 [P3 P4] C M3	*Papio cynocephalus*
	M1 I1 I2 M2 [P4 P3] C M3	*Macaca nemestrina*
	M1 I1 I2 P3 [M2 P4] C M3	Pongid
	M1 I1 I2 C P3 [P4 M2] C M3	*Homo sapiens*
Emergence	M1 I1 I2 M2 [P4 P3] C M3	*Papio cynocephalus*
	M1 I1 I2 M2 [P4 P3] C M3	*Macaca nemestrina*
	M1 I1 I2 [M2 P3 P4] C M3	Pongid
	M1 I1 I2 C P3 [P4 M2] M3	*Homo sapiens*
Apical closure	M1 I1 I2 M2 [P3 P4] C M3	*Papio cynocephalus*
	M1 I1 I2 M2 [P3 P4] C M3	*Macaca nemestrina*
	M1 I1 I2 M2 P3 P4 C M3	Pongid
	M1 I1 I2 C P3 [P4 M2] M3	*Homo sapiens*

[1]Present study.
[2]Present study.
[3]Dean and Wood (1981, includes chimpanzee, gorilla, and orangutan).
[4]Moorrees *et al.* (1963).
Reprinted from Swindler, D. R. and Meekins, D. (1991). Dental development of the permanent mandibular teeth in the baboon, *Papio cynocephalus*. *American Journal of Human Biology* **3**, 571–580. Reprinted with permission of John Wiley & Sons, Inc.

of dental maturational processes between laboratory-born animals and feral animals. There seems to be evidence supporting both sides of the issue as there are data indicating a difference (Phillips-Conroy and Jolly, 1988) and a study suggesting no substantial disparity (Kahumbu and Eley, 1991). And, as Smith *et al.* (1994, p. 203) have said, 'whatever the magnitude of the captive effect, it is fairly certain that it is less than the effect of captivity on skeletal growth, sexual maturation, and body weight.'

Sequence of initial stages of permanent tooth formation

The patterns of initial tooth formation in several primate taxa are presented in Table 3.1. The initial mineralization sequences are the same in all taxa; M1 I1 I2 C P3 [P4 M2] M3. Teeth in the areas of developmental plasticity represent sequence polymorphisms (Garn and Lewis, 1957). By the time of crown completion, the canine has shifted from position four at the beginning of mineralization to position seven in the sequences of all taxa except humans, where it remains the fourth tooth to complete crown formation. It normally remains the fourth tooth to emerge and to complete apical closure in *Homo sapiens.* Early completion of crown formation, early emergence, and early apical closure may be correlated with the phylogenetic and ontogenetic reduction of the canine in *Homo sapiens.*

The premolar region remains ontogenetically variable throughout tooth formation in all stages and in all taxa. Although rare, the P4 M2 emergence pattern is reported in chimpanzees by Clements and Zuckerman (1953) and Krogman (1930) for both chimpanzees and gorillas. On the other hand, the M2 P4 emergence pattern is occasionally observed in humans (Garn and Lewis, 1957). It is interesting then to note that the M2 P4 emergence sequence has always been reported in Old World monkeys and most often in anthropoid apes. Humans, however, can have either P2 M2 or M2 P4 emergence patterns. The data discussed here and presented in Table 3.1 substantiate those of Garn and Lewis (1957), who reminded us that the sequence of emergence cannot always be predicted from the sequence of tooth formation.

The changes in the rates and patterns of dental development presented in Table 3.1 suggest that alterations have occurred in the regulatory system during the phylogeny and ontogeny of primate tooth formation that represent heterochronic changes. Tooth merisms would seem well suited for heterochronic changes, which require only an alteration in the timing of the features already present (Gould, 1977) to produce the sequential changes that occur from initial mineralization to apical closure.

4   *The deciduous dentition*

In 1956 K. D. Jorgensen wrote *The Deciduous Dentition. A Descriptive and Comparative Anatomical Study.* This still remains one of the most thorough and comprehensive studies of the human deciduous dentition. There is nothing, however, in the literature on the deciduous dentition of non-human primates that approaches the magnitude of this work. The present chapter is not intended to fill in the void; this would require a book in itself. Unfortunately, the present data base is generally small for most species since the main purpose of the original collection was to cast the permanent dentition. A few casts, however, were made of the deciduous teeth of several anthropoid species, and considering the limited number of investigations of non-human primate deciduous teeth it was thought appropriate to include this chapter. There were no casts made of the deciduous teeth of the prosimians. In both morphological and metric analyses sexes are combined since in many cases the sex is not known, and there is little if any indication of significant sex differences.

New World monkeys

The Atelidae is the only family represented in the present collection of casts of the deciduous dentition of New World monkeys. These include two genera of the subfamily Atelinae, *Ateles* and *Alouatta*.

Family Atelidae

Subfamily Atelinae

Morphological observations

	n
Ateles fusciceps (brown-headed spider monkey)	5
Ateles belzebuth (long-haired spider monkey)	7
Ateles paniscus (black spider monkey)	3

The deciduous dental formula of all New World monkeys is:

di^2-dc^1-dp^3/ di_2-dc_1-dp_3

Incisors

UPPER. The crown of di^1 is wider mesiodistally than it is thick labiolingually. The labial surface is smooth; the lingual surface is slightly concave and possesses a very narrow cervical marginal ridge. There are no mesial, distal, or incisal marginal ridges. The crowns of the two central incisors slope toward the midline along their mesial borders, whereas the distal borders are more vertical. The morphology of di^2 is similar to that of di^1 but it is about half the size of di^1.

LOWER. The lower central incisor has a narrow, straight crown with slight, lingually elevated distal and mesial marginal ridges outlining a small lingual depression. The lateral incisor is twice as wide mesiodistally as is d_1 and lacks a lingual depression.

Canines

UPPER. A diastema separates the canine from di^2 which receives the lower canine when the teeth are occluded. The canine projects beyond the occlusal plane and is triangular with a slight, narrow cingulum passing completely around the base of the crown. Positioned about one third of the way from the mesial border on the lingual surface is a ridge extending from the lingual cingulum to the tip of the canine. The buccal surface of the crown is convex.

LOWER. There is no diastema between dc_1 and dp_2. The canine extends above the occlusal plane and has convex mesial and distal borders terminating at the crown apex. There is a slight distal heel.

Premolars

UPPER. The dp^2 has a single cusp, the paracone. There is a cingulum passing around the lingual surface of the tooth that creates a narrow ledge that increases the masticating surface of dp^2. The dp^3 is bicuspid, possessing a large paracone that connects with a smaller protocone by an anterior transverse crest, creating a narrow mesial fovea and a broader, deeper distal fovea. The dp^4 usually has three cusps, a large paracone distal to which is a slightly smaller metacone, which is connected with the proto-cone via the crista obliqua. The hypocone may be absent, small or well developed on the distolingual surface of the protocone. When present it is separated from the protocone by a narrow, vertical groove and has an independent apex. The crista obliqua divides the occlusal surface into a broad trigon basin and a much smaller and narrower distal fovea. A small, but distinct hypocone is also present on dp^4 in *A. belzebuth*. Except for their smaller size, the dp^4s are quite similar morphologically to the M^1s in their respective species.

LOWER. The dp_2 has a single cusp with a narrow lingual cingulum along the base of the crown. A protoconid is connected by the protocristid to a more distally positioned metaconid on dp_3 in the three species. In *A. belzebuth* a small entoconid and hypoconid are present on dp_3 but are lacking in the other two species. The dp_4 in *A. belzebuth* possesses five cusps, since a small hypoconulid is present. Interestingly, small hy-poconulids may be present on any permanent lower molar in this genus (see Chapter 6). The dp_4 occlusal surface is divided into a narrow trigonid and a somewhat wider talonid in the three species.

Morphological observations

	n
Alouatta palliata (mantled howler)	13

Incisors (Fig. 4.1)

UPPER. The di^1 is a small, spatulate tooth with a concave lingual surface. The di^2 is about the same size as di^1, but slightly more tapered toward the tip.

LOWER. The di_1 is little more than a narrow, straight-sided enamel crown extending from the root. There is no lingual depression. The di_2 crown is wider than that of di_1 and its distal border inclines mesially toward the cusp tip while the mesial border is vertical. There is a slight lingual depression.

When occluded, the deciduous incisors come together in an edge-to-edge manner as do the permanent incisors in both New and Old World monkeys. It has been found, however, that in leaf-eating monkeys there is often a shift from the edge-to-edge occlusion to an underbite i.e., the lower incisors project mesial to the upper incisors (see Fig. 7.15). The incidence

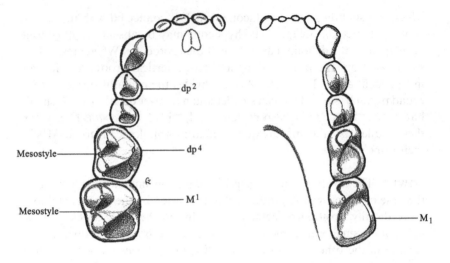

Fig. 4.1. *Alouatta palliata* male, occlusal view of deciduous teeth (27 mm).

of underbite in the deciduous teeth of *Alouatta palliata* is presented in Table 4.1. Schultz (1958, 1960) also reported underbite in juvenile *Alouatta* and *Colobus* monkeys. See Chapters 6 and 7 for more discussion of this condition in adult leaf-eating monkeys.

Canines

UPPER. The dc^1 crown projects below the occlusal plane and both distal and mesial margins taper toward the canine tip, resulting in a triangular crown. From the occlusal view the labial surface is smoothly convex and the lingual surface has a narrow pillar extending from the cusp tip to the lingual cingulum. Distal to the pillar there is a narrow, vertical distolingual groove that passes onto the lingual cingulum and continues distally as a short enamel projection.

LOWER. The crown of dc_1 rises above the occlusal plane and has a convex mesial margin. A convex distal margin projects distally as a short ledge. A narrow lingual cingulum passes from the mesial margin of the tooth distally to join the extended ledge. The labial surface is smooth and rounded mesiodistally. A narrow, vertical groove is present on the distolingual margin of the crown.

Table 4.1. *Percentage frequencies of incisor occlusal variability in juvenile leaf-eating monkeys*

Taxon	Sex	n	Underbite	Edge to edge	Overbite
Colobus polykomos	M	10	50.0	40.0	10.0
	F	11	27.3	72.7	—
Piliocolobus badius	M	6	33.3	66.7	—
	F	9	66.7	33.3	—
Presbytis aygula	M	2	50.0	50.0	—
	F	9	77.8	22.2	—
Presbytis cristatus	M	1	—	100.0	—
	F	2	—	100.0	—
Presbytis pileatus	M	3	66.7	33.3	—
	F	4	100.0	—	—
Nasalis larvatus	M	1	100.0	—	—
	Г	3	100.0		
Rhinopithecus roxellana	M	—	25.0	—	—
	F	5	45.5	100.0	—
Alouatta villosa	M	9	100.0	—	—
	F	7	85.7	14.3	—

Adapted from Swindler (1979).

Premolars

UPPER. The occlusal surface of dp^2 has a large, pointed paracone and a well-formed lingual cingulum, which forms a ledge that increases the masticatory surface of the tooth so that it receives the protoconid of dp_3 when the teeth are in occlusion. About midway along the cingulum there is a minute cusplet that may be the protocone. In their study of the early development of the deciduous dentition of *Alouatta caraya*, Tarrant and Swindler (1973) found that dp^2 initiated calcification from a single site that was located at the apex of a rather caniniform crown, dp^3 had two separate calcification centers, and dp^4 possessed four independent centers of calcification corresponding to the number of cusps present on the erupted teeth. There is a slight development of the ectoloph on the distobuccal surface of the paracone.

The dp^3 has two cusps, the paracone and protocone, that correspond to the two centers of calcification mentioned above. The occlusal surface is also somewhat broader than that of dp^2 and the ectoloph is complete around the buccal surface of the paracone. There are slight enamel thick-enings at the mesial and distal ends of the ectoloph, the parastyle and distostyle respectively. There is no mesostyle.

The dp^4 has four cusps, the paracone, protocone, metacone and hypo-cone, that represent the four centers of calcification mentioned above. The dp^4 is the largest and most molariform of the deciduous molars. The ectoloph is complete around the buccal surface of the tooth. It commences at the parastyle on the mesiobuccal surface of the paracone. Passing distally, the ectoloph joins the mesostyle between the paracone and meta-cone from where the ectoloph terminates as a rather sharp distostyle on the distobuccal surface of the metacone. It is interesting to note that the para-, meso- and distostyles did not have separate calcification centers (Tarrant and Swindler, 1973). The occlusal surface consists of a large deep trigon basin separated from the much smaller talon basin by a well-formed crista obliqua passing obliquely between the protocone and metacone. The hypocone is minuscule, being little more than a slight elevation on the mesiolingual part of the distocrista, where it becomes the entocrista, coalescing with the crista obliqua.

Several of the specimens had M^1s partly emerged; their morphology is very similar to that of dp^4 except for size (Fig. 4.1).

LOWER. The dp_2 has a protoconid projecting above the occlusal plane and is more caniniform than the other premolars. The lingual surface is separated into a vertically oriented shallow, mesial basin and a slightly deeper distolingual basin by a narrow, sharp vertical enamel crest passing from the crown tip to the lingual cingulum. The distolingual basin termin-ates as a slight distal projection. The dp_2 possessed a single calcification center (Tarrant and Swindler, 1973).

The dp_3 is bicuspid, having both a protoconid and a metaconid. Devel-opmentally the protoconid calcified first, followed by the metaconid, and later both cusps were connected by the protocristid (Tarrant and Swindler, 1973). The protoconid is positioned mesial to the metaconid and the two are connected by the protocristid. The protocristid separates the occlusal surface into trigonid and talonid basins. The trigonid basin is slightly larger than the talonid basin.

The largest and most molariform premolar is dp_4. This tooth has four cusps, protoconid, metaconid, hypoconid, and entoconid; this is also the order of cusp calcification (Tarrant and Swindler, 1973). There was no hypoconulid in the present sample but it was found in one of 19 specimens of *Alouatta caraya* (Tarrant and Swindler, 1973). The protoconid is mesial to the metaconid and a protocristid connects the two cusps. In contrast to

dp_3, the talonid basin is larger than the trigonid basin and it is also slightly lower. The cristid obliqua is present only on dp_4.

As with the maxillary dp^4 and M^1, the mandibular dp_4 and M_1 are morphologically very similar. The major difference is that M_1 is a larger, wider tooth resulting in the cusps being further apart on the occlusal surface, particularly in the distance between the protoconid and metaconid (Fig. 4.1).

Old World monkeys

Dental casts were made of several genera of Old World monkeys representing both subfamilies of the Cercopithecidae.

Family Cercopithecidae

Subfamily Cercopithecinae

The deciduous dental formula of Old World monkeys is:

di^2-dc^1-dp^2 /di_2-dc_1-dp_2

Morphological observations: guenons

	n
Cercopithecus mitis (blue monkey)	13
Cercopithecus nictitans (spotnosed guenon)	2
Cercopithecus neglectus (De Brazza's monkey)	2
Cercopithecus lhoesti (l'Hoest's monkey)	1
Chlorocebus aethiops (vervet monkey)	5
Erythrocebus patas (Patas monkey)	3

Incisors (Fig. 4.2)

UPPER. The di^1 is a mesiodistally wide tooth compared with di^2. The labial surface is smooth and rounded mesiodistally; the lingual surface is bordered by mesial, distal and incisal marginal ridges as well as a narrow lingual cingulum. A mesiodistally elongated sulcus lies between the lingual cingulum and the incisal marginal ridge. The lingual sulcus is present in the unworn tooth, but becomes deeper as a consequence of the incisal margins

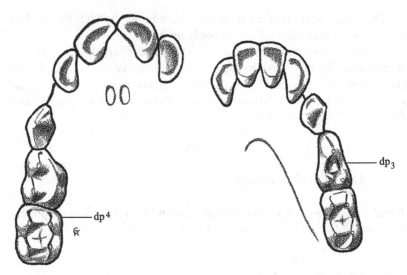

Fig. 4.2. *Cercopithecus mitis* male, occlusal view of deciduous teeth (20 mm).

of the lower incisors wearing against this surface as the teeth occlude. Indeed, the lower incisors pass across the lingual surface of di^1 to rest on the lingual cingulum as the teeth come into full occlusion. A small lingual tubercle rises from near the center of the lingual cingulum and passes as a narrow, vertical ridge across the lingual surface of di^1, terminating prior to reaching the incisal margin. The lingual tubercle quickly disappears with wear. Lingual tubercles and ridges are described by Orlosky (1968) in the di^1s of *Papio anubis, Macaca mulatta* and *Colobus polykomos*. The di^2 crown is more pointed than in di^1. It is more caniniform in *Cercopithecus mitis* and *Chlorocebus aethiops* than in the other guenons. The labial surface is smooth; the lingual surface is slightly concave with a narrow cingulum passing mesiodistally around the lingual surface. There are no lingual tubercles or vertical ridges. There is a diastema between di^2 and dc^1.

LOWER. The labial surface of di_1 is smooth and rounded mesiodistally, whereas the lingual surface is concave. The mesial and distal marginal ridges demarcate the concave lingual surface before becoming continuous with the lingual cingulum.

The distal margin of the crown of di_2 slopes mesially to form a more inclined plane along the incisal border than is present on di_1, resulting in a more caniniform tooth. The mesial marginal ridge is slightly more apparent where it unites with the lingual cingulum. There is a slight lingual sulcus passing transversely along the lingual cingulum. A small ledge may project distally from the distal marginal ridge and is particularly noticeable in *Cercopithecus neglectus.* This projection is reported by Benefit (1994) for *Victoriapithecus*, an early Old World monkey from the middle Miocene of Africa. She also mentions its presence in extant Old World monkeys. In this sample of guenons it is quite variable.

Canines

UPPER. The dc^1 projects above the occlusal plane and is triangular in shape with the base at the cervix. The labial surface is smooth and rounded while the lingual surface is bordered by mesial and distal marginal ridges as well as a slight lingual cingulum. In *C. aethiops* only, the lingual cingulum joins with the distal marginal ridge to form a slight distal projection. There is a narrow but distinct ridge extending from the cusp tip to the lingual cingulum dividing the surface into a rounded mesiolingual portion and a slightly more concave distolingual portion. There may be a narrow, vertical sulcus between the lingual ridge and the mesiolingual marginal ridge. Another vertical sulcus is present on the distolingual portion of dc^1. A distinct lingual tubercle is not present; however, it is reported in *P. anubis* and *M. mulatta* but not in *Colobus polykomos* by Orlosky (1968).

LOWER. The crown tip of dc_1 is only slightly higher than the occlusal plane of the incisors while extending somewhat higher above the occlusal plane of the deciduous premolars. The dc_1 is like that of all cercopithecoids in being long mesiodistally and narrow buccolingually. The labial surface is smooth and rounded mesiodistally. The lingual surface has a crest extending from the cusp tip to the lingual cingulum, separating the surface into a narrow mesiolingual fovea and a wider distal surface. There is a prolongation of the distal marginal ridge as a ledge, or heel. In addition, a narrow, shallow sulcus is present on the distolingual surface of the protoconid. The lingual crest and distolingual sulcus are also reported by Orlosky (1968) and Benefit (1994).

Premolars

UPPER. The dp^3 displays the cercopithecid molar pattern of four cusps: paracone, protocone, metacone and hypocone. In the present sample, the paracone is always mesial to the protocone and connected by an anterior transverse crest; on the other hand, the metacone and hypocone are opposite one another or the metacone is slightly mesial to the hypocone.

Irrespective of the position of the cusps, they are connected by posterior transverse cristae. There are mesial and distal marginal ledges, and a central trigon basin observable from the occlusal view. The dp^3 is slightly elongated mesially. The preparacrista passes mesiolingually from the paracone, forming a slight elevation at the junction with the mesial marginal ridge. The mesial marginal ridge joins the preprotocrista to continue on to the protocone. In his comprehensive study of cercopithecid teeth, Delson (1973, p. 210) noted the similarity between the elevation on the preparacrista and the primitive parastyle but thought that it was 'probably not homologous with this structure.'

The dp^4 is slightly larger than dp^3 in both length and breadth (Appendix 1, Tables 195–220). The occlusal view presents four cusps, the paracone and protocone connected by the anterior transverse cristae and the metacone and hypocone connected by slight posterior transverse cristae. The mesial cusps as well as the distal cusps are positioned opposite each other and it should be noted that the metacone and hypocone are slightly closer together than the paracone and protocone, a condition also found in M^1 (Delson, 1973). In dp^4 the buccal cusps are higher than the lingual in the unworn condition. There are mesial and distal foveae outlined by well-developed mesial and distal marginal ridges and a central trigon basin.

LOWER. The dp_3 is elongated mesiodistally with the protoconid shifted mesial to the metaconid. The cusps may be connected by the protocristid; however, it often passes vertically rather than across to the metaconid. In these cases, the protoconid and metaconid are separated by a narrow central developmental groove that was also reported by Orlosky (1968). A paraconid has been reported by Delson (1973, p. 177) on dp_3s in cercopithecids. He wrote 'the mesial cusp of dp_3 in cercopithecids can safely be termed a paraconid, implying although not proving homology.' In the present sample of guenons and their relatives, there are several individuals among the various species with very low enamel bulges and one specimen of *Chlorocebus aethiops* that has a cuspule at the mesiolingual end of the paralophids. Whether these are paraconids or not, it is difficult, if not impossible, to determine without fossil evidence, or evidence obtained from studies of early dental development. It is interesting to note that Benefit (1994, p. 305) mentions that 'the preprotocristid wraps around the mesial aspect of the crown, being continuous with a mesiolingual ridge' but does not report paraconids on dp_3s in her study of *Victoriapithecus*. Paracristids extend from the protoconids of all dp_3s and outline the buccal border of the somewhat triangular trigonid basin. The mesial extension of the buccal surface of the protoconid and the paracristid extending from it provides a shearing surface for the slightly concave distolingual surface of dc^1, although a honing facet is not discernible on dp_3. The talonid basin is slightly lower than the trigonid basin and is present in all species examined. The distal fovea, however, is reduced and variable in its development among the present sample of guenons. For example, it is present in *C. aethiops* as a nearly circular depression distal to the postcristid, and in *Cercopithecus neglectus* and *C. lhoesti* it is no more than a narrow depression distal to the postcristid. In the other species, the distal fovea is absent.

The dp_4 resembles M_1 in being a truly bilophodont tooth, albeit more narrow than M_1. The protoconid and metaconid are opposite each other and connected by the protocristid, and the postcristid joins the hypoconid and entoconid. The hypoconulid is not present in any of the species studied, including the three *Erythrocebus patas* specimens, although Benefit (1994) found it in one of seven *Erythrocebus patas* dp_4s. The trigonid basin is a well-formed depression bordered by a smooth and continuous marginal ridge from the protoconid to the metaconid in all species. The distal fovea, on the other hand, is a mesiodistally narrow depression never more than a millimeter wide. The mesial and distal cusps are separated by the rather spacious talonid basin, which is accentuated by the buccal and lingual developmental grooves passing between the buccal and lingual cusps, respectively. A similar morphological pattern is present in M_1.

Morphological observations: mangabeys

	n
Lophocebus albigena (gray-cheeked mangabey)	6
Cercocebus torquatus (white-collared mangabey)	3
Lophocebus aterrimus (black mangabey)	2

Incisors (Fig. 4.3, upper teeth)

UPPER. The di^1 presents a smooth labial surface whereas the lingual surface is more concave with mesial and distal marginal ridges and a narrow lingual cingulum. The incisal border is straight in most animals but may also slope distally. The lingual surface wears in a manner similar to that discussed above in the guenons, i.e. the lingual surface becomes deeper, forming a ledge of enamel along the cingulum as the lower incisors come into occlusion on the lingual side of di^1. There is also a slight ledge on the lingual side of di^2. This wear pattern was observed by Baume and Becks (1950, p. 729) in their classic study of the development of the dentition of *Macaca mulatta*. They stated 'The characteristic features of

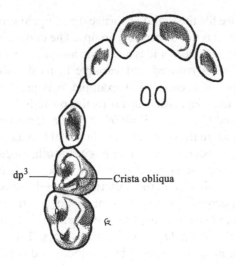

Fig. 4.3. *Lophocebus albigena* male, no lower dentition, occlusal view of deciduous teeth (27 mm).

the occlusal relationship of both deciduous arches consist of interlocking canines, terminal planes forming a mesial step, and a very slight incisal overbite.' The 'mesial step' is the enamel ledge described here, and the 'slight incisal overbite' is characteristic of the majority of the deciduous arches of Old World monkeys, with the notable exception of colobines, where the underbite occlusion is present.

The di^2 is much narrower mesiodistally than di^1 and slopes mesially, terminating in an inclined incisal border. The labial surface is rounded mesiodistally; the lingual surface is bordered by thin mesial, lingual, and distal marginal ridges. There is a diastema separating di^2 from dc^1.

LOWER. The di_1 is a smaller tooth than di_2. The labial surface is convex mesiodistally and the lingual surface is outlined by thin mesial, lingual and distal marginal ridges. There is no distal flange.

Canines

UPPER. The dc^1 projects above the occlusal plane and is triangular, with the base at the cervix. The labial surface is smoothly rounded mesiodistally. The mesial margin is convex toward the crown tip whereas the distal margin inclines mesially toward the tip. There is a narrow pillar dividing the lingual surface from the cusp tip to the lingual cingulum. Distal to the pillar is a slight distolingual depression.

LOWER. The crown tip of dc_1 is slightly lower than the incisors and a little higher than the occlusal plane of the posterior teeth. The tooth is long mesiodistally and narrow buccolingually; this shape seems to be a characteristic of Old World monkey dc_1s. The lingual surface is rounded from just distal to the mesial marginal ridge to the concave distal marginal ridge, at which point the tooth extends distally as a short ledge. There is a narrow, vertical groove on the mesiolingual surface between the mesial marginal ridge and the lingual buttress.

Premolars

UPPER. The dp^3 is somewhat elongated mesiodistally compared with the permanent molars. The mesial two cusps are united by the anterior transverse cristae and the distal two cusps by the posterior transverse cristae. The trigon basin is present as well as the mesial and distal fovea. The position of the paracone and protocone relative to each other is variable in cercopithecids according to Orlosky (1968), who found that these cusps can be opposite each other or that the paracone can be mesial to the protocone. These two patterns are also present in the cercopithecids studied here. The preprotocrista curves buccally around the mesial end of the tooth where it meets the preparacrista, forming the mesial boundary of the mesial fovea. Orlosky (1968) also reported a *Papio anubis* and *Colobus polykomos* with a distal loph that passed mesiolingually from the metacone to the protocone. A similar condition is present on the dp^3 of one *Lophocebus albigena* and one *Cercocebus torquatus*. An obliquely directed crest or loph passes mesiolingually from the metacone to intersect the protocone, forming a crista obliqua, which separates the trigon basin from the distal fovea (talon basin), leaving the hypocone isolated from the other cusps (Fig. 4.3). Benefit (1994) did not find a crista obliqua on dp^3 in *Victoriapithecus* but did find it on dp^4 in 87% of 39 specimens. Moreover, she describes the crista obliqua on both dp^3 and dp^4 of an extant *C. torquatus*, interestingly, the same genus mentioned above with a crista obliqua. In addition, it has been reported as an individual variation in cercopithecids (Colyer, 1936; Remane, 1951). It appears that the crista obliqua, which is normally absent in the deciduous molars of cercopithecids, does occur, albeit rarely. Important studies concerned with crista variation and development are those of Butler (1956) and Hershkovitz (1971). The latter study describes crista diversion and capture that results in new occlusal configurations, whereas the former investigates the ontogenetic processes involved in crista formation. Recently, Swails and Swindler (2002) investigated the early development of occlusal surface structures on the maxillary molars of various primates and concluded that heterochronic differences in the relative rates of growth within the tooth germs and differences in the pattern of initiation of calcification can account for the morphological differences on the occlusal surfaces of primate molars.

As mentioned above, Delson (1973) describes a structure that resembles the primitive parastyle on the dp^3 of cercopithecids. This structure is not present in the guenons examined here, but in *Cercocebus torquatus* a small cusplet is present on the mesiobuccal aspect of the preparacrista that certainly resembles the parastyle. In fact, a slight, narrow developmental groove extends across the preparacrista onto the buccal surface of the paracone just distal to the cusplet (parastyle) which might indicate its independent ontogenetic development.

The buccolingual dimension of dp^4 is wider mesially than distally because the metacone and hypocone cusps are set close together. The mesial and distal cusps are opposite each other and connected by anterior and posterior transverse cristae, forming the characteristic bilophodont pattern of Old World monkey molars. As in dp^3, the trigon basin is flanked by the mesial and distal fovea, and the buccal and lingual developmental grooves are well formed, passing between the mesial and distal moieties of dp^4. As Delson (1973, p. 211) observed, dp^4 and M^1 are so much alike in size and morphology 'that an isolated dp^4 in a small sample of a fossil or recent monkey could easily be identified as an M^1 unless it was abnormally narrow.'

LOWER. The dp_3 is elongated mesiodistally and the protoconid is mesial to the smaller metaconid. The two cusps are set close together and connected by the protocristid, forming the distal wall of the mesial fovea that faces toward the mesiolingual side of the tooth. The mesiobuccal portion of the elongated protoconid provides a shearing surface for the upper canine (Fig. 4.2). There is a slight cuspulid at the mesial end of the preparacristid in *Cercocebus aterrimus* that might be a paraconid. The hypoconid is slightly mesial to the entoconid and the two cusps are connected by the posterior transverse cristid.

The dp_4 is bilophodont, possessing two mesial and two distal pairs of cusps that are opposite each other and connected by transverse lophids. The mesial and distal fovea are narrow; the trigonid basin is somewhat more spacious. Buccal and lingual developmental grooves are present, separating the tooth into mesial and distal moieties. The buccal groove is deeper than the lingual. The dp_4 is similar to M_1 except for being narrower and not as bulbous.

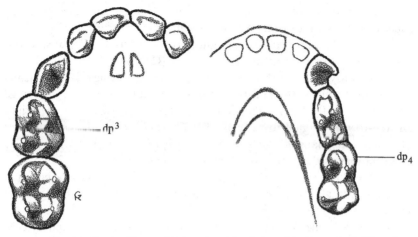

Fig. 4.4. *Trachypithecus cristata* male, occlusal view of deciduous teeth (18 mm).

Subfamily Colobinae

Morphological observations

	n
Piliocolobus badius (red colobus)	10
Trachypithecus cristata (silvered leaf monkey)	18
Rhinopithecus roxellana (golden snub-nosed monkey)	4

Incisors (Fig. 4.4)

UPPER. The di[1]s in these colobines are similar in that they have smooth, mesiodistally rounded labial surfaces. The lingual surface is concave and surrounded by well-formed mesial, occlusal and distal marginal ridges. A lingual cingulum extends across the cervical portion of the tooth. The occlusal marginal ridge is horizontal mesially before sloping downward as it passes distally to become continuous with the distal marginal ridge. There is a narrow pillar on the lingual surface, in unworn teeth, extending from the lingual cingulum to the occlusal marginal ridge in *T. cristata* and *P. badius*. This pillar is also present in *Colobus polykomos* (Orlosky, 1968).

The di[2]s are caniniform in having the mesial and distal marginal ridges incline toward each other to form a point at the apex of the crown. There is a cingulum passing around the lingual surface of the tooth. The lingual surface is slightly convex near the middle of the crown, resulting in narrow concave surfaces on either side. There is a narrow diastema between di[2] and dc[1] in one *T. cristata*; otherwise, this space is absent.

LOWER. The di_1s of *P. badius* are different from those of the other two species in having the distal portion of the incisal border inclined distocervically rather than presenting a straight horizontal surface. This distocervical slope is also present in *C. polykomos* (Orlosky, 1968). The lingual surface is asymmetric: the mesial marginal ridge swings distocervically across the lingual surface of the tooth to intersect the distal marginal ridge at a point slightly distal to the center of the tooth. Orlosky (1968) describes a well-formed lingual fossa in di_1 for *C. polykomos* and Benefit (1994) refers to a lingual sulcus in di_1 for *Victoriapithecus*. A somewhat similar morphology is described for di_1 in *Victoriapithecus* (Benefit, 1994). The lingual surface of di_1 in the other taxa is oriented vertically rather than obliquely. In all taxa, the lingual surface is slightly concave, being a true vertical sulcus in *R. roxellana*.

The di_2 is somewhat caniniform, particularly in *P. badius* and *T. cristata*. In all taxa, the distal margin slopes rather sharply to terminate as a small distal tubercle or flange that is separated from the distal margin of the crown by a narrow labial groove and lingual sulcus. The distal flange is reported by Orlosky (1968) for *C. polykomos*, and Benefit (1994) describes the distal flange in *Victoriapithecus*. Moreover, she considers this a characteristic of the di_2 of extant Old World monkeys. The lingual surface is slightly convex mesially and forms the lingual sulcus distally.

The frequency of incisor occlusal variability in juvenile leaf-eating monkeys is depicted in Table 4.1. The monkeys listed in this table had a complete deciduous dentition or M1 had just emerged above the alveolar bone. According to eruption data from laboratory-raised Old World monkeys, this would indicate that the animals were between six and eighteen months old. This information suggests that monkeys with underbite develop the condition early in life and maintain it as adults (Swindler, 1979). This would seem to suggest a genetic component in the underbite of leaf-eating monkeys.

Canines

UPPER. The crown tip of dc^1 is centrally placed and extends above the occlusal plane. The mesial and distal marginal ridges slope mesially and distally with the distal ridge less inclined and slightly longer than the mesial ridge. The labial surface is convex and smooth; a crest or pillar passes approximately through the middle of the lingual surface from the crown tip to the lingual cingulum. A narrow but distinct vertical groove is present between the mesial marginal ridge and the lingual pillar while the distolingual surface continues smoothly to the distal marginal ridge. The lingual cingulum is a narrow, distinct ledge in all taxa. There is no development of a distal tubercle or flange in these colobines or in *C. polykomos* (Orlosky, 1968).

LOWER. The dc_1 tip is about on the same occlusal plane as the lower incisors but slightly above the lower premolars. The tooth is long mesiodistally and narrow buccolingually; this shape, as mentioned, is characteristic of the dc_1 of extant cercopithecids. In addition the dc_1 of these colobines is similar to that of other cercopithecids in possessing a mesiodistally rounded buccal surface, a lingual pillar extending from the cusp tip to the lingual cingulum, a distolingual sulcus, and a distal marginal extension of a heel or ledge.

Premolars

UPPER. The dp^3 is bilophodont with anterior transverse cristae connecting the paracone and protocone mesially and posterior transverse cristae passing between the metacone and hypocone distally. In general the paracone is mesial to the protocone, but there is some variation regarding the position of these cusps, as indicated by Orlosky (1968). Often by the time M^1 is erupted the protocone is worn down, creating a narrow ledge where the protocone had been. The preparacrista extends mesially, and where it turns lingually to intersect the preprotocrista there may be a slight stylelike process that resembles the primitive parastyle. Among this sample of colobines it is best developed, albeit still small in *T. cristata* and absent in *P. badius*. This cusplet is present among the cercopithecines (see p. 50) and is found among other cercopithecids (Delson, 1973). The size and depth of the mesial fovea varies among these taxa from very narrow and shallow in *P. badius* and *T. cristata* to somewhat larger and deeper in *R. roxellana*. The distal fovea is quite narrow in all taxa. There is a trigon basin in all taxa.

The dp^4 is larger than dp^3 and has a different occlusal morphology, in that the width between the paracone and protocone is greater and the two cusps are always opposite each other in the taxa examined in this study. The mesial and distal fovea as well as the trigon basin are larger than those of dp^3. The median lingual and buccal clefts are slightly wider and deeper than those of dp^3, resulting in a more bilophodont tooth, i.e. a tooth that resembles M^1 more than it does dp^3.

LOWER. The dp_3, as in cercopithecids in general, is a narrow, mesiodistally elongated tooth with the protoconid positioned mesial to the metaconid. The metaconid may be closely approximated to the distolingual surface of the protoconid or separated from it by a narrow developmental groove. The mesial fovea is little more than a slight, triangular depression on the mesiolingual surface of the protoconid. A small cusplid (paraconid) is frequently (80%) present at the mesial end of the paracristid where it intersects the premetacristid. The distal fovea is absent most of the time; when present, it is no more than a slit distal to the posterior transverse cristids. Delson (1973, p. 178) also mentioned the reduction of the distal fovea in the dp_3 of colobines as 'at least a tendency if not a standard feature of the colobine dp_3.'

The dp_4 is bilophodont and resembles M_1 more than it does dp_3. Both mesial and distal pairs of cusps are connected by transverse cristids and separated by a rather spacious talonid basin. The median buccal notch is fairly deep as is the median lingual notch, and both add to the separation between the mesial and distal cusps. The mesial fovea is a small circular depression just distal to the mesial marginal ridge, while the distal fovea is on a lower plane than either the mesial fovea or the talonid basin and appears as a narrow cleft between the posterior transverse cristids and the distal marginal ridge. Indeed, in the sample of *R. roxellana* the distal fovea is absent.

Great apes

Casts were made of several members of the three genera of great apes; unfortunately, no casts were made of the deciduous teeth of hylobatids.

Family Pongidae

Morphological observations

	n
Gorilla gorilla gorilla (western lowland gorilla)	2
Pongo pygmaeus (orangutan)	2
Pan troglodytes (chimpanzee)	8

The deciduous dental formula of the Pongidae is:

di^2-dc^1-dp^2/di_2-dc_1-dp_2

Incisors (Figs. 4.5 and 4.6)

UPPER. Unfortunately, the upper incisors are worn in all specimens; even so, there is some information obtainable. The di^1 is wide mesiodistally,

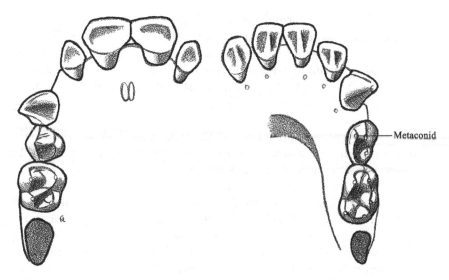

Fig. 4.5. *Pongo pygmaeus* male, occlusal view of deciduous teeth (44 mm).

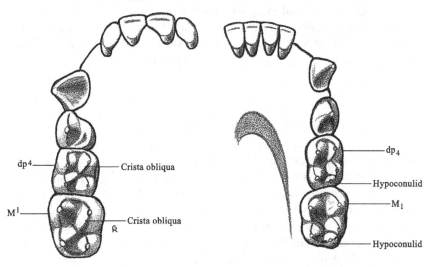

Fig. 4.6. *Gorilla gorilla* male, occlusal view of deciduous teeth (59 mm).

being convex around the labial surface and more concave on the lingual surface. The distal and mesial marginal ridges are present, outlining the lingual surface and passing onto the lingual cingulum. The latter structure is not pronounced but is present in the three taxa. A central tuberculum dentale is present in the unworn condition, passing from the lingual cingulum to almost the incisal margin.

The di^2 is much narrower than di^1 in all animals. In *Pan* and *Gorilla*, however, it is more similar morphologically to di^1 than it is in the orangutan. In the orangutan, di^2 is quite caniniform having inclined mesial and distal marginal surfaces ending in a point at the crown apex. Even allowing for the worn condition of this tooth, it is still possible to see evidence of the mesial, distal and lingual marginal ridges in di^2.

LOWER. Of the two gorilla specimens, one has quite worn incisors and the other has permanent incisors. The di_1 is always narrower than di_2; of the three taxa examined here, the gorilla has the narrowest di_1 even when allowing for the wear mentioned above. The lingual surface of di_1 possesses nearly vertical mesial and distal marginal ridges in *Pongo* and *Pan*. A well-formed cingulum with a single tuberculum dentale extending to the incisal ridge is present in *Pongo*. In *Pan*, the tuberculum dentale is present but not as well developed as in *Pongo*. The lingual surface is too worn in the *Gorilla* to determine whether it is present or not, although a small lingual cingulum is present.

The mesial and distal marginal ridges are less vertical in di_2 and the lingual ridge is inclined upward from distal to mesial. An obvious tuberculum dentale is present in *Pongo* but only slightly discernible in *Pan*. The di_2 of *Gorilla* is too worn to determine this feature.

Canines

UPPER. The dc^1 projects above the occlusal plane in all taxa, particularly in *Gorilla*. The mesial marginal ridge curves slightly distally as it approaches the apex of the cusp and the distal marginal ridge inclines mesially toward the cusp apex, the result being a rather high, pointed cusp apex. There is a narrow lingual ridge passing around the base of the tooth in all taxa. The lingual surface of dc^1 in *Gorilla* has a narrow mesiolingual fossa that may be present in *Pan*, whereas in *Pongo* there is an additional distolingual fossa. The labial surface is smooth and curves mesiodistally. There is a diastema between dc^1 and di^2.

LOWER. The crown of dc_1 projects above the occlusal plane in all taxa. There is a distal ledge from which the distal marginal ridge curves upward to the cusp apex where the mesial marginal ridge passes downward and mesially to contact the lingual ridge that continues around the lingual surface to the distal ledge. The labial surface of dc_1 is smooth and rounded mesiodistally. A diastema is always present (in this sample) between dc_1 and di_2 and may be present between dc_1 and dp_3.

Premolars

UPPER. The dp^3 has two cusps, a large paracone and a smaller protocone. In *Pan* and *Gorilla* these cusps are separated by a deep longitudinal developmental groove. Whereas this groove is present in *Pongo* it is not as deep and there is a transverse crest connecting the paracone and protocone. Jorgensen (1956) described a similar crest in the human upper first deciduous molar and called it the central ridge; later Kraus, Jordan and Abrams (1976) also reported a crest between these two cusps in human upper first deciduous molars and referred to it as the oblique ridge. Jorgensen (1956) reported the presence of the metacone as being variably developed in human upper first deciduous molars as well as the occasional appearance of the parastyle. Neither of these structures is present in this sample, although, in one gorilla, the paracone has a slight transverse groove on its lingual–occlusal surface that just passes over onto the buccal surface of the cusp. This might be an incipient metacone. A slight cingulum is discernible extending along the lingual surface of the protocone in both *Gorilla* specimens.

The dp^4 has four cusps: paracone, protocone, metacone and hypocone. The crista obliqua connects the protocone and metacone and is well developed on all dp^{4}s. Swarts (1988) describes the crista obliqua (post-protocrista) on both dp^3 and dp^4 in his study of orangutan deciduous premolars. The complicated wrinkle pattern so characteristic of the occlusal surface of the permanent molars of orangutans is also present in dp^3 and dp^4. In addition, the occlusal surface presents a rather spacious trigon basin separating a narrow mesial fovea and a somewhat wider distal fovea (talon basin). A lingual cingulum is present in *Gorilla* extending mesially from the hypocone to the mesial surface of the protocone. A cingulum is present in *Pan* only on the lingual surface of the protocone but this structure is absent in *Pongo*. This is essentially the same morphological pattern that is present on M^1 in these three species (Fig. 4.6).

LOWER. The dp$_3$ has four cusps in these taxa, although early attrition tends to wear down the talonid region of the tooth, making it difficult to identify the hypoconid and entoconid. Four cusps are the most common number found in humans, although a rudimentary fifth cusp (hypoconulid) is reported by Jorgensen (1956). The hypoconulid is not present in the taxa examined here. The protoconid is large and connected by the protocristid to the more distally positioned metaconid. The proximity of the cusps and the distal position of the metaconid to the protoconid 'probably represent primitive features in m$_1$ inf' (Jorgensen, 1956, p. 99).

The paracristid extends mesially from the protoconid and continues in a lingual direction to form the mesial border of a small fovea anterior. At this position on the paracristid a small stylid may be present. Jorgensen (1956) has reviewed the literature pertaining to this stylid rather carefully since it has been interpreted differently by various authors through the years. His conclusion is worth quoting in full, since it has broad implications for the field of dental anthropology.

> The present author is of the opinion that no documentation has so far been furnished for the presence of a paraconid on m_1 in any hominids, and that the claims made in the literature to this effect are due to a somewhat too zealous search for primitive features in this tooth.
>
> (Jorgensen, 1956, p. 101)

The buccal surface of the protoconid is slightly convex mesiodistally and serves as a contact surface for the upper canine (distal half) as the teeth come together in occlusion.

The hypoconid is smaller than the protoconid and on a lower plane. The entoconid is the smallest of the cusps. The talonid basin is within the crests connecting these cusps. There is no fovea posterior. A cingulid is present, albeit slight, along the buccal surface of the protoconid in *Gorilla*. There is no indication of cingulids in the other two taxa.

It is well known that a strong morphological resemblance exists between dp_4 and M_1 among primates. Among hominoids, both teeth have five major cusps: protoconid, metaconid, hypoconid, entoconid, and hypoconulid (Fig. 4.6). The hypoconulid may be reduced on these teeth but it is rarely, if ever, completely absent (Jorgensen, 1956; Swarts, 1988; Swindler, Emel and Anemone, 1998). In dp_3 the protoconid is mesial to the metaconid in *Pan* and *Gorilla*, but it is almost directly opposite the metaconid in *Pongo* (Fig. 4.5). In all species, the two cusps are connected by the protocristid. The hypoconid is mesial to the entoconid and the hypoconulid is either centrally placed or is somewhat to the right. The average sequence of cusps according to size is: metaconid = hypoconid > protoconid > entoconid > hypoconulid in *Gorilla*, *Pongo* and *Pan*.

The groove pattern on all dp_4s is the *Dryopithecus* or Y-5 groove and cusp pattern, i.e. contact between the metaconid and hypoconid with a groove passing from the lingual side of the tooth between the metaconid and entoconid to the talonid basin where it bifurcates into a mesial groove passing between the protoconid and hypoconid and a distal groove passing between the hypoconid and hypoconulid (see Chapter 7 for a more complete discussion of the Y-5 pattern). It is interesting to note that the sequence for *Homo* is: entoconid > metaconid > protoconid > hypoconid > hypoconulid (Jorgensen, 1956). The enlargement of the entoconid *pari passu* with the shifting occlusal groove patterns among the cusps ultimately resulted in the formation of the + groove pattern common today on the lower molars of *Homo sapiens* (Gregory and Hellman, 1926).

The trigonid basin is positioned between the protocristid distally and the mesial marginal ridge and is slightly above the larger talonid basin that is situated distally to the mesial cusps surrounded by the hypoconid, entoconid, and hypoconulid. There are no anterior or posterior foveae, although there is a small pit distolingual to the hypoconulid in *Gorilla* (Fig. 4.6). The occlusal surface displays the wrinkle patterns common in the permanent molars of the orangutan. A narrow buccal cingulid extends from the side of the hypoconid across the buccal groove to pass along the side of the protoconid for a short distance in *Gorilla*. There are no buccal cingulids in the other taxa.

As mentioned earlier, there are few investigations of the deciduous dentition of non-human primates, particularly comparative analyses of the deciduous teeth. The study of Swarts (1988) is an exception to this, and interestingly, concludes that orangutans and humans, with respect to their deciduous teeth, share the greatest number of apomorphies among hominoids.

5   *Prosimii*

Superfamily Lemuroidea

Present distribution and habitat

Madagascar is the fourth largest island in the world, situated off the southeast coast of Africa in the southwestern Indian Ocean. Today it is separated from Africa by the 800 km wide Mozambique Channel. The most recent evidence suggests that Madagascar parted from Africa 150–160 million years ago along with Antarctica and India. Much later, Madagascar and India broke apart, about 88 million years ago, leaving India to drift northeastward toward Eurasia while Madagascar moved south-southwestward to reach its present position along the east coast of Africa during the Late Cretaceous, about 70–80 million years ago (Krause, Hartman and Wells, 1997). The question of how and when prosimian primates arrived on Madagascar is still being debated (for excellent discussions of these intriguing questions see Krause *et al.*, 1997; Simons, 1997). The only indigenous primates on the island today are prosimians belonging to the superfamily Lemuroidea. There are two seasons on Madagascar: the hot, wet season and the cooler, drier season. Temperatures generally range from 10 to 29 °C. The island is essentially mountainous with most species living in the lower forested areas, although *Indri* is found from sea level to 1300 m (Wolfheim, 1983).

The prosimians of Madagascar are often compared with Darwin's finches, which have lived and diversified in isolation on the Galápagos Islands for millions of years. The Malagasy prosimians have also lived in relative geographical and ecological isolation, particularly from competition with other primates, for millions of years, during which time they have diversified into several dietary and locomotor habitats. There are five families of prosimians living today on Madagascar and some of the Comoro Islands that stretch northwest toward Africa: Lemuridae, Cheirogaleidae, Indriidae, Lepilemuridae (or Megaladapidae) and Daubentoniidae. There are fourteen genera among these five families, which have locomotor habits ranging from vertical clinging and leaping to arboreal and terrestrial quadrupedalism; their activity rhythms include nocturnal, diurnal and cathemeral species.

The dietary adaptations of prosimians include species that consume various quantities of leaves, bamboo plants, gums, seeds, fruits, and insects. Some have specialized in eating essentially one type of food, e.g. insects, whereas others have included several types of food in their menu. As we survey the anatomy of the teeth of living primates, in this and the following chapters, one emphasis will be on attempting to identify possible associations between tooth morphology and diet. Complications arise since morphology can only set broad limits on what a diet can be, and in many cases, primates living in different regions may differ in diet despite similarities in their morphology (Scheine and Kay, 1982).

At the same time, it is well to remember that the type of food a primate species is eating today is not necessarily the same food the species has been eating throughout its evolutionary history. As Richard (1985, p. 190) so clearly stated 'environments do change, sometimes quite rapidly, and an animal is not necessarily specialized to eat the foods we see it feeding on now: a species can change its diet (within limits, at least) faster than it can change its teeth.' Nevertheless, comparative investigations of primate tooth morphology (Kay, 1975, 1977; Kay, Sussman and Tattersall, 1978; Hylander, 1975; Swindler and Sirianni, 1975; Seligsohn, 1977; Seligsohn and Szalay, 1978; Maier, 1980; Happel, 1988; Lucas and Teaford, 1994; Yamashita, 1998) have demonstrated certain basic features that vary with respect to diet, as we will see in the following chapters.

General dental information

A study by Shellis *et al.* (1998) of molar enamel thickness among primates found that, in prosimians, enamel is thinner for a given tooth size or body mass than it is in anthropoids. The prosimian sample included the upper and lower permanent molars of several taxa of Lemuroidea and Lorisoidea as well as *Tarsius*. A notable exception was *Daubentonia* where M_2 'clearly had extremely thick enamel, although there were too few prosimian M_2 to allow quantification' (Shellis *et al.*, 1998, p. 512). Since *Daubentonia* has small molars and thick enamel Shellis *et al.* (1998, p. 519) suggested that 'Possibly, during the course of its evolution from larger ancestors, only molar size, but not enamel thickness, has become reduced.'

A study of enamel microstructure in Lemuridae (Maas, 1994, p. 221) demonstrated that lemur taxa can have 'all three prism patterns in their

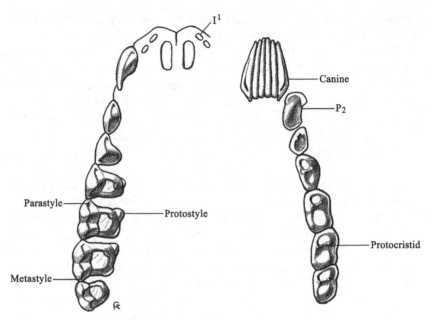

Fig. 5.1. *Eulemur macaco* male, occlusal view of permanent teeth (40 mm).

mature enamel'; further, she cautioned that homoplasy may explain some of the features in the enamel microstructure of lemurids.

The contemporary genera of lemurs, with the exception of the aye-aye (*Daubentonia*), display a unique specialization of the incisors and lower canines. Ankel-Simons (1996), however, has shown that the position of the deciduous incisors of the aye-aye strongly suggests that a tooth comb was also in the ancestry of this lemur. The upper incisors are small with I^1 and I^2 set close together, leaving a wide diastema between the two central incisors that receives the procumbent lower incisors and canines. The mandibular incisors are long, narrow teeth projecting nearly straight forward (procumbent) from the anterior portion of the jaw and, with the procumbent lower incisiform canines, form the dental comb (Fig. 5.1). This is one of the more important features distinguishing lemurs and lorises from other primates. The origin and function of the dental comb has been debated for years. Several early studies have shown that it is used for grooming in living lemurs (Lowther, 1939; Roberts, 1941; Buttner-Janusch and Andrew, 1962). More recently the studies of Szalay and Seligsohn (1977), Rose, Walker and Jacobs (1981), and Rosenberger and Strasser (1985) also point toward grooming as being responsible for the origin of the dental comb. Eaglen (1980, 1986, p. 185) in his studies of the origin and function of the dental comb, came to the following conclusion.

> Although toothcomb size variations among extant strepsirhines are more readily interpreted in terms of gum feeding and bark scraping than they are in terms of grooming, anterior dental morphology as a whole is more easily explained by the grooming hypothesis when existing models of toothcomb origins are considered.

Martin (1975), on the other hand, has pointed out that the dental comb is used by *Phaner, Microcebus* and *Euoticus* to scoop fresh gums and resins from trees. Moreover, he believes that the use of the dental comb for obtaining gum was an important factor in the evolution of this structure and grooming could have been developed as a secondary feature.

Le Gros Clark (1971, p. 93) described the early Miocene genus *Progalago* of East Africa as having a wide space between the upper central incisors and stated that the lower incisors and canines 'were also procumbent.' A fossil contemporary of *Progagalo* is *Komba*, from the early to middle Miocene of western Kenya, which is described by McCrossin (1992, p. 219). He states that the alveoli of I_1, I_2 and the canine are 'mesiodistally compressed and procumbent.' An even earlier fossil prosimian, *Adapis parisiensis* from the late Eocene of France, although not possessing a true dental comb, does have lower canines functioning as incisors and together the six teeth operated together as a single unit or dental scraper (Gingerich, 1975). Gingerich further suggests, as did Martin (1975), that the dental scraper functioned initially to scrape resin or gum from the bark of trees.

As is often the case in evolutionary studies, the original function (adaptation) of a structure (e.g. procumbent lower incisors and canines) may have little if anything to do with the function it presently performs. If, however, the present character is considered within the form–function complex of Bock and von Wahlert (1965, p. 276), where a particular feature is the result of different selective forces and is thus a compromise between these forces (in this case, grooming and scraping), then 'A feature with one or a few utilized functions would be better adapted to each selection force as fewer faculties enter the compromise.' It appears that today these procumbent anterior teeth are used as a tooth scraper both for scraping and tearing bark for resin and gum (nutrition) and as a dental comb for grooming and cleaning one's own fur as well as one's neighbors' fur (social). In this scenario, the repeated use of the lower anterior teeth to scrape or groom was a compromise between these selective forces.

The upper canines are long, trenchant teeth with distally curving crowns. Prosimians have little canine sexual dimorphism (Plavcan and Van Schaik, 1994) that may be attributed to their lifestyles, since males and females compete in similar ways. However, these authors note that there is still much to be learned concerning the variation in prosimian canine size. There is a slight diastema between the upper canines and P^2.

The premolars increase in size and complexity mesiodistally. In both jaws, the second premolars are caniniform, much more so in the upper jaw than in the lower. During mastication the second premolars function somewhat like canines, piercing and tearing the food, and in occlusion, P_2 glides against the distolingual surface of the upper canine. The third premolars are less caniniform, but still essentially unicuspid teeth. The fourth premolar is broadened buccolingually in most genera and has a well-developed crest connecting protocone and paracone. In *Hapalemur* P^4 and P_4 are molariform. The molariform P_4 is unique to *Hapalemur* among extant members of the Lemuridae, but is found in the Galagidae.

The upper molars grade from trituercular to quadritubercular in shape. The lingual cingulum is usually prominent and in some species a protostyle and hypocone are variably developed, a condition also mentioned by Schwartz and Tattersall (1985). The evolutionary history of the hypocone has been considered of some taxonomic importance in primates since the studies of Gregory (1922). The hypocone has been classified either as a true or pseudohypocone depending on whether it develops from the cingulum or the distal surface of the protocone, respectively. Among the notharctines of the North American Eocene the hypocone appears as a pseudohypocone, whereas among adapines it appears as a development from the cingulum, a true hypocone. These differences have been used in the past to include or exclude primate taxa from certain phylogenetic lineages; however, Simpson (1955), Butler (1956) and Hershkovitz (1971) have rejected this distinction, and the hypocone is considered by them to be a derivative of the lingual cingulum independent of the protocone. Moreover, Butler (1992) has traced the history of the hypocone back to early placental mammals of the Cretaceous where it appears as an enlargement of the cingulum distal to the protocone, in such forms as *Gypsonictops* and *Protungulatum*. Butler (1992) also notes that the hypocone has occurred independently in more than one line of early mammals, and Van Valen (1982, p. 306) believes the hypocone 'has evolved separately many times' resulting in parallel evolution. A process, I might add, that is not unusual in the evolution of the mammalian dentition.

A detailed investigation of primate tooth-germ development (Swails, 1993, p. 240) contributes important and appropriate information for this discussion. She states 'The relative positions of the cusps on the crown result, in large part, from the interaction of differential rates of growth and the timing and pattern of initiation of calcification.' Thus, minor changes in any of these factors can alter the relative position of the cusps on the calcified crown and thereby provide insights into how morphological alterations can arise in evolution. As the late Bertram Kraus (1964, p. 206) once said, 'the aspect of the dentition that is critical in evolutionary interpretation is its *morphogenesis* rather than its final morphology.'

The lower molars possess a trigonid and talonid. The paraconid is absent in all genera; the hypoconulid is variable. The protoconid and metaconid are set close together and connected by an oblique transverse crest, the protocristid. In most cases the protoconid is mesial to the metaconid, and buccal cingulids are variably developed.

Seligsohn (1977) presented an extensive investigation of molar adaptations in living and subfossil strepsirhines (22 genera) that is still one of the most complete surveys of the relationship between form and function in the dentition of these primates. He examined the anatomical details of the upper and lower second molars, since they occupy the middle of the molar region where much of the food is chewed, in order to study diet-related molar adaptations. One of several conclusions was that 'The morphological and metrical data appear to demonstrate an often inextricable interrelationship between molar size, heritage and dietary preference as these factors relate to form' (Seligsohn, 1977, p. 105).

Family Lemuridae

Dietary habits

The diet of lemurids is quite varied, consisting of leaves, flowers, fruits, nectar, bark, sap, and the occasional insect (Mittermeier *et al.*, 1994). In most species, the percentages of the different foods eaten vary with the season and habitat. For example, Kay *et al.* (1978) noted that the amount of fruit eaten by *Lemur fulvus rufus* varied from 7% in the dry season to 42% in the wet season. Then there is *Hapalemur*, which eats almost nothing but bamboo plants throughout the year. Indeed, Overdorff, Strait and Telo (1997) found that each of the three species of *Hapalemur* specializes in eating only specific parts of the bamboo plant. These few examples indicate only a few of the complexities facing the field researcher attempting to categorize the diets of primates; however, much progress has been made over the past couple of decades in recording and analyzing dietary information with regard to dental morphology (Yamashita, 1998).

Permanent dentition: I^2-C^1-P^3-M^3 / I_2-C_1-P_3-M_3
Deciduous dentition: di^2-dc^1-dp^3 / di_2-dc_1-dp_3

Morphological observations

	Male	Female
Eulemur macaco (black lemur)	7	8
Eulemur rubriventer (red-bellied lemur)	1	3
Eulemur mongoz (monogoose lemur)	1	2
Lemur catta (ringtailed lemur)	1	2
Varecia variegata (ruffed lemur)	7	1
Hapalemur griseus (gentle lemur)	4	2

Incisors (Figs. 5.1 and 5.2)

UPPER. The upper incisors are quite similar in the species studied. The I^1s are compressed labiolingually and spatulate. The I^2s are smaller than I^1s and there is a wide central diastema between the two I^1s. In *Hapalemur* I^2 is located to the lingual side of the upper canine, which it may touch.

LOWER. The four lower incisors are narrow, long teeth set closely together. They are procumbent and with the canine show the characteristic lemuriform arrangement, the dental comb.

Canines

UPPER. The upper canine is a long, trenchant tooth with a wide base. It curves distally as it rises above the occlusal plane of the other upper teeth. A small distal heel may be present. Rather deep lingual grooves are present on either side of a vertical crest extending from the crown apex to the base.

LOWER. The lower canines are procumbent with the incisors forming the lateral border of the dental comb. They are larger than the incisors and their lateral borders are flared laterally while a narrow, longitudinal groove extends along the occlusal surface.

Premolars

UPPER. The P^2 is the smallest of the three upper premolars in all species examined and may be separated by a diastema from the canine as in *E. macaco* (Fig. 5.1) or set close to the canine as in *H. griseus* (Fig. 5.2). The tooth is compressed buccolingually and is the most caniniform of the three upper premolars. A slight lingual cingulum is present, passing around the base of the paracone. The P^2 of *H. griseus* has an incipient protocone connected to the larger paracone by a narrow occlusal crest. In the present

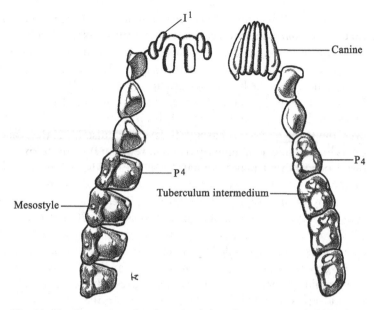

Fig. 5.2. *Hapalemur griseus* female, occlusal view of permanent teeth (30 mm).

sample, *H. griseus* is the only species possessing a protocone, albeit small on P^2. In *E. macaco* P^3 has a prominent paracone with a median lingual ridge extending from the crown tip to the enlarged lingual base. The P^3 is similar in the other lemurs except for *H. griseus* where it has a well-formed protocone. A narrow but perceptible buccal cingulum is present on P^3. The morphology of P^4 is quite variable in these lemurs. In the genera *Eulemur* and *Varecia* there are two cusps, a large paracone and a smaller, lower protocone. In *E. macaco* ($n = 15$) a small but distinct protostyle is observable in 13% of the specimens. A lingual cingulum is present in all species. The P^4 of the gentle lemur *H. griseus* is molariform, possessing, in order of size, paracone, protocone and metacone. Buccal and lingual cingula are prominent structures on P^4 in *H. griseus*. In size, structure, and function P^4 is a molar; indeed, the occlusal surface is nearly as great as that of M^1 and greater than that of M^3 in *H. griseus* (Swindler and Sirianni, 1975). The upper and lower P4s are used to initially puncture-crunch bamboo stems, which are then pulled across these teeth (Petter and Peyrieras, 1970). Thus, as stated by Seligsohn and Szalay (1978, p. 301), it is 'perhaps not surprising that PM4 in *Hapalemur* is highly molariform, and functionally the most derived of the postcanine teeth.'

LOWER. The P_2 is compressed buccolingually and resembles a canine. The tip of P_2 is sharp and projects above the occlusal plane. In occlusion, the tooth glides across the distolingual surface of the upper canine, assuming the function of a true canine. In *H. griseus*, P_2 has a prominent lingual ridge that divides the lingual surface into two facies. P_2 is separated from P_3 by a diastema, which is considered one of several dental synapomorphies linking *Lemur* and *Varecia* (Schwartz and Tattersall, 1985). However, the P_2 of *H. griseus* has a well-developed distal heel that touches P_3 on its lingual side. P_3 is unicuspid and lower than P_2 with sharp mesial and distal borders that may terminate as small stylids. A lingual cingulid is present. P_4 generally has two cusps, the protoconid and metaconid, connected by an interrupted protocristid in *Eulemur* and *Varecia*. There is some variation in cusp number among the species: 28% of *E. mongoz* and 33% of *V. variegatus* have two cusps. In both genera, a small talonid basin extends distally, surrounded by an uninterrupted distal marginal ridge. P_4 in *H. griseus* is molariform as mentioned in the discussion of P^4. The protoconid and metaconid are closely appressed; the hypoconid and entoconid are somewhat separated, creating a rather spacious talonid. A metastylid is connected to the metaconid anteriorly and a tuberculum intermedium is located on the distal slope of the metaconid. The former structure is mentioned by Schwartz and Tattersall (1985) but not the latter cusp. In the present sample, the tuberculum intermedium is present on P_4 in all but one individual. Well-formed trigonid and talonid basins occupy much of the occlusal surface of P_4 in *H. griseus*. In addition, buccal cingulids are limited to the protoconid.

Molars

UPPER. All three molars are essentially tribosphenic, possessing three well-formed cusps, the paracone, protocone, and metacone. Several cristae emanating from these cusps define a shallow trigon basin that receives the hypoconid when the teeth are in occlusion. A prominent lingual cingulum is present on M^{1-2} but absent on M^3. Two variable structures are associated with the cingulum: the protostyle and hypocone. The protostyle is present in all lemurs in the present study except *L. catta*, a condition also found by Schwartz and Tattersall (1985). The hypocone varies from absent in *V. variegatus* and *H. griseus*, to present in the other lemur genera. A similar situation is also reported by Schwarz (1931) and Schwartz and Tattersall (1985). The other species express these traits as follows. In *E. macaco* both hypocone and protostyle are present on M^1; however, on M^2 the protostyle is found but the hypocone is present in only 20% of the sample. Neither characteristic is found on M^3. In *E. mongoz* the protostyle is present on all three molars but the hypocone is found only on M^{1-2}. The protostyle is present on M^{1-2} and the hypocone is present only on M^1 in *E. rubiventer*. The buccal cingulum (ectocingulum) is variably developed in these genera as is the presence of the para-, meso- and metastyles.

LOWER. The lower molars are somewhat compressed buccolingually and are divisible into the trigonid and talonid basins. The trigonid is slightly higher than the talonid; however, the trigonid lacks the paraconid. The paraconid is present in fossil notharctids of the early Eocene (*Pelycodus*), at least on M_{1-2} (Gregory, 1922). The protoconid lies mesially to the metaconid and both cusps are set close together and connected by the protocristid. The hypoconid and entoconid are more separated and not connected by a transverse crest. The talonid basin is shallow and circumscribed by marginal ridges. A distal heel-like process may be present in some species and I believe this to be a true hypoconulid. Schwartz and Tattersall (1985, p. 17), on the other hand, believe that in *H. griseus* 'What appears as a talonid heel on M_3 is, in fact, a large distolingually distended entoconid.' In the present sample of *H. griseus* a tuberculum intermedium is present on M_1 in all specimens; on M_2 it is absent in one animal and on M_3 it is present in only one case. In all molars it is closely associated with the distal surface of the metaconid. In their study of shearing crest development on M_2 in *Lemur fulvus rufus* (primarily folivorous) and *L. fulvus mayottensis* (primarily frugivorous), Kay *et al.* (1978, p. 124) found that, when size differences were corrected for, *L. fulvus rufus* had significantly longer shearing crests than *L. fulvus mayottensis*, and this result led them to conclude that 'there is a general correspondence between the amount of molar shearing and the amount of leaves in the diet.'

Odontometry (Appendix 1, Tables 1 and 2, p. 166)
Sample size is small for each species and only measurements of *Varecia variegatus* are presented in Appendix 1. The upper incisors are small and subequal. The canine is large and blade-like. Of the three premolars, P^4 is wider and longer than the others, although its crown height is slightly less than that of P^3. The upper molar size sequence is $M^1 < M^2 > M^3$.

The lower canine is the largest tooth in the procumbent dental comb. P_2 is caniniform and larger mesiodistally than the others while P_4 is widest buccolingually. The lower molar size formula is $M_1 > M_2 > M_3$. Measurements were not taken of the teeth of *H. griseus* since the sample is small. In molar size relations M1 and M2 are about equal in mesiodistal length and both are longer than M3 in both jaws. A recent paper by Sauther, Cuozzo and Sussman (2001) presents a detailed study of metric and morphological data from a single living wild population of *Lemur catta.*

Family Lepilemuridae

Dietary habits

The diet of *Lepilemur* is folivorous, consisting mainly of leaves, flowers and seeds (Mittermeier *et al.*, 1994). As with other phytophagous primates, the sportive lemur cannot digest cellulose and must depend on intestinal bacteria for digestion. In addition, they are known to reingest their feces, as do rabbits.

Permanent dentition: I^0-C^1-P^3-M^3 / I_2-C_1-P_3-M_3
Deciduous dentition: di^2-dc^1-dp^3 / di_2-dc_1-dp_3

Morphological observations

	Male	Female
Lepilemur mustelinus (sportive lemur)	6	11

Incisors (Fig. 5.3)

UPPER. The permanent incisors are absent, but are present in the deciduous dentition. Permanent incisors are reported in three of 100 specimens of the sportive lemur (Miles and Grigson, 1990).

LOWER. The lower incisors are procumbent.

Canines

UPPER. The upper canines are long and trenchant and curve distally. They are compressed buccolingually. A lingual pillar separates the lingual surface into mesial and lingual grooves. A distal heel or step is in contact with P^2. The apex of the canine is well above the occlusal plane of the upper teeth.

LOWER. The lower canine is in line with the lower procumbent incisors, albeit slightly larger. There is also a slight lateral flare to the canine.

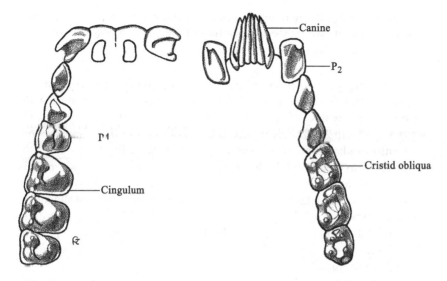

Fig. 5.3. *Lepilemur mustelinus* male, occlusal view of permanent teeth (23 mm).

Premolars

UPPER. P² has a single large paracone that is higher than the crowns of the other two premolars. The tooth is compressed buccolingually, and is therefore caniniform. A complete cingulum passes around the lingual aspect of the tooth. P³⁻⁴ are bicuspid, with the paracone larger than the protocone. A preparacrista passes mesially to the mesial marginal ridge. Both premolars possess buccal and lingual cingula. There may be para- and metastyles present. The paracone–protocone separation is slightly wider in P⁴ than in P³.

LOWER. P₂ is compressed buccolingually with a very high protoconid. A lingual pillar passes down the lingual surface from apex to base. P₂₋₃ have protoconids only; P₄ possesses a small metaconid distal to the protoconid. P₄ has a small buccal cingulid.

Molars

UPPER. The paracone, metacone, and protocone are well developed on M¹⁻³. In the present sample, a hypocone is not present and it is not mentioned by Schwartz and Tattersall (1985), although James (1960) found it present on M¹⁻² but not on M³. The trigon basins are wide and shallow. A lingual cingulum is present, particularly around the distolingual surface of M^{1-2}. A complete buccal cingulum is present on M^{1-3}. Metastyles may also be present on M^{1-3}.

LOWER. M_{1-2} have four cusps and M_3 has a fifth cusp, the hypoconulid. James (1960) describes a fifth cusp for *Lepilemur* without calling it the hypoconulid, whereas Schwartz and Tattersall (1985) do not mention the hypoconulid in *Lepilemur*. In fact, they do not believe that *Lepilemur* has an entoconid; rather, they identify the cusp immediately distal to the metaconid as a metastylid. I have examined the 17 *Lepilemur* specimens in this sample and believe the cusp separated by an obvious developmental groove from the metaconid is a true entoconid and the cusp distolingual to it is the hypoconulid. Seligsohn and Szalay (1978), although not mentioning a hypoconulid, show an excellent illustration (their Fig. 1) of the dentition of *Lepilemur* that, in my opinion, depicts a hypoconulid projecting from the distal end of M_3.

The protoconid and metaconid are set close together and connected by the protocristid. The protoconid is larger and always mesial to the metaconid. The protoconid may be connected by the paracristid to a slight elevation of the cingulid as it passes onto the mesial surface of the tooth, the mesiostylid. A cristid obliqua is present as the talonid part of the centrocristid between the two buccal cusps. M_2 has well developed shearing crests as found in folivorous primates (Kay *et al.*, 1978). When viewed from above, the outline of the buccal half of each molar appears W-shaped.

Odontometry (Appendix 1, Tables 3–6, pp. 167–8)

The upper canines are about equal in size in both sexes. In females they are slightly wider in the buccolingual dimension, although the difference is not significant at the 0.05 level of probability. Of the maxillary premolars P^2 has the largest mesiodistal dimension, while P^4 has the greatest breadth diameter. The size relation of the upper molars is: $M^1 > M^2 > M^3$.

In the mandible, P_3 is the longest and narrowest of the series while P_2 and P_4 are about subequal in their dimensions. The molar formula is: $M_1 < M_2 < M_3$.

It is obvious from Tables 3–6 that the size difference of all teeth between the sexes is small. *t*-Tests show that there were only two cases where sexual dimorphism was statistically significant; the buccolingual diameters of M^1 ($p < 0.04$) and M^2 ($p < 0.04$). In both cases females were larger than males and of the 26 measurements, 17 had slightly higher means in females suggesting, on average, that the teeth of females were slightly larger than those of males.

Family Cheirogaleidae

Dietary habits

The cheirogaleines are primarily gummivorous or frugivorous; however, mouse lemurs (*Microcebus*) are mostly faunivorous (Hladik, Charles-Dominique and Peter, 1980). *Microcebus* eats invertebrates (a variety of insects) in addition to small vertebrates, e.g. frogs. *Cheirogaleus* is mainly a fruit eater that also consumes large amounts of nectar during the dry season (Wright and Martin, 1995). The forked-marked lemur (*Phaner furcifer*) is a specialized gum eater but also feeds on nectar and insect secretions (Nowak, 1999).

Permanent dentition: I^2-C^1-P^3-M^3 / I_2-C_1-P_3-M_3
Deciduous dentition: di^2-dc^1-dp^3 / di_2-dc_1-dp_3

Morphological observations

	Male	Female
Cheirogaleus major (greater dwarf lemur)	1	—
Microcebus murinus (gray mouse lemur)	3	2
Phaner furcifer (forked-marked lemur)	2	—

Incisors (Figs. 5.4–5.6)

UPPER. I^{1-2} are set close together and I^1 is larger than I^2. I^2 is separated from the canine by a small diastema. In *Phaner* I^1 is much longer than I^2 and inclines mesially, a condition not present in other lemurs. The diastema between the two central incisors is wide.

LOWER. I_{1-2} are narrow, procumbent teeth, slightly longer than the lower canine that forms the lateral border of the dental comb. In *Phaner* these teeth are quite elongated and narrow.

Canines

UPPER. The upper canine is a rather robust tooth that curves distally well above the occlusal plane. Its greatest dimension is mesiodistal and there is a slight heel projecting distally. There is a slight lingual cingulum, which is better developed in *Microcebus* than in the other species. In *Microcebus* it terminates as a distal stylid. The canine in *Phaner* is long and dagger-like, curving distally near the tip.

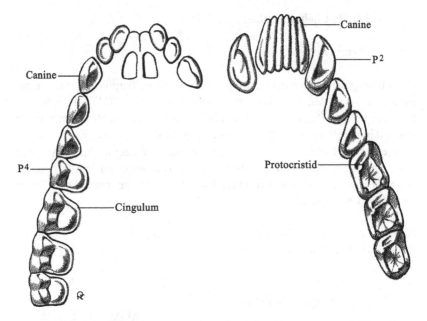

Fig. 5.4. *Cheirogaleus major* male, occlusal view of permanent teeth (20 mm).

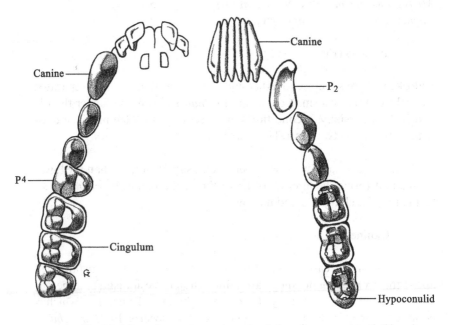

Fig. 5.5. *Microcebus murinus* male, occlusal view of permanent teeth (20 mm).

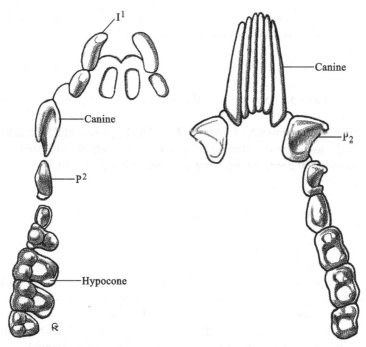

Fig. 5.6. *Phaner furcifer* male, occlusal view of permanent teeth (21 mm).

LOWER. The lower proclivous canine is slightly shorter and wider than the lower incisors. There is a slight longitudinal keel on the lingual surface of each tooth in the dental comb of *Cheirogaleus*; in *Microcebus* there is a slight distal marginal ridge. In *Phaner*, the canine is more robust than the incisors and collectively, this is a very long dental comb.

Premolars

UPPER. P^2 is compressed buccolingually and is caniniform. A slight lingual cingulum is present around the base of the tooth. The crown of P^2 is higher than it is on P^{3-4} in these species; however, in *Phaner* it is much higher than P^{3-4} and is the most caniniform among extant lemurs. Indeed, this unique combination of the long I^1, long, robust upper canine, long stout P^2, and long procumbent dental comb are ideally suited for scraping the bark off of trees to stimulate the flow of gum and then licking it with their long tongues (Hladik, 1979; Fleagle, 1999). P^3 has a single cusp with a vertical pillar on the lingual surface that contacts the lingual cingulum passing around the base of the tooth. P^4 has two cusps, a large paracone and a much smaller protocone. A lingual cingulum skirts around the base of P^4, and para- and distostyles are present.

LOWER. P_2 has a single cusp that is large and caniniform. The apex of the cusp is higher than the occlusal plane of the teeth distal to it. A slight depression occupies the distolingual surface of P_2. P_3 has a single cusp that is lower than the crown of P_2; there is also a small distal heel projecting toward P_4 in *Cheirogaleus*. The P_4 protoconid is larger than it is in P_3, as is the distal heel. Cingulids are present on all three lower premolars in the three species but are usually better developed in *Microcebus* and *Phaner*.

Molars

UPPER. The upper molars are tribosphenic in *Cheirogaleus* and *Microcebus*, and lack any development of the hypocone in the present sample. However, James (1960) and Schwartz and Tattersall (1985) report hypocones in *Microcebus* (M^{1-2}). The hypocone is present in *Phaner* on M^{1-2} in this sample as well as those reported by James (1960) and Schwartz and Tattersall (1985). These findings suggest some degree of variability in the hypocone in extant cheirogaleines. Lingual and buccal cingula are present on M^{1-3} and the buccal cingulum is particularly well developed in *Phaner*. The buccal stylar region is better formed on M^{1-2} than on M^3. Para- and metastyles are present in *Microcebus* on M^{1-2} but only a metastyle is present on M^3. *Cheirogaleus* lacks styles and only the parastyle is present on M^1 in *Phaner*. In addition, however, *Phaner* may have a minuscule protoconule and metaconule on M^{1-2}.

LOWER. The protoconid and metaconid are close together and connected by the protocristid on M_{1-3}. The hypoconid and entoconid are approximately the same size and separated by a longitudinal developmental groove. These cusps are bunodont in *Cheirogaleus*, as reported by Schwartz and Tattersall (1985). Of the three cheirogaleid species, the hypoconulid is present on M_3 only in *Microcebus*, as is also reported by James (1960) and Schwartz and Tattersall (1985). The trigonids are rather spacious on all three molars even though M_3 is shorter than M_{1-2}, a condition also discussed by Schwartz and Tattersall (1985). A buccal cingulid is present on the three molars but better developed on M_{1-2}. Kay et al. (1978) pointed out that cheirogaleids have relatively short shearing blades and low cusps on M_2, a condition generally associated with a primarily frugivorous–gummivorous diet.

Odontometry

The teeth were not measured in the cheirogaleid specimens owing to the small samples. The molar size formulas are presented for the mesiodistal dimensions.

Cheirogaleus: $M_1 > M_2 > M_3 / M^1 = M^2 > M^3$.
Microcebus: $M_1 = M_2 < M_3 / M^1 = M^2 > M^3$.
Phaner: $M_1 = M_2 > M_3 / M^1 = M^2 > M^3$.

Family Indriidae

Dietary habits

Avahi is a vegetarian, subsisting almost exclusively on leaves (Ganzhorn *et al*, 1985) *Indri* also eats leaves, preferably young leaves and shoots, but will add fruits on occasion. Both of these preferences seem to depend on the season. Like *Avahi*, they appear to avoid leaves containing alkaloids (Fleagle, 1999). *Propithecus*, like the other two species, eats leaves or fruit depending on the season.

Permanent dentition: I^2-C^1-P^2-M^3 / I_2-C_0-P_2-M_3
Deciduous dentition: di^2-dc^1-dp^3 / di_2-dc_1-dp_3

There has been debate through the years concerning the dental formula of the deciduous and permanent dentition of indriids. The question is, what is the most lateral tooth in both the deciduous and permanent procumbent tooth comb of these primates, incisor or canine? From his study of juvenile indriid dentitions, Schwartz (1974) claimed the tooth to be a canine. However, in his study of dental variation in Indriidae, Gingerich (1977) found convincing evidence that the lateral permanent tooth is an incisor and that the lower permanent canine has been lost in all genera of Indriidae. Further, he showed that the deciduous lateral tooth is 'almost certainly' a deciduous canine. The deciduous and permanent dental formulas presented here are those suggested by Gingerich (1977).

Morphological observations

	Male	Female
Propithecus verreauxi (sifaka)	3	8
Indri indri (indri)	1	—
Avahi laniger (eastern woolly lemur)	2	—

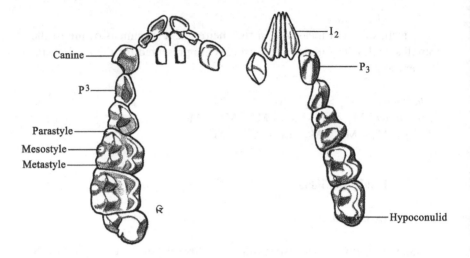

Fig. 5.7. *Propithecus verreauxi* male, occlusal view of permanent teeth (40 mm).

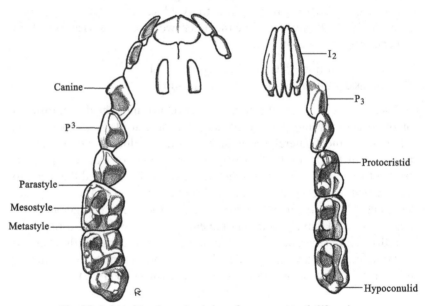

Fig. 5.8. *Indri indri* male, occlusal view of permanent teeth (48 mm).

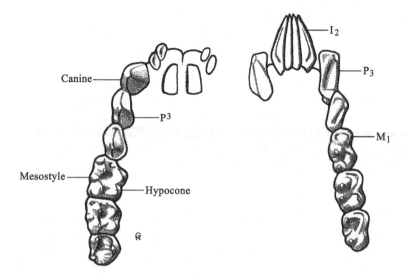

Fig. 5.9. *Avahi laniger* male, occlusal view of permanent teeth (23 mm).

Incisors (Figs. 5.7–5.9)

UPPER. The upper central incisors are separated by a wide diastema except in *Propithecus* where they lean mesially, which reduces the width between the central incisors. In *Propithecus* I^1 has a slightly concave surface. I^2 is narrower than I^1 except in *Avahi* where it is slightly larger.

LOWER. The lower incisors are procumbent as in the Lemuridae. The difference between the Indriidae and Lemuridae is that in the former there are four incisors and no canines in the dental comb whereas in the latter there are four incisors and two canines forming the dental comb (see Gingerich, 1977). The incisors have narrow longitudinal grooves on their labial surfaces.

Canines

UPPER. The canines are large, stout teeth extending above the occlusal plane. In *Indri* a distal stylid passes to the lingual side of P^3 and there is a diastema between the canine and I^2. The canine is compressed buccolingually in *Avahi*, so that it is blade-like with a flange that touches the lingual surface of P^3. There is a mesiolingual groove on the dagger-like canine of *Propithecus*.

LOWER. The permanent lower canines are absent in Indriidae (Gingerich, 1977).

Premolars

UPPER. P^3 has a single, large paracone in the three taxa. In *Propithecus*, P^3 has a paracone crest that separates the occlusal surface into a mesial and distal basin. There is also a complete lingual cingulum. P^3 is more caniniform in *Indri* than in the other two genera. There is a rather feeble buccal and lingual cingulum on the P^3 of *Indri*. P^3 is blade-like in *Avahi*, being quite compressed buccolingually.

LOWER. P_3 is rather large, single-cusped and compressed buccolingually, forming a blade-like caniniform tooth in all indriids. There is usually a narrow distolingual cingulid terminating as a distal ledge. P_3 is separated from I_2 by a diastema in indriids. P_4 is also a single-cusped tooth, and although it is elongated mesiodistally, it is not as compressed buccolingually as P_3.

Molars

UPPER. The molars are quadritubercular with M^3 being slightly inset lingually and always the smallest of the series; in fact in *Indri* M^3 lacks a hypocone. The four cusps are well formed, rather sharp and connected by a pattern of crests between the mesial and distal cusp pairs. Indeed, Maier (1977a,b, p. 307) referred to the molars of the Indriidae as the only bilophodont molar type that evolved in primates 'independently from the Cercopithecoidea.' The crista obliqua is present on M^{1-2}, passing between protocone and metacone; it is especially well developed in *Propithecus*, slightly less so in *Avahi*, and absent in *Indri*. This crest is also present in certain Cebidae and variably developed in Hominoidea. Functionally, the crista obliqua occludes between the hypoconid of one lower tooth and the protoconid of the tooth immediately distal so it fills in the slight embrasure between adjacent lower teeth, thereby actively participating in chewing the bolus of food. Buccal cingula are present on M^{1-3}, and para-, meso- and metastyles are usually well developed on M^{1-2}. In *Avahi* occlusobuccal crests interconnect the apices of the styles with those of the para- and metaconules to form a dilambdodont pattern.

LOWER. M_{1-2} have well-formed protoconids, metaconids, hypoconids and entoconids connected by crests (the protoconid is connected to a parastylid by a paracristid and the cristid obliqua interlocks the distal surface of the metaconid to the hypoconid). When viewed from the buccal aspect a W-pattern is formed in *Avahi* and *Propithecus*. This crown pattern is less obvious in *Indri*. The hypoconulid is present on M_3 in all indriids. In *Propithecus* the protoconid and metaconid are widely separated and lie opposite each other, as do the hypoconid and entoconid of M_3. M_3 is longer than either M_1 or M_2. The trigonid and talonid basins are present on M_{1-3}. A well-developed tuberculum intermedium is present on the distal surface of the metaconid of M_{1-3} in both specimens of *Avahi*. Bennejeant (1936) mentions that the tuberculum intermedium is present frequently in *Avahi*; Schwartz and Tattersall (1985, p. 33) refer to a 'distinct but low metastylid (largest on M_1 and smallest on M_3) that is the terminus of thick, lingually deflected postmetacristids' in *Avahi*. The metastylid is another name for the tuberculum intermedium (see Hershkovitz, 1971).

Odontometry (Appendix 1, Tables 7–10, pp. 169–70)
Measurements were made only of the teeth of *Propithecus* owing to the small samples of *Avahi* and *Indri*. There is no sexual dimorphism in the anterior teeth, and the only case in the premolar region is P^3 ($p < 0.03$), which is significantly larger in males. The male molars are significantly larger than the female molars in the following manner: M_2 is significantly longer mesiodistally ($p < 0.009$) whereas the same dimension is less for M_3 ($p < 0.02$). The males are also larger in the buccolingual direction, M_1 ($p < 0.02$), M_2 ($p < 0.01$), and M_3 ($p < 0.03$). The difference in M_1 is the talonid whereas in M_{2-3} it resides in the trigonids. Two upper molars are significantly different mesiodistally, M^2 ($p < 0.002$) and M^3 ($p < 0.001$); all three are significantly wider, M^1 ($p < 0.007$), M^2 ($p < 0.003$) and M^3 ($p < 0.002$).

Family Daubentoniidae

Dietary habits

Aye-ayes are nocturnal prosimians that have many anatomical specializations in addition to their unusual dentition. Their diet consists mainly of insect larvae, seeds, leaves, and fruits. They are known to tap on logs with their extended middle finger, listen for sounds within, and then gnaw the wood with their rodent-like incisors to extract insects. It has been sugges-

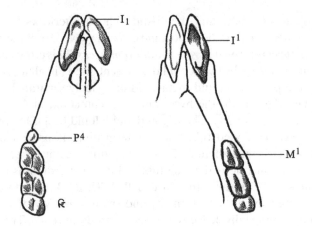

Fig. 5.10. *Daubentonia madagascariensis* male, occlusal view of permanent teeth (30 mm).

ted that the aye-aye occupies the woodpecker niche on Madagascar since there are no woodpeckers on the island (Cartmill, 1972, 1974).

Permanent dentition: I^1-C^0-P^1-M^3 / I_1-C_0-P_0-M_3
Deciduous dentition: d^1-dc^1-dp^2 / di_1-dc_1-dp_2

Morphological observations

	Male	Female
Daubentonia madagascariensis (aye-aye)	1	—

The deciduous dental formula presented here is that suggested by Ankel-Simons (1996). This formula is based on her review of the literature back to 1866 and the use of radiographs and dissections of two newborn specimens at the Duke University Primate Center. The permanent formula presented here is generally accepted by most scholars.

Incisors (Fig. 5.10)

UPPER. The upper incisors are robust, mesiodistally compressed teeth that are continuously growing. The outer surface is enamel and the lingual surface is softer dentine that wears away, resulting in the lingual concavity (James, 1960). They are chisel-like and used for gnawing.

LOWER. The lower incisors are also robust, compressed teeth that grow throughout the life of the animal. As in the upper incisors, enamel covers the outer surface and dentine is on the lingual surface, resulting in a chisel-like tooth (Rosenberger, 1977). James (1960, Fig. 13e, p. 77) presents the inner side of a mandible that has been cut away to show the long incisor root below the molars that extends well distal to the third molar. It is much like the long mandibular incisor root of beavers.

Canines

There are no upper or lower permanent canines.

Premolars

UPPER. The P^4 is the only premolar present. It is a small, peg-like tooth that is separated from I^1 by a wide diastema. The occlusal surface is smooth.

LOWER. There are no lower premolars.

Molars

UPPER. M^{1-2} are small and have four rounded cusps, two buccal and two lingual. The buccal cusps are positioned slightly mesial to the lingual ones and separated from them by a mesiodistally coursing developmental groove. M^3 is somewhat oval and has three cusps; there is no hypocone.

LOWER. The lower molars are separated by a large diastema from I_1. There are four cusps on M_{1-2} and, as in the upper molars, there is a developmental groove between the buccal and lingual cusps. There is no hypoconulid on M_3; however, the entoconid is positioned centrally. The occlusal surfaces of both upper and lower molars are worn, a condition also reported by Gregory (1922) and James (1960).

Odontometry

No measurements were taken of the single specimen. The molar size formula is the same for both uppers and lowers: $M_1 < M_2 > M_3$.

Superfamily Lorisoidea

Present distribution and habitat

There is a radiation of prosimians living in Africa composed of galagos and lorises and a radiation in Asia made up of only lorises. At the present time there is much debate concerning the systematics of the two groups. I have followed Fleagle (1999) in separating the superfamily Lorisoidea into two families Galagidae and Lorisidae, and kept this spelling (Schwartz, 1996) rather than Galagonidae and Loridae (Jenkins, 1987). The Galagidae have four genera, *Galago, Otolemur, Galagoides* and *Euoticus*; there are five genera in Lorisidae, *Loris, Nycticebus, Arctocebus, Pseudopotto* and *Perodicticus*. All members of the Lorisoidea have a procumbent dental comb.

They are all arboreal and for the most part, prefer tropical rain forests, although they are occasionally found in montane forests. Their locomotor behavior is quite different, in that the galagos are primarily leapers whereas the lorises are slow climbers.

Family Galagidae

Dietary habits

Their diet consists of a wide range of foods including insects, fruits, leaves, gum, and birds' eggs. In many species, the diet varies with respect to locality as well as seasonally. In *Otolemur crassicaudatus* the diet seems to be mainly gums, insects, and fruit (Nash, 1986).

Morphological observations

	Male	Female
Otolemur crassicaudatus (thick-tailed bushbaby)	17	14
Galago senegalensis (Senegal bushbaby)	22	17

Permanent dentition: I^2-C^1-P^3-M^3 / I_2-C_1-P_3-M_3
Deciduous dentition: di^2-dc^1-dp^3 / di_2-dc_1-dp_3

Incisors (Fig. 5.11)

UPPER. The upper central incisors are separated by a wide diastema and both I^1 and I^2 are rather tall, slender teeth with slight marginal ridges.

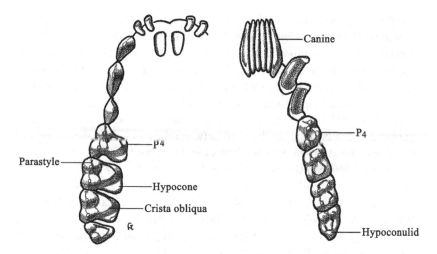

Fig. 5.11. *Otolemur crassicaudatus* male, occlusal view of permanent teeth (24 mm).

LOWER. I_{1-2} are long, narrow procumbent teeth forming the central portion of the dental comb. They possess slight marginal cristids.

Canines

UPPER. The upper canine is a tall, well-developed tooth that stands much higher than the occlusal plane. Mesial and distal marginal ridges pass along their respective borders to merge with a faint lingual cingulum along the base of the tooth. There is an indication of an enamel spur distally.

LOWER. The procumbent lower canines are in line with the lower incisors and form the lateral tooth of the dental comb. They flare slightly laterally and are more robust than the incisors.

Premolars

UPPER. P^2 has a single paracone that is compressed buccolingually, giving the impression of a canine. Buccal and lingual cingula are present, terminating as sharp para- and distostyles. These styles are quite prominent and are discussed in some detail by Schwartz and Tattersall (1985). P^3 is more variable than P^2 in that it may have a small, pointed protocone just distal to the middle of the tooth. The protocone is also described on P^3 in *Otolemur crassicaudatus* by Schwartz and Tattersall (1985). In the present sample, the protocone is variable, being found in 11.8% of male and 7.1% of female *O. crassicaudatus* and in 27.7% of male and 0% of female *Galago senegalensis*. Interestingly, Schwartz and Tattersall (1985) did not mention the protocone in *G. senegalensis*. P^4 is quite molariform as it possesses a rather bulbous paracone, a well-formed, but smaller protocone, a distinct metacone and a low, broad hypocone. A large trigon basin occupies the space bordered by the paracone, protocone and metacone; a smaller, narrower talon basin is sandwiched between the base of the metacone and the hypocone. A crista obliqua forms the distal border of the trigon basin as it passes between the protocone and the base of the lingual side of the metacone. Para- and metastyles are present as well as several occlusal cristae in addition to the crista obliqua. The only other extant prosimians that approach such an abrupt change in premolar morphology are *Hapalemur* and *Arctocebus* (Figs. 5.2 and 5.15). Indeed, in their concluding remarks about the P^4 of *Otolemur crassicaudatus*, Schwartz and Tattersall (1985, p. 62) wrote 'Without a doubt, the posterior premolar is a strange and unique tooth.'

LOWER. P_2 is caniniform and the crown rises well above the occlusal plane of all distal teeth. A complete lingual cingulid sweeps around the base of the lingual surface to end as a distostylid. The distal surface of P_2 is slightly concave whereas the mesial surface is convex. P_3 is morphologically similar to P_2 except that it is smaller. As in P_2, a marginal ridge passes down the mesial border of the tooth, swings lingually to form the lingual cingulid and terminates as the distostylid. P_4 is molariform, possessing four cusps, several occlusal cristids, and a rather spacious talonid basin. The large protoconid and smaller metaconid are set close together; indeed, the metaconid is on the distolingual surface of the protoconid, a position aptly referred to as 'almost entirely melded' by Schwartz and Tattersall (1985, p. 70). A sharp paracristid passes mesially from the protoconid to form a small stylid at the base of the metaconid. This stylid is in the position of a paraconid and Schwartz and Tattersall (1985) believe that it may indeed be a small paraconid. In the present study, the metaconid is variably expressed: in *O. crassicaudatus* it is absent in 52.9% of males and 64.2% of females and in *G. senegalensis* it is absent in 22.7% of males and 23.5% of females. Thus there is a rather significant difference in the development of the metaconid between these two species.

Molars

UPPER. M^{1-2} have four cusps; M^3 lacks the hypocone. The hypocones appear to arise from the cingulum and are on a lower plane than the protocone. The hypocone is normally larger on M^2 than on M^1. The trigon basin is rather spacious and bounded distally by the crista obliqua passing between the protocone and metacone. The para- and metastyles are present on M^{1-2}; however, if a style is present on M^3 it is the parastyle.

LOWER. M_{1-2} possess four cusps and M_3 has a hypoconulid. The protoconid is mesial to the metaconid and the cusps are connected by the protocristid. The trigonid basin lies mesial to the protocristid and is present on the three molars. The hypoconid and entoconid are farther apart and not connected. The talonid basin is rather spacious and becomes narrower from M_1 to M_3. The cristid obliqua is present on all lower molars but it is better developed on M_{1-2}. Regarding the presence of the M_3 hypoconulid, this cusp is not mentioned by Schwartz and Tattersall (1985, p. 72), but they do say that there is a 'moderately developed heel' on M_3 and this is clearly shown in their Fig. 33, p. 82. In addition, M_3 is not the smallest of the three lower molars as stated by Schwartz and Tattersall (1985), rather it is the longest of the three (Appendix 1, Tables 20 and 22). The cusps of both the upper and lower molars are considered to be more adapted to compression and grinding than to shearing during mastication (Seligsohn, 1977; Maier, 1980). In her study of enamel microstructure in *Otolemur crassicaudatus* Maas (1993, p. 217) found 'that galago occlusal enamel is organized so as to resist abrasion of different functional regions' which she points out may have important implications for maintaining dental functional efficiency in the molars of *Otolemur*.

Odontometry (Appendix 1, Tables 19–26, pp. 175–8)
Of the two species studied, *O. crassicaudatus* is the larger animal and it also has slightly larger teeth. There is little sexual dimorphism according to the *t*-test in either species, although there are a few more sex differences in the teeth of *O. crassicaudatus* than in those of *G. senegalensis*. The upper canine is the only tooth consistently different between the sexes.

The molar lengths for *O. crassicaudatus* are: $M^1 > M^2 > M^3$ / $M_1 = M_2 < M_3$.

The molar lengths for *G. senegalensis* are: $M^1 > M^2 > M^3$ / $M_1 = M_2 < M_3$.

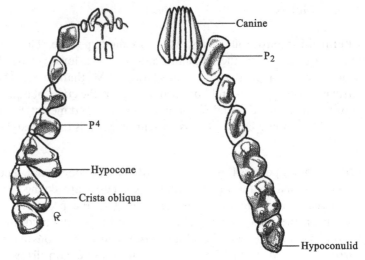

Fig. 5.12. *Loris tardigradus* male, occlusal view of permanent teeth (23 mm).

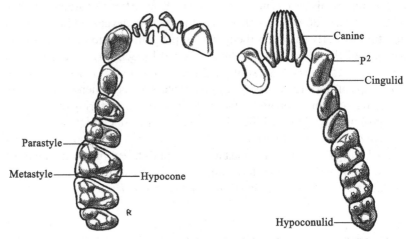

Fig. 5.13. *Nycticebus coucang* male, occlusal view of permanent teeth (25 mm).

Family Lorisidae

Dietary habits

The diet of the lorisids is similar to that of the galagids, consisting of fruits, animal matter and gums. For example, the slow loris (*Nycticebus coucang*) is frugivorous whereas the angwantibo (*Arctocebus calabarensis*) eats mainly insects and seems to prefer noxious caterpillars (Fleagle, 1999).

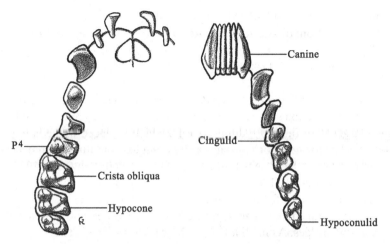

Fig. 5.14. *Perodicticus potto* female, occlusal view of permanent teeth (22 mm).

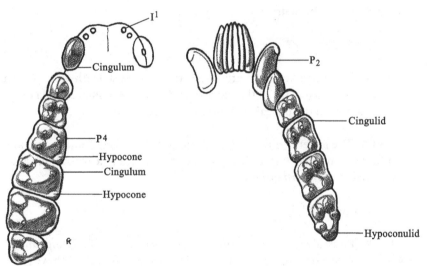

Fig. 5.15. *Arctocebus calabarensis* male, occlusal view of permanent teeth (17 mm).

Morphological Observations

	Male	Female
Loris tardigradus (slender loris)	4	1
Nycticebus coucang (slow loris)	18	12
Perodicticus potto (potto)	11	9
Arctocebus calabarensis (angwantibo)	1	—

Permanent dentition: I^2-C^1-P^3-M^3 / I_2-C_1-P_3-M_3
Deciduous dentition: di^2-dc^1-dp^3 / di_1-dc_1-dp_3

Incisors (Figs. 5.12–5.15)

UPPER. I^{1-2} are small and peg-shaped in *L. tardigradus* and *A. calabarensis* and a wide diastema separates the two central incisors. In *N. coucang* I^1 is much larger than the more diminutive I^2, which is placed immediately distal to I^1. The former tooth also has a slight lingual cingulum. I^2 may be shed early in life or be absent (Hill, 1953; James, 1960; Schwartz, 1974; Schwartz and Tattersall, 1985). In the present sample, I^2 is absent in 27.7% of males and 41.6% of females. There is a diastema between I^{1-1}. In *P. potto* I^{1-2} are about equal in size, separated from each other, and both have slight marginal ridges. As in all lorisoids there is a diastema between I^{1-1}.

LOWER. The lower incisors are procumbent as in all lorisids, however they are not quite as procumbent as in lemurs.

Canines

UPPER. The upper canines are large, rather robust teeth that project above the occlusal plane of the other teeth. There are lingual cingula on all upper canines and a buccal cingulum only in *A. calabarensis*.

LOWER. The lower canine forms the lateral tooth of the dental comb and is always somewhat larger than the incisors. There is a mesial marginal ridge except in *L. tardigradus*.

Premolars

UPPER. P^2 is always caniniform and possesses a complete lingual cingulum, which is also present on P^{3-4}. In lorisids, P^{2-4} frequently have para- and distostyles, which are generally better developed in *N. coucang*. P^3 is often smaller than P^2, the reverse of what is found in lemurs. The presence of a protocone, however, is reported on P^3 in *N. coucang* (Gregory, 1922; Schwartz and Tattersall, 1985). A protocone is present on P^3 in this sample of *N. coucang* in 61.1% of males and 66.6% of females and this cusp is always present on P^4. In addition to the protocone, P^4 in *A. calabarensis* has a metacone and hypocone, making it the only molariform P^4 among the lorisids studied here.

LOWER. P_2 is always caniniform, possessing a single high cusp with a lingual cingulid. P_2 is separated from the proclivous lower canine by a diastema. P_3 has a single protoconid with a lingual cingulid that may terminate as a small distal heel. P_4 is variable in lorisids. In *L. tardigradus* there is a metaconid attached to the distolingual surface of the protoconid with its apex slightly removed from the protoconid. In *P. potto* there is a very small metaconid extending above the cingulid and in *A. calabarensis* there are four cusps, protoconid and metaconid close together, and the hypoconid and entoconid more separate, forming the boundary of the talonid basin. There is also a cristid obliqua and buccal and lingual cingulids on P_4 in *A. calabarensis*.

Molars

UPPER. M^{1-2} generally have four high cusps; although the hypocone is variable it is present on these teeth in *P. potto* in the present sample and Schwartz and Tattersall (1985) report its presence in their study of this lorisid. M^3 is the smallest of the series and the hypocone is frequently absent. In this sample of *N. coucang* the hypocone is present on M^3 in only 22.2% of males and is not present in the females. The crista obliqua is present on M^{1-2} as well as other occlusal cristae. Para- and metastyles are generally present on M^{1-2} but only the parastyle is present on M^3. Buccal and lingual cingula are often present on M^{1-3}.

LOWER. The lower molars have four well-developed cusps and M_3 has the additional hypoconulid. The protoconid is mesial to the metaconid and connected by the protocristid that forms the distal border of the trigonid basin. The hypoconid and entoconid are not connected and are positioned on either side of a rather spacious talonid basin. The lower cusps are rather sharp and the mesial two are slightly higher than the distal cusps. Seligsohn (1977) mentioned that, of all lorisines, the molar cusp relief and acuity as well as the length and sharpness of crests is greatest in *A. calabarensis*, an insectivorous molar pattern. The trigonid basin is narrower on M_{1-3} in *P. potto* than in the other genera. A well-formed cristid obliqua passes between the hypoconid and metaconid, and in *A. calabarensis* a paracristid extends down the protoconid to end as a mesiostylid. A buccal cingulid is present on the protoconid and may extend more distally onto the hypoconid. Lingual cingulids are often observed passing around the metaconid and entoconid.

Odontometry (Appendix 1, Tables 11–18, pp. 171–4)
In *N. coucang* $I^1 > I^2$ in both dimensions. P_2 is the largest lower premolar,

whereas P^4 is the largest upper premolar. The upper molar mesiodistal formula is $M^1 > M^2 > M^3$ and the lower is $M_1 < M_2 < M_3$. t-Tests reveal only one instance of sexual dimorphism: the mesiodistal diameters of the upper canines are significantly larger ($p < 0.03$) in males. In *P. potto* the only significant sex difference is in the buccolingual dimension of the upper canine ($p < 0.04$). The lengths of the lower molars are $M_1 < M_2 > M_3$ and M^2 is usually the longest upper molar. Measurements were not taken of *L. tardigradus* or *A. calabarensis* because of small sample size.

Superfamily Tarsioidea

Family Tarsiidae

Present distribution and habitat

Tarsiers (*Tarsius* spp.) are found today only in Southeast Asia, although their earliest antecedents have been discovered in Eocene formations in both North America and Europe (Kay, Ross and Williams, 1997). Simons and Bown (1985) have also described a fossil tarsier from the Oligocene Fayum formations of Egypt. They live on many of the islands of the Malay Archipelago, from Sumatra in the west to the southern Philippines in the north, and as far east as Sulawesi. They have not been found on the mainland of Asia. Tarsiers are among the smallest of primates and have a mixture of prosimian and anthropoid morphological characteristics (Fleagle, 1999; Swindler, 1998).

Tarsiers live in lowland tropical rain forests and are rarely seen much above sea level. They frequent coastal areas and some species, e.g. *T. syrichta*, seem to prefer regions of secondary growth (Oxnard, Crompton and Lieberman, 1990). They have a highly specialized vertical clinging and leaping mode of locomotion (Napier and Walker, 1967) and have been observed to leap up to 3 m, although most leaps are 1.5 m or less (Oxnard *et al.*, 1990). Tarsiers are nocturnal, possessing extremely large eyes but lacking the reflective tapetum that is present in other nocturnal primates.

Dietary habits

Tarsiers are insectivorous although perhaps a better term is 'carnivorous' since they prey on a wide variety of insects and small vertebrates. For

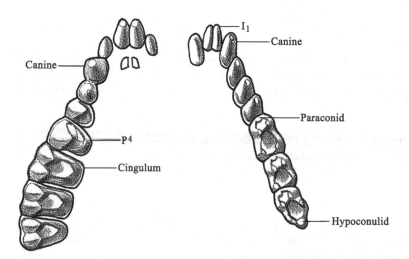

Fig. 5.16. *Tarsius bancanus* female, occlusal view of permanent teeth (15 mm).

example, a detailed study of *T. bancanus* by Crompton and Andau (1986) has shown that these animals eat beetles, crickets, grasshoppers, cockroaches, and occasionally crabs and frogs. Indeed, Niemitz (1984, p. 59) has stated

'The Tarsiers are, in fact, the only true carnivorous Primates in existence.'

General dental information

Permanent dentition: I^2-C^1-P^3-M^3 / I_1-C_1-P_3-M_3
Deciduous dentition: di^0-dc^1-dp^2 / di_0-dc_1-dp_2

The permanent dentition of tarsiers differs from that of other prosimians as follows. Tarsiers have three premolars and one lower incisor, whereas other prosimians either have two premolars and two lower incisors (indriids), or three premolars and two lower incisors (lorisids and lemurs). Of course, *Daubentonia* and *Lepilemur* have additional numerical differences. Tarsiers lack the procumbent lower dental comb of other prosimians.

Morphological observations

	Male	Female
Tarsius spectrum (spectral tarsier)	1	5
Tarsius bancanus (Bornean tarsier)	1	1
Tarsius syrichta (Philippine tarsier)	7	9

Incisors (Fig. 5.16)

UPPER. I^{1-2} are vertically implanted and I^1 is larger than I^2. Both teeth are conical and set close together, i.e. there is no diastema separating them. A complete cingulum encircles the base of I^1 and there is little space between I^2 and the canine.

LOWER. I_1, the only lower incisor, is a vertically implanted conical tooth with a slight cingulid that passes around the base.

Canines

UPPER. The upper canine has a sharp crown tip that is barely higher than the premolars. The tooth is girdled by a cingulum that develops into a distal flange or slight heel.

LOWER. The lower canine is more robust than the incisors or premolars and extends above the occlusal surface of these teeth. It has a cingulid passing around the base of the tooth that is well developed and U-shaped on the buccal surface of the tooth.

Premolars

UPPER. P^2 is conical and resembles the canine except it is smaller. A narrow, sharp crest passes down the lingual surface of the tooth from the crown tip to the lingual cingulum dividing it into two nearly equal parts. P^3 is similar to P^2 except that it is larger and the lingual cingulum is more pronounced. P^4 has two cusps, a large paracone and a smaller protocone that arises from the cingulum and has an independent apex. There is no crest passing between the paracone and protocone. P^4 has a complete cingulum, occasionally incipient para-and distostyles as well as an ento-style on the distolingual portion of the cingulum. In fact, a cingulum passes completely around the base of each premolar and, as noted above, it is more prominent on the lingual aspect of each tooth.

LOWER. P_{2-4} are rather plain, uniform teeth possessing a single, high protoconid encircled by a cingulid. P_2 is very small compared with the large, caniniform P_2 of lemurs. $P_{3,4}$ lack any development of the trigonid or talonid, but there is a cingulid encircling the base of each tooth. The cingulid is slightly more developed on the buccal surface and terminates on the distal surface of P_4 as a narrow, distal ledge. Maier (1984, p. 55) believed the upper and lower premolars were 'probably adapted to puncturing, cutting and pulping food; some phase II- grinding may occur in the last two premolars.' Certainly, the surface between the paracone and protocone of P^4 would receive the trigonid complex of M_1 as the lower teeth slide across the uppers, permitting some chewing or grinding (Mills, 1963).

Molars

UPPER. M^{1-3} are tribosphenic molars with three well-formed cusps: paracone, protocone and metacone. The cusps are not particularly high or steep, in contrast to the situation in insectivorous primates. A lingual cingulum is slightly thickened in the region of the hypocone but the cusp is absent. The trigon basin is complete and rather spacious. A crista obliqua and buccal cingula are present on all molars, and para- and metastyles may be present; however, the mesostyle is absent. Protoconules and metaconules are variable on M^{1-3} in the present sample. Various transversely and anteromedially directed wear striations have been identified on the grinding surfaces of both upper and lower molars in *T. bancanus*. These support earlier investigations that demonstrated that the masticatory mechanism of tarsiers has changed considerably from the scissor-like shearing characteristics of insectivorous primates (Maier, 1984; Jablonski and Crompton, 1994).

LOWER. M_{1-3} are tribosphenic with a trigonid basin consisting of paraconid, protoconid and metaconid and a distal talonid basin with the hypoconid and entoconid. A hypoconulid is present on M_3. A cristid obliqua is present as well as a V-shaped protocristid. A narrow but distinct buccal cingulid is present on M_{1-3}, there is no lingual cingulid.

Odontometry (Appendix 1, Tables 27–30, pp. 179–80)
Sexual dimorphism in tooth size was studied in tarsiers by pooling the three species. A single difference was found: the buccolingual dimension of the trigonid of M_3 is significantly larger in the male ($p < 0.02$), otherwise there are no sex differences in tooth size. Molar size relations are: $M^1 < M^2 > M^3 / M_1 < M_2 < M_3$.

6   *Ceboidea*

Family Cebidae

General dental information

Permanent dentition: I^2-C^1-P^3-M^3 / I_2-C_1-P_3-M_3

Deciduous dentition: di^2-dc^1-dp^3 / di_2-dc_1-dp_3

The above dental formulas are the same for all New World monkeys with the exception of the callitrichines that have only two permanent molars. *All* New World monkeys have three upper and lower permanent premolars. The dental formulas are not repeated for each of the subfamilies.

Subfamily Callitrichinae

Present distribution and habitat

Members of the Callitrichinae are the smallest of the New World primates and are made up of the marmosets, tamarins, and Goeldi's marmoset (*Callimico*). The marmosets include two genera, *Callithrix* and *Cebuella*, and there are also two genera in the tamarins, *Saguinus* and *Leontopithecus*. The callitrichines live in the tropical and montane rain forests of Central and South America. *Saguinus* is found as far north as Panama but the other genera are limited to South America.

All callitrichines are arboreal, spending the majority of their lives in trees. Their locomotion is squirrel-like, consisting of quick, jerky movements across the substrate, described as scansorial leaping and clinging (Rosenberger, 1992). Because their digits possess claws, except the big toe, they are able to cling to the sides of trees to feed on gums and insects. There are various feeding patterns within the callitrichines, but Garber (1992, p. 469) was able to state 'Large-branch feeding and the use of vertical clinging postures appear to be a primary adaptation among virtually all callitrichines, distinguishing them ecologically from other platyrrhine taxa.' All members of the subfamily are diurnal.

Dietary habits

Callitrichines eat a variety of foods: fruits, insects, exudates, and flowers. According to Ramirez (1989) reported in Rosenberger (1992), *Saguinus mystax* is the only taxon that spends any time foraging on leaves, although about twice as much time is spent eating fruits. Here, as with many primates, the diet is rather eclectic. The Rosenberger (1992) contribution to platyrrhine feeding adaptations is an excellent investigation and analysis of the range of problems involved in correctly identifying any primate dietary category and the associated dental specialization. This paper is part of a symposium, *Feeding Adaptations in New World Primates: An Evolutionary Perspective*, edited by P. A. Garber and the late W. G. Kinzey (1992). This symposium is an invaluable contribution to the evolutionary biology of New World primates.

General dental information

Permanent dentition: I^2-C^1-P^3-M^2 / I_2-C_1-P_3-M_2 (callitrichines)
Permanent dentition: I^2-C^1-P^3-M^3 / I_2-C_1-P_3-M_3 (*Callimico*)
Deciduous dentition: di^2-dc^1-dp^3 / di_2-dc_1-dp_3

The number of permanent molars differs between *Callimico*, which possesses three permanent molars, and all other callitrichines, which possess two permanent molars. The lower incisors separate the tamarins from the marmosets. In tamarins, the lower canines are much higher than the occlusal level of the lower incisors, hence the name long-tusked tamarins. In marmosets, the lower incisors and lower canines are on the same occlusal level, hence the name short-tusked marmosets (Fig. 6.6a,b). In addition, marmosets lack enamel on the lingual surface of their lower incisors (Rosenberger, 1977).

Morphological observations

	Male	Female
Callithrix penicillata (black tufted-ear marmoset)	3	2
Cebuella pygmaea (pygmy marmoset)	2	1
Saguinus geoffroyi (Geoffroy's tamarin)	12	15
Leontopithecus rosalia (golden lion tamarin)	—	2
Callimico goeldii (Goeldi's marmoset)	3	—

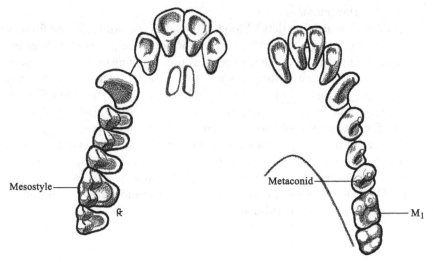

Fig. 6.1. *Callithrix penicillata* male, occlusal view of permanent teeth (13 mm).

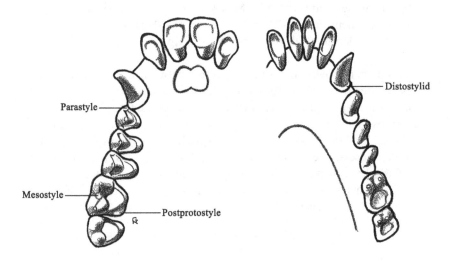

Fig. 6.2. *Cebuella pygmaea* male, occlusal view of permanent teeth (11 mm).

Incisors (Figs. 6.1–6.5)

UPPER. I^1 is a rather broad, spatulate tooth with a lingual cingulum and mesial and distal marginal ridges. The labial surface is smooth and slightly convex mesiodistally. I^2 is narrower than I^1 with a somewhat pointed

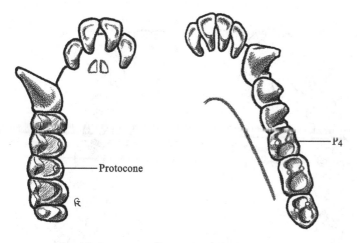

Fig. 6.3. *Saguinus geoffroyi* male, occlusal view of permanent teeth (16 mm).

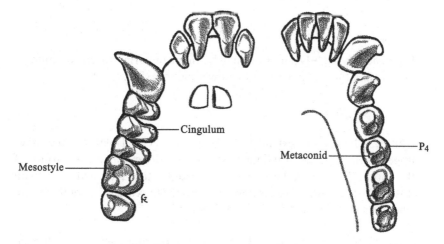

Fig. 6.4. *Leontopithecus rosalia* female, occlusal view of permanent teeth (18 mm).

incisal border, particularly in *L. rosalia*. I^2 has a lingual cingulum and mesial and distal marginal ridges.

LOWER. I_{1-2} are about equal in height and are as high as the lower canines in the marmosets (Fig. 6.6a). These enlarged incisors and canines are used to gouge holes in the bark of trees to elicit the flow of gums and resins. There are slightly developed mesial and distal marginal ridges on I_{1-2}; and

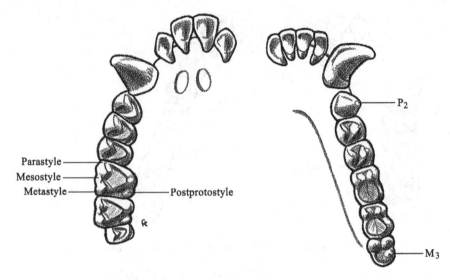

Fig. 6.5. *Callimico goeldi* male, occlusal view of permanent teeth (17 mm).

I_1 is generally slightly narrower than I_2. In the tamarins, the lower canine is higher than the incisal level of the lower incisors (Fig. 6.6b).

Canines

UPPER. The canines extend well above the occlusal plane of the other upper teeth and a longitudinal groove passes down the mesial surface of the tooth. The canine has a lingual cingulum that may end as a small distostyle; the cingulum does not pass onto the buccal surface of the tooth. There is a diastema between the canine and I^2.

LOWER. The canines are robust teeth, possessing well-formed lingual cingulids that terminate as distostylids. In tamarins, the canines stand well above the lower incisors, whereas they are about on the same occlusal level as the incisors in the marmosets. In a study of canine sexual dimorphism in New World Monkeys, Kay *et al.* (1988) found that in these taxa, which included several species of callitrichine, canine dimorphism increased as male–male competition for females increased.

Premolars

UPPER. P^{2-4} have two cusps, paracone and protocone, except in *Leontopithecus* where the protocone is absent. The paracone is always the larger

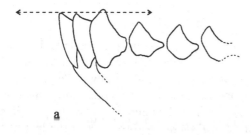

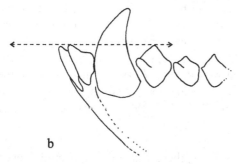

Fig. 6.6. Canine–incisor relations in the Callitrichinae. (a) Short-tusked marmoset, b) long-tusked tamarin. Reprinted from Swindler, D. R. (1976) *Dentition of Living Primates*, with the permission of Academic Press.

of the two cusps, and para- and distostyles are variably developed. For example, Kinzey (1973) reports that para- and distostyles are always present in *S. mystax* and *S. oedipus*, a condition found in the present sample of *S. geoffroyi*. In addition, entostyles may also be present on the upper premolars in these three species; in *S. geoffroyi* they are present 85% of the time on P⁴. Buccal cingula are present in *C. penicillata* as well as para- and distostyles on P$^{2-4}$. Kinzey (1977) found buccal cingula only on P⁴ in 6% of 31 specimens of *C. argentata* and a mesostyle in 7%. He also reports the presence of lingual cingula on P$^{3-4}$, none on P². A similar condition is found in *C. penicillata*.

LOWER. P$_2$ is caniniform with an obvious lingual cingulid around its base in each taxon. P$_3$ also has a single cusp in all species except *C. goeldii* and *S. geoffroyi*, where the metaconid is connected by a protocristid to the protoconid. P$_4$ has well-developed protoconids and metaconids; small talonid basins are present in *C. penicillata*, *L. rosalia* and *C. goeldii*. Stylids are variably developed in all taxa, as are buccal cingulids. Kinzey (1973) reports protostylids and distostylids as variable on P$_{2-4}$ in *C. argentata*. In *C. penicillata* only distostylids are present.

Molars

UPPER. The M^1 has three cusps, paracone, protocone and metacone, in these taxa, although hypocones have been reported on M^1 in *S. mystax* (16%) by Kinzey (1973). Rosenberger and Kinzey (1976) also found that *C. argentata* lacks a hypocone, and the protocone is opposite the buccal notch and is rimmed by the lingual cingulum, a condition similar to the crown morphology of *C. penicillata*. With respect to the loss of the hypocone in the callitrichines, Plavcan and Gomez (1993) believed that the absence of the hypocone might be functionally related to their insectivorous–gummivorous diet. Para- and metastyles are frequently present in these species as well as mesostyles. M^2 has three cusps but is smaller than M^1. M^3 is absent in all genera except in *C. goeldii* where it is small and possesses two to three cusps. Lucas *et al.* (1986a) suggested that the loss of M^3 in these taxa may be a dietary specialization since they eat very few leaves; however, Ford and Bargielski (1985) attributed the loss of M_3 to dwarfism.

LOWER. M_{1-2} have four cusps; there is also no hypoconulid, even on the M_3 in *Callimico*. The mesial two cusps are connected by a protocristid that defines the distal border of a small trigonid relative to the talonid. There is no paraconid. Buccal cingulids are present on the protoconids of M_{1-2}, except in *Callimico* and *C. penicillata*, although Kinzey (1973) reported them in *C. argentata* on M_{1-2}. In general, the molar cusps of callitrichines are more sharply pointed than the bunodont cusps of the molars of the other subfamilies of the Cebidae. Indeed, the molars of callitrichines have been characterized as a functional complex that is 'well suited to processing foods of high "impact strength"' which is 'the amount of energy it takes to break a unit volume of material' (Rosenberger and Kinzey, 1976, p. 286).

Odontometry (Appendix 1, Tables 31–34, pp. 181–2)

There is little odontometric information on the callitrichines: only the data presented here on *Saguinus geoffroyi* (the only genus measured, owing to small sample size) and those of Hanihara and Natori (1987), who presented complete odontometrics on six species of *Saguinus*. Their findings are similar to those presented here for *S. geoffroyi*. In the present study and that of Hanihara and Natori (1987), the teeth are nearly the same size between the sexes, and when not, the males are usually slightly larger, although there are several instances when a female tooth (and this occurs in all species) is slightly larger than the male. This seems to be particularly true in *S. midas*, although the question of the larger female teeth is not discussed by Hanihara and Natori (1988).

In addition to the above odontometric information on the genus *Saguinus*, Kanazawa and Rosenberger (1988) studied upper and lower molar reduction in marmosets. These four teeth were measured in all species, except *Callimico*. Length and breadth measurements were made, and since no sex differences were determined, combined mean values were presented. They found that *L. rosalia* had the largest molars for its body size and *Saguinus* possessed relatively small molars. Currently, more comparative odontometric investigations of the Callitrichinae are needed, a suggestion put forward by Hanihara and Natori (1988).

Subfamily Cebinae

Present distribution and habitat
Members of this subfamily range throughout the neotropical forests from Mexico to Argentina and from the east coast of South America to the Peruvian Andes. The species are found from sea level to approximately 2000 m. They are all arboreal and possess anatomies well adapted for life in the trees. *Cebus*, the capuchin or organ-grinder's monkey, is perhaps the best known of New World monkeys.

Dietary habits
Cebus and *Saimiri* are frugivorous taxa feeding on fruits, nectar, seeds, leaves, berries and a variety of insects. According to Janson (1986) *C. apella* eats 96 varieties of fruit and during the dry season it can get along quite well eating pith from *Scheelea* palm fronds. In addition, of all New World primates of similar body size, the capuchins (*Cebus*) depend most on faunivory for protein (Janson and Boinski, 1992). It is also interesting to note that Shellis *et al.* (1998) found that, among the different anthropoid primates they examined, *C. apella* had the thickest molar enamel when scaled with tooth size and body mass.

Morphological observations

	Male	Female
Cebus apella (tufted capuchin)	33	16
Saimiri sciureus (squirrel monkey)	20	6
Saimiri oerstedii (red squirrel monkey)	11	6

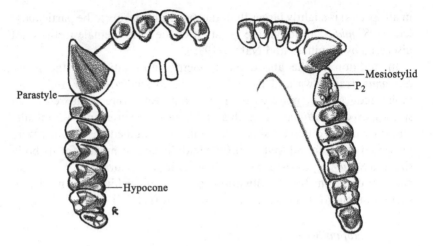

Fig. 6.7. *Cebus apella* male, occlusal view of permanent teeth (33 mm).

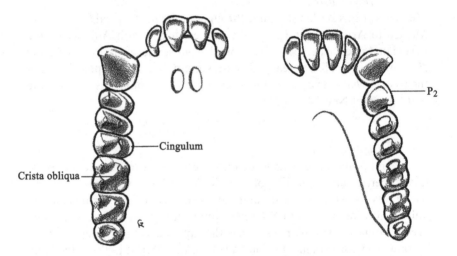

Fig. 6.8. *Saimiri sciureus* male, occlusal view of permanent teeth (19 mm).

Incisors (Figs. 6.7 and 6.8)

UPPER. I^1 is a broad, spatulate tooth; I^2 is more pointed. I^{1-2} are heteromorphic. Marginal ridges are well formed on both teeth, and in *Saimiri* there is a shallow lingual fossa present on both teeth.

LOWER. In *Cebus* I_{1-2} are vertically implanted and their crowns are relatively flat. Marginal ridges are present on I_{1-2} and the incisal border of

I_2 slopes distally. Incidentally, I_2 is larger than I_1, a condition present in New World monkeys, whereas in Old World monkeys and hominoids it is generally the reverse. Eaglen (1984, p. 263) pointed out that platyrrhine monkeys possess small incisors that have a negative allometric relationship with body mass, and in addition he suggested that small incisors represent a 'plausible primitive condition' in these monkeys. It should be mentioned, however, that Anapol and Lee (1994) considered the fruit- and fauna-eating (omnivores, their classification) *Cebus apella, Saimiri sciureus* and *Saguinus midas* to possess relatively wide incisors that are usually associated with frugivory (they used the condylobasal length of the skull to represent body size).

Canines

UPPER. The upper canine is a large, robust tooth projecting well above the occlusal plane in both of these genera, particularly in *Cebus*. Indeed, Anapol and Lee (1994) found both the upper and lower canines of *Cebus* to be exceptionally robust teeth equaling those of *Chiropotes satanas*, and that both taxa use these teeth to open the hard outer covering of palm nuts. The development of lingual cingula is more prominent in females than in males, a condition first reported by Orlosky (1973). A longitudinal groove passes down the mesial surface of the tooth in *Saimiri* and a mesial crest passes from near the cervix to the tip of the tooth in *Cebus*. Canine sexual dimorphism is well known in most living primates, as discussed in the last chapter. Incidentally, most primate skulls can be sexed by simply looking at the canine teeth. In platyrrhines, the male canine is larger than it is in females with the notable exception of *Callicebus moloch*, where no canine sexual dimorphism is present (Kinzey, 1972; Orlosky, 1973). Kinzey (1972) suggests that the lack of canine sexual dimorphism may be due to the low level of aggression in this species (see *Callicebus* below). Another explanation of canine sexual dimorphism in platyrrhines is offered by Kay *et al.* (1988, p. 385); they concluded 'that intermale competition among platyrrhine species is the most important factor explaining variations in canine dimorphism.' Both of these suggestions indicate the important role that social organization has on the development of canine sexual dimorphism in New World primates.

LOWER. The canine is a large, stout tooth that rises above the occlusal plane of the other teeth. There is no diastema between the canine and I_2 or P_2. Orlosky (1973) discusses non-metric sexual differences in the canine of *Saimiri*, consisting of well-formed mesiolingual grooves and a partial

106   *Ceboidea*

lingual cingulid in males. In fact, of 68 females and 69 males, only one male varied from this pattern.

Premolars

UPPER. P² is bicuspid in *Cebus*, but more caniniform in *Saimiri* where a protocone is present in less than 10% of the present sample and lacks any development of a lingual cingulum, a condition also reported by Kinzey (1973) as well as by Orlosky (1973). P³⁻⁴ are bicuspid and have lingual cingula in both taxa. In the present sample of *Saimiri* these appear in over 95% of P³⁻⁴, particularly the type that Kinzey (1973) calls the postprotostyle. Para- and distostyles represent remnants of the buccal cingulum that may occur on all three premolars of both taxa. The parastyle is more common than the distostyle, being present on P² 52% of the time in *C. capucinus*, 15% of the time in *C. apella* and completely absent in *C. albifrons* (Kinzey, 1973). In both species of squirrel monkey the parastyle occurs 60–65% of the time and the distostyle occurs 20–25% of the time.

LOWER. P₂ is somewhat caniniform even though there is a small metaconid attached to the distolingual surface of the protoconid in *Cebus*. In *Saimiri* the metaconid is only present about 15% of the time. The tooth is incipiently sectorial in that it is mesiodistally elongated, forming a mesiobuccal honing surface for the distolingual surface of the upper canine. A mesiostylid is present in *Cebus* on P₂ 55% of the time in this sample; Kinzey (1973) reports it in 46% of *C. apella*. P₃₋₄ are bicuspid and connected by a protocristid that separates the larger trigonid from a narrow talonid. Distostylids are present in *Cebus* as follows: P₂, 58%; P₃, 30%; P₄, 10%; these are approximately the incidences reported by Kinzey (1973) for this taxon. Both mesio- and distostylids are common on P₂₋₄ in *Saimiri sciureus*; Kinzey's (1973) findings suggest that this species has the highest frequencies of these structures among extant cebids.

Molars

UPPER. M¹⁻² have four rather bulbous cusps and M³ frequently lacks the hypocone. Kinzey (1973) reports M³ hypocones 82% and 23% of the time in *C. capucinus* and *C. albifrons*, respectively. M¹⁻² tend to become separated into mesial and distal moieties resembling somewhat the bilophodont molars of Old World monkeys that will be discussed later. Rosenberger and Kinzey (1976, p. 289) drew attention to this morphological similarity, but noted that it was 'certainly not a functional one,' rather, they thought the molars were more 'suited for crushing-grinding mode of mastication.' Indeed, over eighty years ago Gregory (1922, p. 225) stated that the upper and lower molars of *Cebus* 'may be a progressive adaptation to insectivorous and frugivorous habits,' a diet we know today that *Cebus* prefers. A true crista obliqua is not always present in *Cebus* (40% of present sample), since the crista obliqua and entocrista merge to form a transverse crest between the metacone and hypocone. A metaconule may be present. A well-formed crista obliqua is present on M^{1-2} and forms the distal border of the rather spacious trigon basin in *Saimiri*. Buccal and lingual styles are variable on M^{1-3}.

LOWER. M_{1-3} have four cusps; the hypoconulid is absent in Cebinae. A sharp protocristid connects the protoconid and metaconid, forming the distal wall of the trigonid basin. The talonid basin is on a slightly lower level than the trigonid. Lingual cingulids are absent, but buccal cingulids are variably developed on M_{1-3} in this sample and similar findings are reported by Kinzey (1973) and Orlosky (1973).

Odontometry (Appendix 1, Tables 45–56, pp. 188–93)
t-Tests revealed that, in the great majority of tooth dimensions, male *C. apella* were significantly larger than females. The canines were appreciably larger in males ($p < 0.01$). The significant differences are few in the squirrel monkey, but when they do occur they are in the dimensions of the maxillary and mandibular canines and P_2. In males, the mesiodistal diameter of the upper canine surpasses that of all other teeth.

Subfamily Aotinae

Dietary habits
Aotus is the only nocturnal anthropoid primate. Originally the genus contained eight to ten subspecies; today these subspecies have been elevated to species mainly on the basis of chromosomal evidence (Groves, 1993). They are small primates living in a variety of forest habitats from Panama to Argentina. Their diet consists mainly of fruit, supplemented by leaves and insects.

Morphological observations

	Male	Female
Aotus trivirgatus (night monkey)	9	15

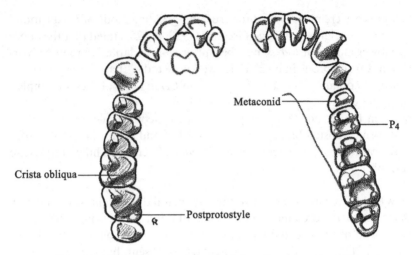

Fig. 6.9. *Aotus trivirgatus* male, occlusal view of permanent teeth (22 mm).

Incisors (Fig. 6.9)

UPPER. I^1 is quite wide with well-developed marginal ridges circum-navigating the rather flat lingual surface of the tooth. I^2 is narrow and pointed compared with I^1, thus I^{1-2} are heteromorphic.

LOWER. The incisal border of I_2 slopes distally rather sharply compared with the more level incisal border of I_1. Mesial and distal marginal ridges outline the concave lingual surfaces of I_{1-2}, resulting in what Rosenberger (1992, p. 537) refers to as 'a broad scoop-like battery of all four lowers combined.'

Canines

UPPER. The canine is rather large and projects well above the occlusal plane of the other teeth. There is an obvious lingual cingulum skirting the base of the tooth.

LOWER. The canine is small but does stand above the occlusal plane of the other teeth. The crown is pointed and there is a small distostylid that occupies the space between the canine and P_2. Among monogamous platyrrhines, *A. trivirgatus* has the least amount of canine sexual dimorphism (Kay *et al.*, 1988).

Premolars

UPPER. P^{2-4} have two cusps: a large paracone and a smaller protocone. The protocone is mesial to the paracone. Lingual cingula are not present on P^2 but are on P^{3-4}. Para- and distostyles are present on P^{2-4} about 75% of the time, a finding not dissimilar from that of Kinzey (1973).

LOWER. P_2 is more caniniform than P_{3-4} and rises above the other premolars. P_{3-4} can have identifiable metaconids, hypoconids and entoconids, particularly P_4. Thus, there is a molariform gradient from P_3 to P_4.

Molars

UPPER. M^{1-2} have four well-formed cusps and are quadrangular in occlusal view. M^3 usually has two cusps: the paracone and protocone. The crista obliqua is present on M^{1-2} but rarely present on M^3. Buccal cingula are not common on the upper molars of *Aotus*, but when present, they can be found on M^{1-2} (Orlosky, 1973). Lingual cingula, particularly the post-protostyle, are present in over 30% of the present sample on M^{1-2}, less on M^3 (15%). Kinzey (1973) found a higher incidence in his sample of *Aotus trivirgatus*: thus M^1 (100%), M^2 (88%), and M^3 (74%).

LOWER. M_{1-3} have four cusps; there is no hypoconulid on M_3. Protocristids are present on M_{1-3} and separate the occlusal surface into smaller trigonid basins and larger, more spacious talonid basins. Buccal cingulids (distostylids) are not present in these specimens; however Kinzey (1973) reports their presence on M_1 (18%), and M_{2-3} (5%) for *A. trivirgatus*.

Odontometry (Appendix 1, Tables 35–38, pp. 183–4)
In the present sample of *Aotus*, sexual differences are found only in the buccolingual dimension of the upper canine, M^2 and M^3 ($p < 0.01$, $p < 0.03$ and $p < 0.01$, respectively). M_3 also shows a significant difference mesiodistally ($p < 0.02$).

Family Atelidae

Present distribution and habitat

The geographic distribution of members of the Atelidae is similar to that of the Cebidae. The atelines are the largest platyrrhines, and they are also the

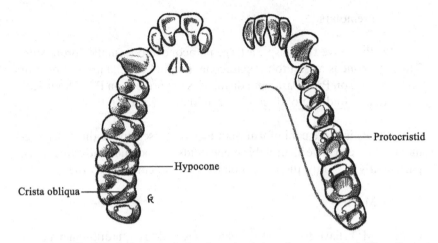

Fig. 6.10. *Callicebus moloch* male, occlusal view of permanent teeth (22 mm).

monkeys with prehensile tails. Indeed, the underside of the tail has dermal ridges (Midlo and Cummins, 1942) that function like fingerprints by increasing the friction when the animal is suspending itself from a branch of a tree or swinging outward to nibble a tender leaf at the far end of a branch.

Subfamily Callicebinae

Dietary habits

The diet of *Callicebus* monkeys (there are now considered to be several species of callicebines (Hershkovitz, 1990)) consists mainly of fruits, leaves, and insects (Rosenberger, 1992).

Morphological observations

	Male	Female
Callicebus moloch	3	1

Incisors (Fig. 6.10)

UPPER. I^{1-2} are different in that I^1 is broad and the incisal border is relatively horizontal whereas I^2 is narrow and has a more pointed crown. Marginal ridges are present on both teeth and the lingual surface is rather concave. A lingual cingulum is present on both teeth.

LOWER. I_{1-2} are relatively tall, slender teeth and subequal in size. The lingual surfaces are concave and the labial surfaces are convex. Marginal ridges are present; the distal marginal ridge is better developed than the others.

Canines

UPPER. The upper canine has a stout base from which it tapers to end above the occlusal plane. The distolingual surface is slightly concave and a noticeable lingual cingulum swings around the cervical portion of the tooth. No mesial groove is present.

LOWER. The lower canine curves slightly mesially and then distally to extend above the occlusal plane. A lingual cingulid is present, which terminates distally as a small heel. It is important to note that, as early as 1977, Kinzey demonstrated that *Callicebus* uses I_{1-2} and the lower canine to scrape and open the tough covers of certain fruits and the hard husk of the palm nut (Kinzey, 1977).

Premolars

UPPER. P^{2-4} have two cusps: a large paracone and a smaller protocone that is mesial to, and connected with, the paracone as in *Aotus*. Cingular remnants are present on the lingual surfaces of P^{2-4}; Kinzey (1973) found that the postprotostyle increased in frequency from P^2 53%, P^3 79%, to P^4 100% in his sample of 40 *Callicebus torquatus*. He also found the parastyle present 13% on P^{2-3}, and 8% on P^4. The distostyle frequency is P^2 18% and P^{3-4} 23%. Thus, the incidence of buccal and lingual cingula is reversed in *Aotus* and *Callicebus*.

LOWER. P_2 is caniniform and the protoconid is large and usually higher than it is on the bicuspid P_{3-4}. On these teeth the metaconid is small and almost directly opposite the protoconid and the two cusps are connected by a protocristid. Although the present sample is too small for statistical analysis, Kinzey (1973) found distostylids on P_2 24%, P_3 19%, and P_4 8% of his sample of *C. torquatus*. Orlosky (1973) found entoconids (91%) and hypoconids (77%) on P_4 in *C. moloch*.

Molars

UPPER. M^{1-2} have four cusps and are square from the occlusal view. The hypocones are relatively large and connected to the protocones by the entocrista. The hypocone is very small on M^3 and may be absent. A lingual cingulum is present on M^{1-3} and may extend distally around the protocone to become continuous with the hypocone. It is always better developed on M^{1-2}. Buccal cingula are not as common as lingual cingula, indeed the mesostyle is the only remaining part of the buccal cingulum. Kinzey (1973) found it present in *C. torquatus* on M^1 95% and M^2 53% of the time in his sample of 40 animals. It is present in 70% of this small sample on both M^1 and M^2.

LOWER. M_{1-2} have four cusps and the hypoconulid occurs on M_3. A protocristid separates the occlusal surface into a small, narrow trigonid basin and a larger, more spacious talonid basin. On M_{1-3} the metaconid is distolingual to the protoconid. A cristid obliqua is well defined only on M_{1-2}. A distostylid is present on M_1 56%, M_2 83%, and M_3 10% of 40 *C. torquatus* (Kinzey, 1973). It is also more frequent on M_2 in the present sample.

Odontometry

No measurements were taken of the present small sample. Orlosky (1973) did not report any sexual dimorphism in tooth size in *C. moloch*. As mentioned previously Kinzey (1972), in his study of canine development in *C. moloch*, attributed the lack of sexual dimorphism to the low level of aggression in the species.

The mesiodistal molar relations are $M^1 > M^2 > M^3 / M_1 < M_2 > M_3$.

Subfamily Pitheciinae

Dietary habits

The three genera belonging to the Pitheciinae, *Pithecia, Chiropotes* and *Cacajao*, are frugivorous primates that eat considerable amounts of seeds and unripe fruit with hard pericarps (Ayers, 1989; van Roosmalen, Mittermeier and Fleagle, 1988: Kinzey and Norconk, 1990). For example, *Chiropotes satanas* feeds predominately on immature seeds and ripe fruit and seems particularly fond of the members of the Brazil nut family, Lecythidaceae, which have very hard, tough seed pods (van Roosmalen *et al.*, 1988, p. 32). In fact, *C. satanas* is referred to as 'a neotropical seed predator' by these investigators.

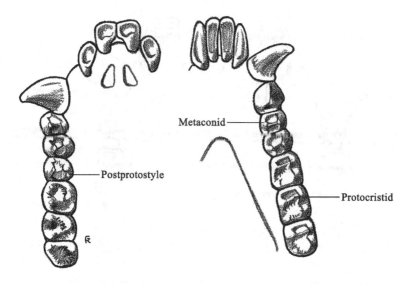

Fig. 6.11. *Pithecia pithecia* male, occlusal view of permanent teeth (28 mm).

Morphological observations

	Male	Female
Pithecia pithecia (white-faced saki)	2	1
Chiropotes satanas (black-bearded saki)	1	1
Cacajao calvus (bald uakara)	3	3

Incisors (Figs. 6.11–6.13)

UPPER. I^1 is larger than I^2 and both have a well-defined lingual fossa. Marginal ridges outline the lingual surfaces of both teeth and a well-formed lingual tubercle is present on I^1 (Orlosky, 1973; Kinzey, 1992). Orlosky (1973) found this tubercle or cusplet to be present on I^1 60% of the time in 11 specimens of *P. monachus*. Lingual tubercles are lacking on I^2. There is a wide diastema between I^2 and the upper canine.

LOWER. I_{1-2} are elongated and I_2 is larger than I_1. Both upper and, particularly, the lower incisors are somewhat procumbent (inclined slightly mesially from root to crown tip). It has been suggested by Kinzey (1992) that this incisor arrangement is well adapted for nipping or cropping while the animal is feeding. Incidentally, *P. pithecia* is apparently the most

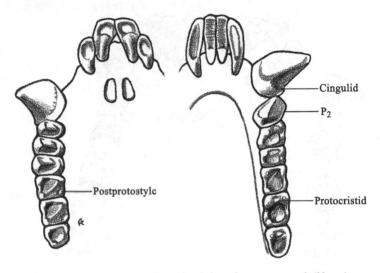

Fig. 6.12. *Chiropotes satanas* male, occlusal view of permanent teeth (29 mm).

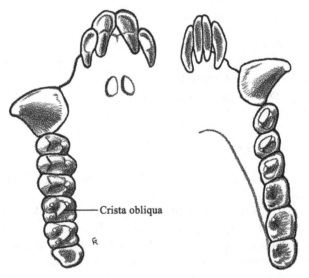

Fig. 6.13. *Cacajo calvus* male, occlusal view of permanent teeth (37 mm).

folivorous of the pitheciines. In addition, *Chiropotes* has been observed to bite a hole in fruit near the operculum and then use the 'procumbent incisors rather like a can opener to pop off the operculum and gain access to the seeds inside' (van Roosmalen *et al.*, 1988, p. 26).

Canines

UPPER. The upper canine is a large, robust tooth that splays laterally and becomes buccolingually tapered as it rises well above the occlusal plane. This is true, to different degrees, for both upper and lower canines in the pitheciines (Hershkovitz, 1985). There is a groove on the mesial surface of the tooth that passes from the cervical region to the crown tip.

LOWER. The lower canine is also large, tilts towards the buccal side of the tooth row and is separated from I_2 by a wide diastema. It has three surfaces and three borders, of which the distolingual border is rather sharp. A lingual cingulid passes around the cervical region and terminates distally as a distostylid. The position of the canines outside the dental arcade and separated from the lateral incisors permits the canine to operate with more force when biting into hard food objects (Kinzey, 1992). In addition, there is little sexual dimorphism in the canines, which led Kinzey (1992) to suggest that the large canines are more likely associated with feeding habits rather than with aggressive habits. Lucas *et al.* (1986b) mention that the shape of the canines as well as the large amount of projection in both sexes is unusual among anthropoids.

Premolars

UPPER. P^{2-4} are bicuspid and the paracone is larger than the protocone, although the latter cusp is almost as large as the paracone on P^{3-4} in *P. pithecia* and *C. satanas*. Postprotostyles as well as mesostyles are variably developed on P^{2-4}, being more frequent on P^{3-4} according to Kinzey (1973). The occlusal surfaces of unworn P^{3-4} have narrow, delicate crenulations within the boundaries of the marginal ridges. P^4 is the largest of the series and can be termed molariform.

LOWER. P_2 is slightly elongated mesiodistally with a high protoconid whose paracristid may end as a mesiostylid. Kinzey (1973) found mesiostylids on P_2 31% of the time in *C. rubicundus*. A small metaconid is variable. The elongated buccal surface of P_2 is somewhat similar to the sectorial P_3 of Old World monkeys. Although P_2 is not as elongated as the P_3 of Old World monkeys, it does pass along the more cervical distolingual portion of the upper canine, which has a vertical depression that receives the protoconid when the teeth are occluded. In the present small sample of pitheciines this functional relationship is better developed in *C. calvus*. P_{3-4} are bicuspid and frequently have hypoconids and entoconids, indeed, P_4 is molariform as is P^4. Crenulations are present on the occlusal surfaces of P_{3-4}, again as in P^{3-4}.

Molars

UPPER. M^{1-3} have four rather low cusps and the outline of the molars is rectangular in occlusal view. In *C. calvus* the hypocone may be absent. A crista obliqua is present on the molars, as are crenulations. Trigon and talon basins are present on M^{1-2}, but more variable on M^3. Mesostyles, protostyles, postprotostyles and metastyles are variable in pitheciines and have been thoroughly discussed by Kinzey (1973) and Orlosky (1973).

LOWER. M_{1-3} have four low cusps; there is no hypoconulid on M_3 in the pitheciines. The protoconid is mesial to the metaconid, and the two are connected by the protocristid. The trigonid basin is narrow and the talonid basin is more spacious. Crenulations are present on the molars of the pitheciines and appear to be particularly well developed in *P. pithecia*.

The combinations of molariform P^4 and P_4 and crenulations on P^{3-4} and P_{3-4} as well as M^{1-3} and M_{1-3} are considered hallmarks of the pitheciine dentition as well as derived features (Kinzey, 1992). At one time it was believed that the dentition of pitheciines represented a specialization of their posterior teeth related to chewing hard foods; however, recent field studies (Kinzey and Norconk, 1990, 1993) have found that after opening the hard outer pod cover, the seeds are rather soft, pliable and easy to masticate. Earlier, Lucas and Luke (1984) suggested that the crenulations and low cusp relief in pitheciines probably assist in the secondary reduction of seed particles during mastication. This suggestion would still seem to be correct irrespective of whether the seeds are hard or soft, since the low cusp relief and crenulations should continue to grind and crunch the seeds in the spacious talonid basins.

Odontometry
No measurements were made owing to small sample sizes; however, Orlosky (1973) reports no significant sex differences for any tooth dimension of *Pithecia monachus*.

Subfamily Atelinae

Dietary habits
The atelins are the largest of the New World monkeys and all have

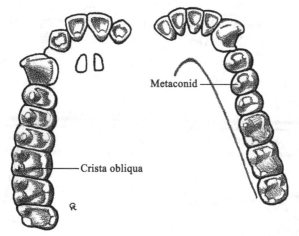

Fig. 6.14. *Ateles geoffroyi* female, occlusal view of permanent teeth (37 mm).

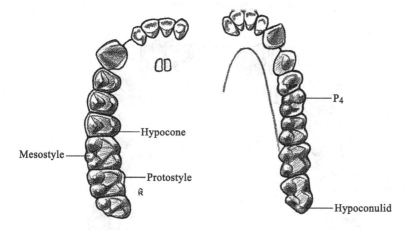

Fig. 6.15. *Brachyteles arachnoides* male, occlusal view of permanent teeth (44 mm).

prehensile tails from which they may suspend themselves from a branch while eating, particularly *Ateles*. Their diet consists mainly of leaves, fruits and flowers, although the proportions consumed during a year may vary a great deal among the four genera (Strier, 1992). Howler monkeys are the most folivorous of the New World monkeys and although there is some seasonal variation in their diets, leaves still make up half or more of their yearly diet (Fleagle, 1999). In addition, as with most primates, there is much seasonal variation in what they eat; for example, insects are eaten almost exclusively at certain times of the year by *Lagothrix*.

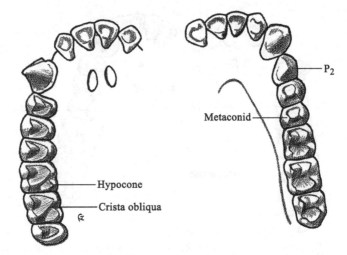

Fig. 6.16. *Lagothrix lagothricha* female, occlusal view of permanent teeth (40 mm).

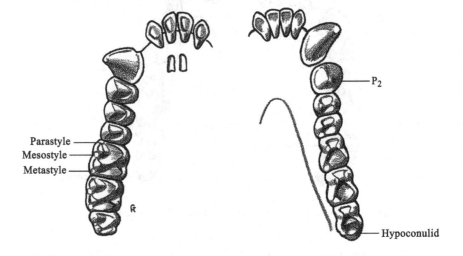

Fig. 6.17. *Alouatta palliata* male, occlusal view of permanent teeth (47 mm).

Morphological observations

	Males	Females
Ateles geoffroyi (black-handed spider monkey)	13	12
Ateles belzebuth (long-haired spider monkey)	11	8
Brachyteles arachnoides (woolly spider monkey)	1	—
Lagothrix lagothricha (common woolly monkey)	3	3

Alouatta palliata (mantled howler)	15	42
Alouatta seniculus (red howler)	18	18
Alouatta belzebul (black-and-red howler)	11	10

Incisors (Figs. 6.14–6.17)

UPPER. I^{1-2} are heteromorphic, i.e. I^1 is larger and the incisal border wider than in I^2, except in *Alouatta* where the two teeth are more homomorphic. Marginal ridges are present on both incisors. A lingual tubercle may be present on I^1 but is not present on I^2 in the present sample. Regarding the incisors of New World monkeys, Eaglen (1984) was able to demonstrate that these teeth were smaller, relative to body size, than they were in Old World monkeys, regardless of diet. In addition, Rosenberger and Strier (1989, p. 732) suggested that the larger I^1 relative to I^2 of *Ateles* is an 'automorphic adaptation reflecting a strong commitment to frugivory.'

LOWER. I_{1-2} are of similar size and morphology in *Ateles* and *Lagothrix* with only a slight suggestion of marginal ridges. I_{1-2} in *Brachyteles* are somewhat heteromorphic: I_1 is narrow with a flat incisal border while I_2 is more pointed and larger than I_1. *Brachyteles* has a rather high incidence of mandibular underbite, though not as high as in *Alouatta* (Miles and Grigson, 1990; Serra, 1951; Zingeser, 1973).

I_{1-2} of *Alouatta* are similar although I_2 is more caniniform than I_1. Incisor occlusion is frequently different from that of other atelines as well as that of the other platyrrhines: instead of the edge-to-edge incisor occlusion that is the normal condition in most monkeys — both New World and Old World species — the howler monkey has a high incidence of mandibular incisor protrusion, i.e. underbite or epharmosis. Colyer (1936) was one of the first students to record the presence of underbite in non-human primates (Miles and Grigson, 1990). He found 9.5% in *A. seniculus* and 50% in *A. villosa (palliata)*. Later Serra (1951) recorded an incidence of 43.5% for the species and Schultz (1960) found only two instances of edge-to-edge occlusion in 493 skulls of this species. Zingeser (1968, 1973) after finding an incidence of 100% underbite in *A. caraya,* a species living on islands in the Rio Parana in northern Argentina, posited that the underbite was the result of their herbaceous diet. In the present sample of howler monkeys the incidence of underbite is 92.8% in *A. palliata*, 47.1% in *A. seniculus* and 75% in *A. belzebul*. There is no significant sex difference in underbite in these three genera of howler monkeys (Swindler, 1979). When underbite is not present in howler monkeys the edge-to-edge bite is almost always present, although overbite and openbite do occur (Swindler, 1979). Further discussion of the high incidence of underbite in leaf-eating monkeys appears in Chapter 7.

Canines

UPPER. The canines are large, stout teeth that project well above the occlusal plane in atelines. A lingual cingulum is usually well developed, although it is rather feeble in *Alouatta* where an occasional small buccal cingulum may be found. Sexual dimorphism obtains in the development of the lingual cingulum in *Ateles*. In females, the cingulum is much more prominent than it is in males (Orlosky, 1973).

LOWER. The canine is rather robust and projects well above the occlusal plane, except in *B. arachnoides* where it is barely higher than adjacent teeth and a diastema separates it from P_2. The distal portion of the canine in *B. arachnoides* forms a ledge that is part of the honing mechanism described by Zingeser (1968, 1973). A lingual cingulum passes around the cervical region of the tooth and may form a small distal heel.

Premolars

UPPER. P^{2-4} generally have two cusps, paracone and protocone, and the former is larger than the latter in *Ateles* and *Lagothrix*. A small hypocone is frequently present on P^{3-4} in *A. belzebuth*, but is lacking in *A. geoffroyi*. Moreover, P^{3-4} in *A. belzebuth* have a greater frequency of para- and distostyles ($p < 0.05$) than in *A. geoffroyi* (Orlosky, 1973). Kinzey (1973) reported a complete absence of buccal and lingual remnants in both *A. geoffroyi* and *A. paniscus*. The *Brachyteles* P^4 may have a hypocone as reported here and by Zingeser (1973). The hypocone also has a variable expression on P^4 in *L. lagothricha*. P^2 lacks a protocone in 38% of the present sample of spider monkeys but the lingual cingulum is large, probably functioning as a bicuspid tooth. Mesostyles are not present in the atelines examined here but Kinzey (1973) reports mesostyles on P^2 in 7% of 29 specimens of *L. lagothricha*. In the present sample of *Alouatta*, para- and distostyles are always present on P^{3-4} but variable on P^2 (65% present for both styles).

LOWER. In atelines P_2 is unicuspid; if present, the metaconid is small. P_{3-4} are bicuspid and if P_4 has three cusps, the third is the hypoconid. In *L. lagothricha* P_4 has a large talonid basin with the hypoconid and entoconid on either side of the basin. A constricted trigonid basin is mesial to the protocristid. A protocristid connects the protoconid and metaconid and if the metaconid is absent, the protocristid passes distolingually to merge with the marginal ridge. Orlosky (1973) reports paraconids (mesiostylids) on P_2 in *A. belzebuth* whereas they are present on P_4 in *B. arachnoides* at the end of the paracristid (Zingeser, 1973). Buccal cingulids are present on P_{2-4} in howler monkeys, variable in spider monkeys and absent in the present small sample of woolly monkeys, although Kinzey (1973) reports them in his study of woolly monkeys.

Molars

UPPER. M^{1-3} usually have four cusps; however, the hypocone is smaller than the other cusps and may be completely lacking on M^3 (50%). The hypocone is absent in *A. geoffroyi* in the present sample; Kinzey (1973) reports its absence in 62% of *A. geoffroyi* and 55% of *A. paniscus*. The hypocone may also be absent on M^3 in *L. lagothricha*. The postprotostyle is small and may be present on M^{1-3} in *Ateles* but is less common in the other taxa. On the other hand, the crista obliqua is present on M^{1-2} in all atelines in the present sample. It is less common on M^3. Frequently associated with the crista obliqua in *Alouatta* is a slight enlargement near the base of the metacone, the metaconule. The free crest of the metaconule is sharp and in occlusion occupies the re-entrant space between the postmetacristid of one tooth and the paracristid of the next lower tooth. In turn, the metaconule may be associated with the transverse widening and lengthening of the talonid that enhances the shearing action of the molars, a functional mechanism recognized by Serra (1952) that he associated with the leaf-eating habits of the howler monkey. In their comprehensive study of ateline primates Rosenberger and Strier (1989, p. 730) discussed in detail the functional relationships between the upper and lower molars of *Alouatta* and emphasized 'the essence of *Alouatta* occlusion is buccal shear.'

Mesostyles are present in atelines, but they are not common except in *Alouatta* where they are present 100% of the time on M^{1-2} but are rare on M^3. Para- and metastyles are also very common in *Alouatta* and, as with mesostyles, they are quite rare in the other ateline species.

LOWER. M_{1-2} have four cusps in atelines and hypoconulids are present on M_3 in *Brachyteles, Ateles* and *Alouatta*. Indeed, *Ateles* may have small hypoconulids on all three lower molars (Orlosky, 1973). A protocristid is present on M_{1-3}, separating the occlusal surface into a trigonid basin and a larger talonid basin, particularly in *Alouatta* and *Lagothrix* where it is quite wide. The trigonid is slightly higher than the talonid.

It is interesting that Remane (1960) compared the upper and lower 'bunodont' molars of *Brachyteles* with artiodactyl molars and coined the term 'selenobuntodonty' to describe them. Later Zingeser (1973) noted the similarity of the lower molars, especially M_1, to the bilophodont molars of the Old World colobine (leaf-eating) monkeys (see Chapter 7).

The M_{1-3} of *Alouatta* have four large cusps and the hypuconulid is present on M_3. The protocristid connecting the protoconid and metaconid is sharp. The trigonid is wider on M_2 than on M_1. The talonid basin is spacious and receives the protocone of the maxillary molar. Both the paracristid and cristid obliqua are well formed and sharp, and play important roles during mastication. M_3 is somewhat different in morphology in that (a) the protoconid and metaconid are opposite each other, (b) the trigonid is as wide as the talonid and slightly elongated, (c) the cristid obliqua is not as oblique, and (d) both the hypoconulid and tuberculum intermedium may be present. In fact, the hypoconulid is present 67% of the time in *A. palliata*, 88% in *A. seniculus*, and 75% in *A. belzebul*. The tuberculum intermedium is present 24% of the time in *A. belzebul*, 14% in *A. seniculus*, and 1% in *A. palliata*. It is interesting to note that small paraconids have been described on the lower molars by Le Gros Clark (1971). These were not present in this sample.

Odontometry (Appendix 1, Tables 39–44 and 57–64, pp. 185–7 and 194–7)

In several *Ateles geoffroyi* the female teeth are larger than those of the male (Tables 39–42). The measurements of *Brachyteles* were made by Zingeser (1973) and are presented here with his permission. Sexes were not separated and the sample consisted of 21 skulls. *Lagothrix* was not measured owing to the small number of specimens. Tables 57–64 present the dental sexual dimorphism of these two howler species; in all cases, males are larger than females. The greatest length and breadth differences occur in the upper and lower canines of the three species. *Alouatta* males have quite robust canines and Kay *et al.* (1988) report that the males of *A. belzebul* had the greatest degree of sexual dimorphism of the many platyrrhine species they examined, which they attributed to intermale competition.

The lower molars of *Alouatta* increase mesiodistally from M_1 to M_3, a progression probably first reported by Pocock (1925, p. 32) who wrote, '*Alouatta* stands alone among the Platyrrhini in having the last molar longer than the first.'

7   *Cercopithecidae*

Present distribution and habitat

The family Cercopithecidae is divided into two subfamilies, the Cercopithecinae and Colobinae, which are distributed throughout Asia and Africa between latitudes 35°N and 30°S. The single exception to this is the Japanese macaque (*Macaca fuscata*), which lives as far north as 41° on the island of Honshu. The cercopithecines have cheek pouches, non-sacculated stomachs, and are essentially non-leaf-eating, whereas the colobines do not have cheek pouches, their stomachs are sacculated, and they are leaf-eating. The cercopithecines have a wider geographic distribution than the colobines, ranging from North Africa in the west to Japan in the east; moreover, the cercopithecines are both terrestrial and arboreal whereas the colobines are predominately arboreal. There is one genus of Colobinae in Africa; all the other members of the subfamily live in India, China, Malaysia and parts of Indonesia.

Members of both subfamilies range from sea level to the snow line. In fact, some langurs live in excess of 3400 m in the Himalayas and one species of guenon, *Cercopithecus mitis*, is found at 3300 m in Africa (Rowe, 1996). The predominant mode of locomotion of all Old World monkeys is quadrupedal, whether on the ground or in the trees. Both subfamilies are diurnal.

Dietary habits

Old World monkeys are, for the most part, vegetarians: they eat mostly fruit, leaves, bark, grass, nuts, seeds, and flowers. The cercopithecines may also add insects, small vertebrates, birds and eggs to their diets. Baboons are known to eat gazelle and other mammals on occasion. And there is the crab-eating macaque, *Macaca fascicularis* which really eats crabs. As we have mentioned earlier, the diet of many primates varies seasonally and it is the same with the Old World monkeys.

123

General dental information

Permanent dentition: I^2-C^1-P^2-M^3 / I_2-C_1-P_2-M_3
Deciduous dentition: di^2-dc^1-dp^2 / di_2-dc_1-dp_2

Studies of enamel thickness in Old World monkeys have shown that in general, the enamel of molars is thicker in frugivorous primates that eat harder foods than in the more folivorous forms that eat softer foods, suggesting that thick enamel might be an adaptation to eating tough or hard food items. Kay (1981, pp. 145, 147) was able to show that the 'relative thickness of enamel gives no information regarding whether an animal forages primarily on the ground or in the trees.' He did conclude, however, 'that thick enamel is routinely seen among species that eat very hard foods.' It is interesting to recall that Shellis *et al.* (1998) found that among all the anthropoids they investigated *Cebus apella*, an inveterate eater of fruits with hard covering, possessed the thickest molar enamel. In addition, Shellis *et al.* (1998) also demonstrated that the scaling of enamel thickness with tooth size and body mass tended to show positive allometry among anthropoid primates.

Perhaps the most characteristic or hallmark feature of the dentition of Old World monkeys is their bilophodont upper and lower molars (Fig. 7.1). Their anatomy consists of a rather high crown with four cusps at the margins connected by two transverse crests or lophs that separate the occlusal surface of each molar into three fovea. On the lower molars these fovea or basins are from mesial to distal the trigonid basin, the talonid basin, and the distal fovea. On the upper molars the fovea are from mesial to distal the mesial fovea, the trigon basin, and the distal fovea or talon basin. In addition, the crown is constricted between the mesial and distal pairs of cusps.

In occlusion, the transverse lophs on the lower teeth fit into corresponding embrasures on the upper teeth and vice versa. Thus, shearing and crushing actions are performed each time the molars come together. In addition, these molars have higher cusps, longer shearing blades, and somewhat larger crushing surfaces relative to body mass (Kay, 1975). Indeed, it has been suggested that bilophodonty suggests a somewhat folivorous ancestry for the lineage (Kay and Hylander, 1978), which is not too dissimilar from the suggestion of Happel (1988, p. 324) that the evolution of bilophodont molars 'may reflect an increasing reliance on seeds.' It is well known that the molars of some other primates that consume large amounts of leaves approach the bilophodont condition, e.g. *Propithecus* and *Alouatta* (see Figs. 5.7 and 6.17). In addition, over eighty years ago

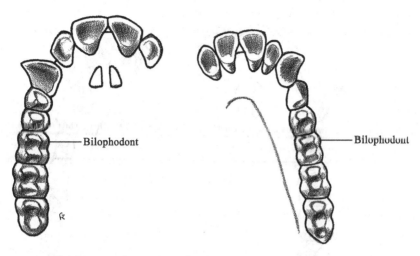

Fig. 7.1. *Miopithecus talapoin* female, occlusal view of permanent teeth (25 mm).

Gregory (1922) noted the parallel development of bilophodont molars in other leaf-eating mammals, such as tapirs and kangaroos. More recently, Jablonski (1994) published an insightful study of the evolution of bilophodont molars in two distant lineages of vegetarians, the macropodine marsupials and the cercopithecine primate *Theropithecus gelada*.

Delson (1973, 1975) discussed the differences in the degree of crown flare in bilophodont molars from the cusp tips to the cervical region (more apparent on the buccal surface of lowers and the lingual surface of uppers) in the two subfamilies. The flare was greatest in papionins, particularly in *Cercocebus* and *Papio*, and least in *Theropithecus*. In general, molar flare is less in the other cercopithecines and least in colobines.

The lower P_3 is sectorial in all Old World monkeys, and as we will see later, in the lesser and great apes (Fig. 7.2). The sectorial P_3 has a single cusp (protoconid) that is compressed buccolingually to form an oblique cutting edge that shears against (sharpens) the lingual surface of the upper canine. This C^1–P_3 honing mechanism of cercopithecoids is described by Zingeser (1968) and Walker (1984) discussed the function of this important mechanism for maintaining the sharpness of the male upper canine in Old World monkeys, using the male baboon as the model.

Studies of various enamel hypoplasias in the teeth of non-human primates were not common until about the mid-1980s (Guatelli-Steinberg and Lukacs, 1999) as discussed in Chapter 2. For the most part, these studies have been of catarrhine primates except for Newell's (1998; cited by Guatelli-Steinberg and Lukacs, 1999) investigation of sex differences in

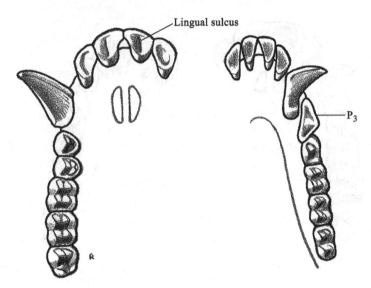

Fig. 7.2. *Erythrocebus patas* male, occlusal view of permanent teeth (55 mm).

linear enamel hypoplasia (LEH), which included several species of platy-rrhine. In their study of enamel hypoplasia in cercopithecids Vitzthum and Wikander (1988) report an incidence of 1.5–3% in *Chlorocebus aethiops, Cercopithecus mitis, Papio* and *Mandrillus*. They found hypoplasias more often in the anterior dentition than in the molars of these animals. Guatelli-Steinberg and Skinner (2000) reported more variability in LEH in cercopithecids than previously recognized; for example, the incidence of LEH ranges from 0% in *Cercocebus atys* and *C. galeritus* to 42% in *C. torquatus* and 54% in *Piliocolobus badius*. In addition, they posited that the monkey–ape dichotomy (that apes have a greater incidence of LEH) is not as clear-cut as some research suggests, and what differences there are, are probably due to differences in the timing of crown formation between the two taxa. In addition, they did not rule out differential environmental stress as a potential contributing factor.

Subfamily Cercopithecinae

The morphology sections in this large subfamily are separated into three groups for ease of presentation and discussion.

Morphological observations: guenons

	Male	Female
Cercopithecus cephus (moustached monkey)	9	12
Cercopithecus nictitans (spot-nosed monkey)	11	7
Cercopithecus mona (mona monkey)	26	16
Cercopithecus mitis (blue monkey)	30	21
Cercopithecus lhoesti (l'Hoest's monkey)	3	7
Cercopithecus neglectus (De Brazza's monkey)	13	7
Cercopithecus ascanius (red-tailed monkey)	30	18
Chlorocebus aethiops (grivet monkey)	35	20
Miopithecus talapoin (talapoin monkey)	2	2
Erythrocebus patas (patas monkey)	5	2

Incisors (Fig. 7.2)

UPPER. I^{1-2} tend to vary a great deal among the species examined here. I^1 is spatulate, having a rectangular labial surface and a rather triangular lingual surface. Mesial and distal marginal ridges are present as well as a lingual ridge or cingulum. The lingual ridge varies in its construction among cercopithecines and may be absent, V-shaped or U-shaped (Sirianni, 1974). For example, the lingual ridge is absent in over 25% of *Chlorocebus aethiops* and V-shaped in the remainder, whereas in *Cercopithecus ascanius* it is U-shaped (Fig. 7.8). Medial lingual sulci (Fig. 7.2) are variably developed in the cercopithecines, although it approaches more the configuration of a fossa in some taxa.

I^2 is much narrower mesiodistally than I^1 and frequently truncated and pointed. Marginal ridges are present but are narrower than those on I^1. A lingual tubercle is absent on I^{1-2} in the guenon species studied. In their discussion of the morphology of the lingual surface of I^1 in primates Henke and Rothe (1997) describe and illustrate features very similar to those presented here for the cercopithecids.

LOWER. I_1 is wider than I_2 in the guenons studied here. The lingual surfaces of both I_1 and I_2 are outlined by marginal ridges and this surface is somewhat triangular whereas the labial surface is convex mesiodistally.

Walker (1976), in his investigation of the lingual surfaces of both I^{1-2} and I_{1-2} in Old World monkeys, found that wear striations in cercopithecines tend to be oriented labiolingually, i.e. perpendicular to the incisal borders, whereas in colobines the striations are oriented in a predominately mesiodistal direction, i.e. parallel to the incisal edge. Walker (1976, p. 306) opined that these findings offered important information 'concerning habitat preferences and feeding adaptations of these animals.' In their investigation of enamel distribution on the lingual surfaces of I_1 in several Old World monkeys Shellis and Hiiemae (1986) found a significant difference in enamel thickness between cercopithecines and colobines, as did Strasser and Delson (1987). In the former taxon, which is mainly frugivorous or omnivorous, the enamel is virtually absent, resulting in a sharp incisal border; in the latter taxon, which is essentially folivorous, the enamel is thicker on both the labial and lingual surfaces, resulting in blunt incisal edges. Shellis and Hiiemae (1986, p. 103) offered a functional explanation. 'It is suggested that colobine incisors are used mainly in gripping or tearing leaves, whereas cercopithecine incisors are better adapted to cutting and scraping.' Hylander (1975), in an extensive comparative study of incisor size in primates, found that cercopithecines had relatively larger incisors than colobines. He also attributed the difference to incisor use, i.e. cercopithecines use their incisors for opening the tough outer coverings of fruits in order to get at the softer inner parts.

Canines

UPPER. The canines are large, robust teeth that project well above the occlusal plane and have been described as 'large stabbing weapons' by Delson (1975, p. 175). The base is well formed and the crown tapers off to end as a sharp point, especially in males. The mesial surface is wider than the more blade-like distal surface, which is maintained by occlusion with P_3. The functional relationship or honing mechanism between the cercopithecid C^1 and P_3 is thoroughly discussed by Zingeser (1968; see also Delson, 1973); later studies by Greenfield and Washburn (1991, 1992) are concerned with the polymorphic aspects existing in the maxillary canine and lower P_3 of a large sample of anthropoid primates. The upper canines show a high degree of sexual dimorphism in shape and size that is associated with intermale competition, and as demonstrated by Plavcan (1993) this canine size and shape difference is not related to diet among cercopithecids, although some anthropoid species, e.g. the pithecines, have canines specialized for dietary purposes.

LOWER. The crown splays distolabially as it passes above the occlusal plane, and indeed, the tooth is somewhat twisted so that the long axis is set slightly obliquely to the mesiodistal line of P_3–M_3. A narrow mesial longitudinal groove is present in most species, although it is absent in *C. ascanius*. A heel-like projection is present in all species and in *Chlorocebus aethiops* and *Erythrocebus patas* it is quite well developed, particularly in males. The lower canine is largest in males.

Premolars

UPPER. P^3 always has a large paracone; however, the number of lingual cusps varies from none to two. The protocone is absent on P^3 in *Cercopithecus neglectus* but is present in *Chlorocebus aethiops* 65% of the time (Sirianni, 1974). In the majority of species examined here a protocone is absent in less than 25% of the sample. A narrow anterior transverse crista connects the paracone and protocone; when the protocone is absent, the crista ends on the lingual cingulum. In either case, the occlusal surface is separated into a narrow mesial fovea and a more expanded distal trigon basin. P^4 has two cusps in the majority of guenons, and as in P^3, a mesial crista connects the two cusps, dividing the occlusal surface into a narrow mesial fovea and a larger distal trigon.

LOWER. P_3 is sectorial, i.e. it is compressed buccolingually forming a mesiodistally elongated surface consisting of one cusp (protoconid) that shears against the upper canine during mastication. As mentioned earlier, in the discussion of the upper canine, the C–P_3 honing mechanism has been considered in detail by various students. This honing whetstone is variably developed among Old World monkeys as well as in the lesser and great apes. There is generally as much or even more sexual dimorphism in P_3 as in the canine (see Appendix 1). P_4 has two cusps, the protoconid and metaconid, and there may be an entoconid or hypoconid present, as in two of the seven specimens of *Erythropithecus*. A protocristid connects the protoconid and metaconid, thereby dividing the occlusal surface into narrow trigonid and slightly expanded talonid basins.

Molars

UPPER. The molars are bilophodont. The cusps are high with well-defined shearing blades. In all species of guenon studied, M^{1-2} has four cusps, whereas M^3 often has fewer cusps; the missing cusp is almost always the hypocone. Sirianni (1974) in her study of cercopithecid teeth found the hypocone absent most frequently (46.2%) in *Cercopithecus ascanius*. Remnants of a lingual cingulum (protostyle) are frequently present on M^1 (80%) in *C. neglectus*, although they may occur on all three molars (Sirianni, 1974). The protostyle is least common on *C. cephus* (2% on M^2). A distoconulus is present on M^3 in one *Erythrocebus patas*.

LOWER. The molars are bilophodont. M_{1-2} have four cusps whereas M_3 lacks the hypoconulid, as it does in all members of the genus *Cercopithecus* as well as in *Erythrocebus patas*. A protostylid is present 85% of the time on the lower molars of *Chlorocebus aethiops* and on all lower molars of *Miopithecus talapoin*.

Odontometry (Appendix 1, Tables 65–92, pp. 198–211)
The degree of sexual dimorphism varied among the taxa. In all species measured, the upper canine and P_3 were always significantly different in their mesiodistal dimensions. Males were always larger than females. In addition, molar size relations express a great deal of sequential variability in guenons, with no general pattern emerging (Swindler and Olshan, 1982). In general, variations of size and structure among the molars of Old World monkeys are attributable to differences in diet (Kay, 1978; Kay and Hylander, 1978; Lucas *et al.*, 1986a). Cope (1993, p. 236) measured the teeth of 13 *Cercopithecus* species in his study of dental variation within and between these species, and further, he tested the efficacy of the coefficient of variation (CV) for identifying multiple species in fossil samples. He concluded that if the CV is consistently high for several dental dimensions 'this is compelling evidence that more than one [sic] species is represented.'

Morphological observations: mangabeys

	Male	Female
Cercocebus torquatus (white-collared mangabey)	14	10
C. galeritus (Tana River mangabey)	10	10
Lophocebus albigena (gray-cheeked mangabey)	30	29

Incisors (Fig. 7.3)

UPPER. I^1 is a broad, spatulate tooth that is much larger than I^2. Mesial and distal marginal ridges are prominent. Within the marginal ridges, well-formed enamel elevations are separated by a rather deep median lingual sulcus that is also present in *Papio* and *Macaca* (Swindler, 1968). I^2 shows little definition on its lingual surface and there is no median lingual sulcus.

LOWER. I_1 has a horizontal incisal border whereas this border inclines distally on I_2. Narrow marginal ridges delineate smooth, slightly concave lingual surfaces on both I_1 and I_2. The lingual surfaces of I_{1-2}, as in frugivores in general, has little or no enamel (Shellis and Hiiemae, 1986).

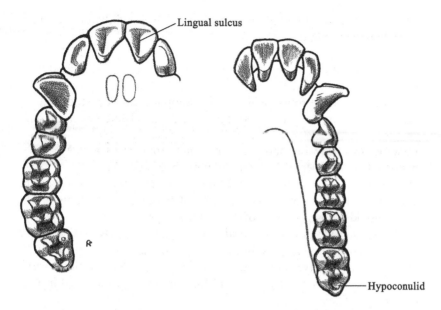

Fig. 7.3. *Lophocebus albigena* female, occlusal view of permanent teeth (53 mm).

Canines

UPPER. The upper canine is large and projects well above the occlusal plane. It is narrow distally and broad mesially. A mesial groove is present that is wider and deeper in males. A diastema separates the canine from I^2. The lingual surface of the upper canine hones across the lateral surface of P_3 during mastication.

LOWER. The lower canine is robust and inclines labially. There is a well-developed heel.

Premolars

UPPER. P^{3-4} have two cusps; indeed, P^4 has two lingual cusps in 15% of the present sample. In both teeth an anterior transverse crista passes between the paracone and protocone, dividing the occlusal surface into a narrow mesial basin and a larger trigon basin.

LOWER. P_3 is sectoral and P_4 is bicuspid and often has an entoconid and hypoconid on the talonid. A protocristid connects the protoconid and metaconid, thereby separating the occlusal surface into a narrow trigonid basin and a larger talonid basin. The talonid is on a slightly lower occlusal level than the trigonid.

Molars

UPPER. M^{1-3} have four cusps and are bilophodont. A distoconulus is present on M^3 in 10% of the present sample; no sexual difference is present.

LOWER. M_{1-3} are bilophodont. M_{1-2} have four cusps and M_3 has a hypoconulid. A tuberculum intermedium is present on M_1 (7%), M_2 (9%) and M_3 (21% of the time). A tuberculum sextum is present between the entoconid and hypoconulid on M_3 20% of the time in this sample. The hypoconulid is absent in 13% of these mangabeys.

The lingual side of cercopithecid lower molars is nearly vertical to the alveolar plane; however, the buccal side, as discussed by Delson (1973, 1975), may flare outward, resulting in a different width between the cervical and apical dimensions. He found the flare most pronounced in *Cercocebus* and *Papio* and least in colobines.

Odontometry (Appendix 1, Tables 93–104, pp. 212–17)

Mangabeys display considerable sexual dimorphism in tooth size. Of the species examined, *Lophocebus albigena* is significantly different in almost all teeth whereas *Cercocebus torquatus* has the least amount of sexual dimorphism in its teeth. In all taxa, the mesiodistal diameters of the upper canine and P_3 are significantly different ($p < 0.01$).

Morphological observations: papionins

	Males	Females
Macaca mulatta (rhesus macaque)	90	83
Macaca fascicularis (crab-eating macaque)	60	55
Macaca nemestrina (pigtailed macaque)	85	50
Macaca nigra (Celebes black macaque)	10	7
Papio cynocephalus (yellow baboon)	50	53
Mandrillus sphinx (mandrill)	1	1
Theropithecus gelada (gelada)	3	3

Incisors (Figs. 7.4–7.7)

UPPER. Swindler (1968) studied the lingual surfaces of I^{1-2} in several taxa of Old World monkeys and described various morphological differences,

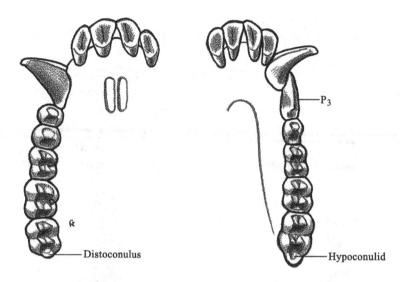

Fig. 7.4. *Macaca nemestrina* male, occlusal view of permanent teeth (64 mm).

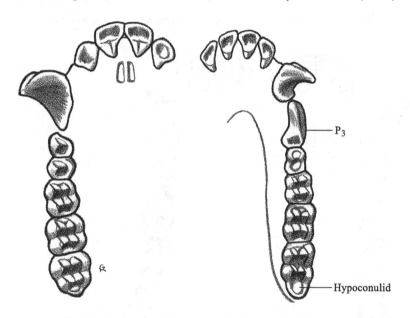

Fig. 7.5. *Papio cynocephalus* male, occlusal view of permanent teeth (85 mm).

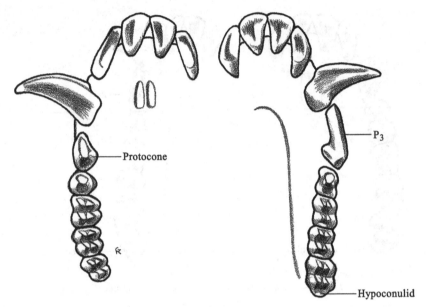

Fig. 7.6. *Mandrillus sphinx* male, occlusal view of permanent teeth (98 mm).

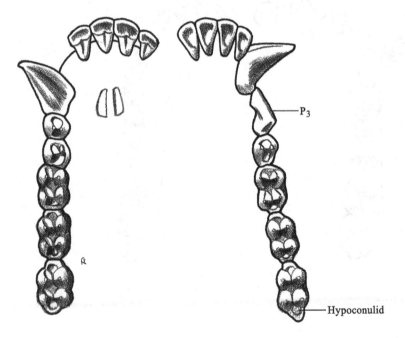

Fig. 7.7. *Theropithecus gelada* male, occlusal view of permanent teeth (78 mm).

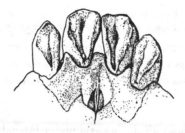

Papio anubis

Macaca mulatta

Cercocebus albigena

Cercopithecus ascanius

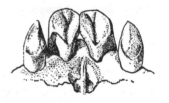

Cercopithecus aethiops

Fig. 7.8. Lingual surface variation of the upper incisors in several taxa of cercopithecids. Not drawn to scale. Adapted from Swindler (1968).

which are depicted in Fig. 7.8. I^1 is a broad, spatulate tooth that has a quadrilateral labial surface and a rather triangular lingual surface. Mesial and distal marginal ridges are present and the distal ridge is generally better developed in baboons and macaques than in the other taxa. A median lingual sulcus is present in all species and is particularly deep in *Papio*. Enamel elevations converge toward the cervical end of I^1 in baboons and macaques to form a V-shaped contour. A lingual tubercle is present at the base of the V in 43% of the macaques and it has a similar distribution in the two sexes (Swindler, 1968). The tubercle is not present in the other species, although a pit is frequently present at the junction of the ridges in *Papio*. I^2 has a median sulcus but it is not as well developed as it is on I^1. The median sulcus is absent on I^2 in the two specimens examined of *Mandrillus*.

LOWER. I_{1-2} have smooth, rather triangular lingual surfaces with occasional median lingual sulci. The labial surfaces are mesiodistally convex. Noble (1969), in a histological study of the lingual surfaces of the lower incisors of *Papio*, determined that enamel was thin or absent on these surfaces; this was an adaptation for developing the sharp edges of these teeth, not unlike the situation in the lower incisors of rodents. Later Delson (1973, 1975, p. 103) opined that the absence or thinning of enamel on the lingual surfaces of the lower incisors resulted in their sharp incisal edges and was 'characteristic of Papionini among the cercopithecines.' As mentioned previously, Shellis and Hiiemae (1986) found little or no enamel on the lingual surface of the lower incisors of papionins and cercopithecines. A recent study of Papionini lower incisors by Gantt, Strickland and Rafter (1998), using scanning electron microscopic (SEM) analysis, demonstrated the lack of enamel on the lingual surfaces of the lower incisors of these primates. As suggested by Gantt *et al.* (1998), these findings indicate a major change in tooth morphogenesis between the two subfamilies Colobinae and Cercopithecinae.

Canines

UPPER. The upper canines in Papionini are large, robust teeth that project well above the occlusal plane. They are sexually dimorphic, males always larger than females, even though female canines are large. The mesial groove is present in both sexes, but it is better developed in males. As in all cercopithecids, the C^1–P_3 honing mechanism is present. The studies of Greenfield and Washburn (1991, 1992) and Plavcan (1993) on the canines of Old World monkeys were discussed previously.

LOWER. The lower canine is large and projects labiodistally as it rises above the occlusal plane. A mesial groove is present and in *Mandrillus* another groove curves up along the lingual surface. A distal heel is present in all species. In *Papio* a lingual cingulid is discernible as it passes distally to become a well-formed heel.

Premolars

UPPER. P^{3-4} are bicuspid and the paracone is larger than the protocone. The two cusps are connected by the anterior transverse cristae, dividing the occlusal surface into a mesial fovea and a larger trigon basin. In *Mandrillus* P^3 has a greater extension of the crown onto the mesiobuccal root (Fig. 7.6). This feature is present to various degrees in cercopithecids and Delson (1973) discusses its reliability in distinguishing isolated P^3s from P^4s. In several taxa the protocone is almost as large as the paracone on P^4. In addition, P^{3-4} in *Theropithecus* has a mesiolingual cleft, which is not usually present in other taxa (Delson, 1973).

LOWER. P_3 is sectorial and its elongated protoconid shears against the upper canine as in all Old World monkeys. A fossa is usually present on the inclined distal surface of P_3 that Delson (1973) believes is homoplastic but probably not homologous with the talonid basin of the lower molars. The study of Walker (1984) is important here since it demonstrated that the honing surface of P_3 has a thicker layer of enamel than the corresponding surface of the upper canine, at least in *Papio cynocephalus*. He therefore concluded that the differential of enamel between the two surfaces was sufficient to protect P_3. P_4 is always bicuspid and may have small entoconids and hypoconids on the talonid rim. P_4 is molariform in *Papio* and *Mandrillus*. In *Papio* the metaconid is wider and higher than the protoconid in 100% of 21 animals studied by Hornbeck and Swindler (1967). Because of wear many animals in the sample could not be examined. The protoconid and metaconid are connected by the protocristid. There are often small cusplids along the talonid rim in addition to the major cusps, resulting in a rather complex occlusal surface in P_4.

Molars

UPPER. M^{1-3} have four cusps and are bilophodont. The lingual developmental groove is usually wider and deeper than the buccal groove. In macaques, an interconulus (a cingular remnant) may be present at the base of the lingual groove. It appears to be more common in *Macaca fuscata* (67% in M^3) than in other species of macaque (Saheki, 1966). It is present in *M. nemestrina* (38% in M^3) and in *M. mulatta* (4% in M^3). A distoconulus (on the distal occlusal surface of M^3) also appears more often in *M. fuscata* (38%) (Saheki, 1966); in the present sample, it is present 10% of the time in *M. nemestrina* and 4% of the time in *M. fascicularis*. It is absent in *M. mulatta*. These accessory cusps are not found in the other papionins studied. As is typical of *Papio*, the molars flare between the cusp tips and the cervix. *Theropithecus* has high-crowned molars with deep and widely separated basins with long shearing blades, and the molars are quite long. Of the three basins, the middle or trigon basin is particularly wide and deep and the developmental groove passing transversely through it forms clefts in the buccal and lingual surfaces, especially the latter. The mesial and distal basins are also somewhat elongated. These high-crowned molars are much higher than the lower-crowned molars, with their rounded cusps, seen in *Mandrillus*.

LOWER. The lower molars are bilophodont, rectangular, and have four cusps, except M_3, which usually has a well-formed hypoconulid. Molar flare is more pronounced on the buccal aspect than on the lingual; indeed, the lingual side is nearly vertical. The trigonid, talonid and distal basins are present on the occlusal surface. The tuberculum intermedium and tuberculum sextum are variably developed in these species, and are more common in macaques and baboons. In *Macaca fuscata* the tuberculum intermedium is more frequent on M_2 (38%), than in the samples of *M. nemestrina* (M_2 11%), *M. fascicularis* (M_2 7%), and *M. mulatta* (M_2 3%). The tuberculum sextum varies on M_3 from 61% in *M. nemestrina*, 68% in *M. fascicularis*, 26% in *M. mulatta* to 39% in *M. fuscata* reported by Saheki (1966). The tuberculum intermedium is present in *Papio* on M_1 in 56.8% of males and 71.7% of females, on M_2 in 75% of males and 75% of females, and on M_3 in 70.8% of males and 77% of females. The tuberculum sextum is about twice as common in male baboons as in females, 56.7% compared with 25.9%. These tubercles are apparently rare in mandrills and geladas, although the sample size is too small to make a definitive statement.

Theropithecus gelada is a predominantly grass-eating cercopithecine primate (Jolly, 1970; Teaford, 1993). Much has been written about the various dental modifications made by grass-feeding mammals. These usually include increase in the surface area of the teeth, evolution of complex enamel patterns, and increase in crown height. These dental strategies have been considered in detail for *Theropithecus* by Jolly (1972), Meikle (1977) and Jablonski (1993, 1994). It has been shown that the unworn molars of geladas present a complex pattern of enamel ridges (loph(id)s) for separating plant material. Perhaps more importantly, owing to the high crowns (hypsodont) and hard enamel crests alternating with softer dentine depressions, the occlusal surfaces of *Theropithecus* molars retain their ability to cut, incise, and reduce the grasses eaten by the aging animals (Meikle, 1977; Jablonski, 1994). It is well known that a shelf or cusp extends from the distal surface of M_{1-3} and in M_3 it is the hypoconulid (Figs. 7.7 and 7.9). Swindler (1983) suggested that the distal cusps

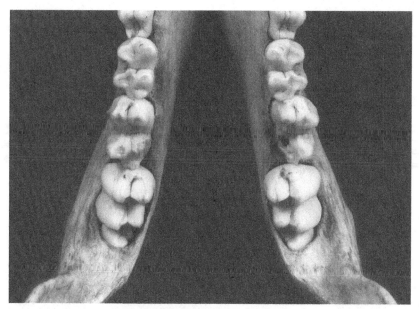

Fig. 7.9. Lower jaw of *Theropithecus gelada* showing the development of a heel of enamel on the distal surfaces of M_{1-2} and the hypoconulid of M_3. Courtesy of N. G. Jablonski.

on M_{1-2} were serially homologous with the hypoconulid of M_3 since they are topographically and functionally similar to the hypoconulid on M_3. Certainly, they are part of the dental pattern referred to by Jablonski (1994, p. 39) as 'structurally and functionally unique among primates' since they increase the occlusal length of lower molars and help to maintain the contact of these teeth during interstitial wear and mesial drift (Swindler and Beynon, 1993).

Odontometry (Appendix 1, Tables 105–128, pp. 218–29)
These animals present a great deal of dental sexual dimorphism: males are always larger than females. In most cercopithecoids, the greatest sex difference is centered around the C^1–P_3 complex. Sequential molar relations are variable among cercopithecoids and few clear patterns are discernible (Swindler and Olshan, 1982). Canine dimorphism has been widely studied through the years. Garn, Kerewsky and Swindler (1966) suggested that a canine 'field' of sexual dimorphism in tooth size appears to characterize the permanent teeth of many primates, and that adjacent teeth are more influenced by the morphogenetic field than are the dental elements twice removed. This is supported by the present data.

Subfamily Colobinae

Morphological observations

	Male	Female
Colobus polykomos (king colobus)	48	30
Piliocolobus badius (red colobus)	26	26
Nasalis larvatus (proboscis monkey)	26	15
Simias concolor (pig-tailed langur)	1	6
Pygathrix nemaeus (douc langur)	7	4
Rhinopithecus roxellana (golden snub-nosed langur)	6	11
Trachypithecus pileatus (capped leaf monkey)	18	8
Trachypithecus cristata (silver leaf monkey)	24	34
Trachypithecus phayrei (Phyre's leaf monkey)	4	5
Presbytis comata (Javan leaf monkey)	15	18
Kasi johni (John's langur)	3	2

The geographical distribution and major dietary habits of the colobine monkeys were discussed in conjunction with those of the cercopithecines. Many of the general dental features of Old World monkeys were also considered, so here it is only necessary to add a few general comments on their teeth as well as certain of their dental specializations.

Incisors (Figs. 7.10–7.14)

UPPER. I^{1-2} are heteromorphic: I^1 is larger with a more horizontal incisal border, and I^2 is smaller and more pointed. Mesial and distal marginal ridges are present, outlining the lingual surfaces of each tooth. A lingual cingulum is present on I^{1-2} in all species, and in the African colobines and *Nasalis* a lingual tubercle is present on I^1 5.8% of the time in the former and 87% of the time in the latter. A median lingual sulcus is present on I^1 in all taxa and it is usually deeper and wider in *Simias concolor*. It is variably developed on I^2 since a median ridge is more often present on this tooth.

LOWER. The lower incisors are more or less similar in colobines, i.e. with rather narrow marginal ridges that outline slightly concave lingual surfaces, and with I_2 more narrow and pointed than I_1. Shellis and Hiiemae (1986) found a substantial layer of enamel on both lingual and labial surfaces of I_{1-2} in colobine monkeys and attributed it to their predominantly folivorous diet. Hylander (1975) demonstrated that, relative to body size, leaf-eating colobines have smaller incisors than the more fruit-eating cercopithecines.

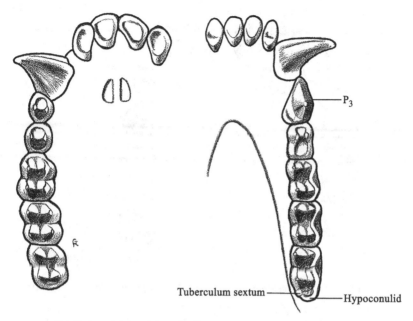

Fig. 7.10. *Colobus polykomos* male, occlusal view of permanent teeth (48 mm).

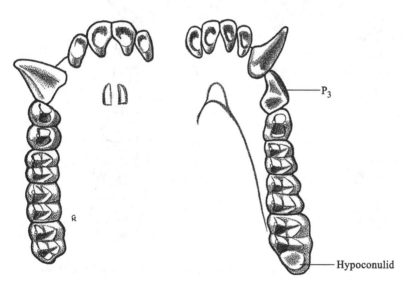

Fig. 7.11. *Nasalis larvatus* male, occlusal view of permanent teeth (50 mm).

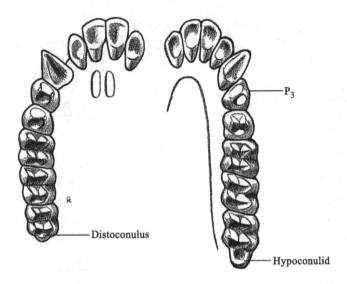

Fig. 7.12. *Simias concolor* female, occlusal view of permanent teeth (38 mm).

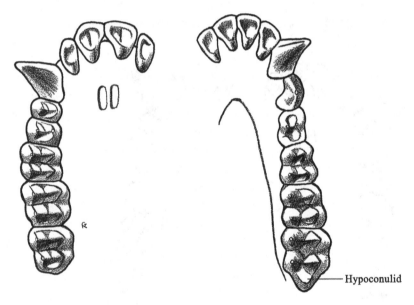

Fig. 7.13. *Rhinopithecus roxellana* female, occlusal view of permanent teeth (48 mm).

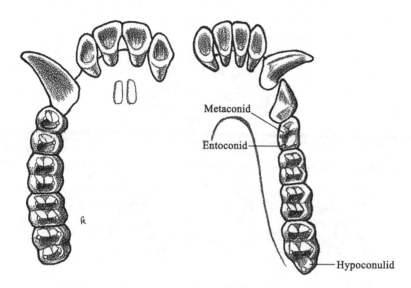

Fig. 7.14. *Trachypithecus cristata* male, occlusal view of permanent teeth (37 mm).

The normal, or at least the most common, occlusion of the incisors in both New and Old World monkeys is edge-to-edge; an exception is found among predominantly leaf-eating monkeys where a high incidence of underbite or mandibular protrusion occurs (Fig. 7.15) (see also Chapter 6). Colyer (1936) as mentioned earlier, was one of the earliest students to associate leaf-eating with the underbite occlusion among monkeys. Since then several investigators have studied the occlusion of leaf-eating monkeys (Serra, 1951; Schultz, 1958, 1960; Zingeser, 1973; Swindler, 1979; Emel and Swindler, 1992) and all have reported some degree of underbite in these species. Table 7.1 presents the incisor occlusal variability of various species of leaf-eating monkey. There are differences in frequencies between genera as well as within species of the same genus. To my knowledge, underbite is extremely rare in non-leaf-eating monkeys. The conditions of overbite and openbite are rare in monkeys from natural populations but do occur in laboratory monkeys, particularly thumb-suckers (Swindler and Sassouni, 1962; Moore, McNeill and D'Anna, 1972). The incidence of underbite is negatively correlated with maxillary incisor size and palate length (Sirianni, 1974; Swindler, 1979). In their study of underbite in colobine monkeys, however, Emel and Swindler (1992, p. 188) concluded that 'The high frequency of underbite seen in colobines is more likely to be the product of the scaling of facial dimensions in combination with apparent relaxed selection on the anterior dentition, rather than a

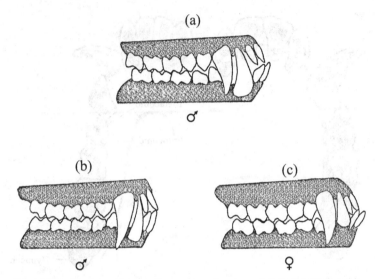

Fig. 7.15. Lateral view showing the permanent molar and incisor occlusion: (a) openbite, (b) edge-to-edge and (c) underbite. Adapted from Swindler (1979).

direct adaptation to folivory.' It is interesting to note that underbite was found in a 2 million year old colobine skull from Ethiopia indicating that the underbite condition has a long history in colobine monkeys (Swindler, 1979).

Canines

UPPER. Colobine canines are not unlike those of other cercopithecids in that they project beyond the occlusal plane and the characteristic cercopithecid mesial groove is present. There is a cingulum skirting around the lingual base of the tooth that varies in thickness among these genera. The distolingual surface of these canines may be quite sharp.

LOWER. The lower canine is large, projecting well above the occlusal plane. A lingual cingulid passes distally to form a heel that becomes more obvious with wear (Lucas and Teaford, 1994).

Premolars

UPPER. P^{3-4} conform to the upper premolar cercopithecid morphotype i.e. a small mesial fovea, paracone and protocone, and a rather spacious trigon basin. The variability pertains to the formation of the protocone on P^3. The cusp varies from absent 93% of the time in male *Colobus*

Table 7.1. *Percentage frequency of incisor occlusal variability in leaf-eating monkeys*

Taxon	Sex	n	Underbite	Edge to edge	Overbite	Openbite
Colobus polykomos	M	61	29.5	54.1	9.8	6.6
	F	33	51.5	42.4	6.1	—
Piliocolobus badius	M	36	41.7	52.7	5.6	—
	F	34	47.1	50.0	—	2.9
Presbytis aygula	M	14	92.9	7.1	—	—
	F	18	95.5	5.5	—	—
Presbytis cristatus	M	22	13.6	81.8	—	4.6
	F	32	56.3	40.6	—	3.1
Presbytis pileatus	M	24	41.7	58.3	—	—
	F	14	78.6	21.4	—	—
Nasalis larvatus	M	26	53.8	46.2	—	—
	F	18	83.3	11.1	5.6	—
Rhinopithecus roxellana	M	4	25.0	75.0	—	—
	F	11	45.5	54.5	—	—
Alouatta villosa	M	18	66.7	27.8	5.5	—
	F	41	87.8	12.2	—	—
Alouatta belzebul	M	9	77.8	22.2	—	—
	F	3	66.7	33.3	—	—
Alouatta seniculus	M	24	54.2	33.3	4.2	8.3
	F	18	44.4	33.3	5.6	16.7

Adapted from Swindler (1979).

polykomos to always present in female *Nasalis larvatus* (Swindler and Orlosky, 1974). P^3 in *Pygathrix* is always bicuspid, and in *Trachypithecus* and *Presbytis* it may have as many as three cusps, the third cusp being the metacone. P^4 has two cusps, the paracone and protocone.

LOWER. As in all cercopithecids, the lower premolars are heteromorphic. P_3 is sectorial with a large mesiobuccal flange upon which the upper canine is honed. There may be a small to large metaconid on P_3 in *Colobus* and *Trachypithecus* but it is never as large as the protoconid. When present, the metaconid is connected by the protocristid to the protoconid. The distal basin, if present, is always quite small. P_4 is a molariform tooth, always with a protoconid and metaconid connected by a protocristid that separates the occlusal surface into a small trigonid and a somewhat larger talonid basin. A major difference between the P_4 of cercopithecines and colobines is that in the latter taxon there is a more prominent mesiobuccal flange present on the tooth (Zingeser, 1968; Delson, 1973). In over 90% of the present sample of *Nasalis larvatus* a third cusp is present on P_4, the entoconid; in *Rhinopithecus* a fourth cusp, the hypoconid, is present 87% of the time.

Molars

UPPER. M^{1-3} are typical bilophodont molars with three occlusal basins and two transverse crests. In contrast to cercopithecine molars, those of the colobines have relatively high cusps that are positioned near the tooth margins. This results in longer molar crests, which are associated with the more folivorous diets of colobine monkeys (Kay, 1977; Kay and Hylander, 1978). In general, colobine molars are larger, relative to body mass, than they are in cercopithecines (Kay, 1975).

LOWER. M_{1-2} have four cusps, M_3 five, and of course they are bi-lophodont. The position of the hypoconulid is variable, being situated either centrally or buccally; in most species, it is more often situated buccally. In one specimen of *Nasalis larvatus* and in 2% of *Colobus polykomos* the hypoconulid is absent. The trigonid basin is narrow and shallow and there is more buccal flaring compared with the lower molars of cercopithecines. A tuberculum intermedium is present on M_{1-3} in *Rhinopithecus* M_1 (7%), M_2 (62%), and M_3 (46%) and in *Pygathrix* M_1 (9%) and M_3 (11%). The tuberculum sextum is quite variable on M_3 in all species. The discussion concerning crown height and the development of crests in the upper molars can be equally applied to the lower molars. Moreover, Lucas *et al.* (1986a) have also shown that the breadth: length ratio of M_3 is highly correlated with the percentage of leaves in the diet.

Odontometry (Appendix 1, Tables 129–164, pp. 230–47)
The upper $C-P_3$ honing complex is significantly different in size between the sexes in all colobine taxa. Among colobines, however, it appears that the degree of sexual dimorphism vacillates between genera; for example, *Colobus polykomos* and *Nasalis larvatus* have more differences than any of the *Presbytis* or *Trachypithecus* species. Differences also exist among close-ly related genera: *Colobus polykomos* has more sexual dimorphism than *Piliocolobus badius*. Finally, it might be noted that M^3 is often smaller than the other molars, because its distal portion is reduced.

8 *Hylobatidae*

Present distribution and habitat

The Hylobatidae consists of a single genus, *Hylobates*, which currently has 11 species. The genus *Symphalangus*, the siamang, is now included with the gibbons and known as *Hylobates syndactylus*. Gibbons are the smallest and the most diverse of living apes. They live in monogamous family groups, but see Falk (2000) for different current views on the subject. The males and females of the same species are about the same size and have canine teeth of about equal length.

Gibbons live throughout the evergreen rain forests of Southeast Asia, China, Burma, Sumatra, Borneo and Java, as well as on several of the smaller islands. They are arboreal brachiating primates, preferring the middle to the upper canopy in the forests. In the trees gibbons are the most acrobatic of all primates, with the spider monkey of the New World a close second. When on the ground, they walk bipedally with arms raised above their heads for balance.

Dietary habits

The gibbon diet is mainly fruit (50–70%), although the siamangs apparently eat more leaves (up to 59%) than other gibbons (Curtin and Chivers, 1978). In addition, gibbons eat termites, caterpillars and animal prey. As with most primate species, the diet varies seasonally as well as from locality to locality.

General dental information

Permanent dentition: I^2-C^1-P^2-M^3 / I_2-C_1-P_2-M_3
Deciduous dentition: di^2-dc^1-dp^2 / di_2-dc_1-dp_2

There are several differences in the morphology of the teeth of hominoids

147

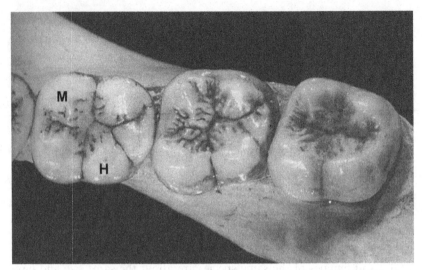

Fig. 8.1. The *Dryopithecus* Y-pattern of the lower molars of hominoids. M, metaconid; H, hypoconid. In the Y-pattern the metaconid and hypoconid are in contact on M_{1-3} as in this chimpanzee. The lower side of the teeth is buccal and the upper is lingual. Reprinted from Swindler, D. R. (1998) *Introduction to the Primates*, with the permission of the University of Washington Press.

and those of the non-human primates so far discussed. These will be discussed later in the chapter. For now, a major difference in the morphology of the lower molars will be considered: the presence of five cusps on all three molars and the arrangement of the sulci separating these cusps to form the *Dryopithecus* or Y-5 pattern (Fig. 8.1). This pattern was initially described by Gregory (1916) for the lower molars of the dryopithecine apes (hence the name) found in European and Indian Miocene formations. Later it was realized that this arrangement of cusps and grooves on the occlusal surface of the lower molars of *Dryopithecus* is also present in humans. Thus, the evolutionary importance of the pattern was early recognized: it is characteristic of all Hominoidea, and therefore useful for establishing phylogenetic relationships.

A brief definition of the pattern is as follows. The metaconid and hypoconid are in contact in the talonid basin; the lingual groove between the metaconid and entoconid forms the stem, and the grooves between the protoconid and hypoconid and the hypoconid and hypoconulid form the fork, of the Y. Therefore, when viewed from the lingual side the Y-5 pattern is observed (Fig. 8.1). During the course of hominoid evolution the Y-5 pattern has been altered, so that today various modifications are observed. There is the + 5 pattern (the protoconid and entoconid contact, resulting in a + pattern of sulci); the Y-4 pattern (the hypoconulid is absent but cusp contact is still between the metaconid and the hypoconid); and the + 4 pattern (the hypoconulid is absent but the cusps have shifted and cusp contact is as in the + 5 pattern). In agreement with others, I believe that the contact between the metaconid and hypoconid (see, for example, Erd-brink, 1965; Frisch, 1965; Jorgensen, 1956; Johanson, 1979) is one of the principal characteristics of the pattern. In other words, if there is a point contact or a clear contact between the protoconid and entoconid the pattern is not a Y. The incidence of the lower molar patterns in living hominoids is presented in Table 9.1 (p. 162).

Martin (1985) studied enamel thickness in a sample of hominoids and concluded that gibbons have thin enamel. Shellis *et al.* (1998, p. 520), however, in their study of enamel thickness in $M_1–M_2$ regressed as a measure of tooth size on body mass among anthropoids, found that 'in terms of tooth size, *Hylobates* has average enamel thickness.' Part of the problem of interpretation has been the use of different methods by investigators in measuring enamel thickness. This is changing now that many researchers are using similar and more refined techniques.

Guatelli-Steinberg and Lukacs (1999) and Guatelli-Steinberg and Skinner (2000) examined a series of *Hylobates lar* for enamel defects and found that their lower canines had more evidence of stress episodes than monkeys, but fewer than recorded for great apes. There was no sex difference in enamel defects; they suggested that this might be related to the lack of canine sexual dimorphism in gibbons, since the canines in both sexes take about the same amount of time to develop.

Morphological observations

	Male	Female
Hylobates klossi (Kloss's gibbon)	18	15
Hylobates moloch (silvery gibbon)	5	4
Hylobates lar (white-handed gibbon)	5	2
Hylobates syndactylus (siamang)	5	—

Incisors (Fig. 8.2)

UPPER. I^{1-2} are heteromorphic. I^1 is broad and the lingual surface is concave. A well-developed lingual cingulum is ledge-like as it passes around the base of the tooth. I^2 is narrow and pointed with a lingual cingulum. It also has a median ridge on the lingual surface, passing from the cervix to the incisal border.

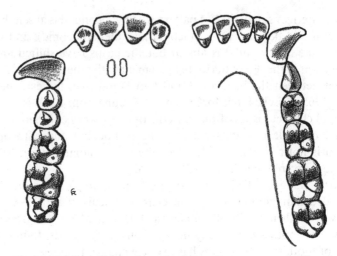

Fig. 8.2. *Hylobates klossi* male, occlusal view of permanent teeth (38 mm).

LOWER. I_{1-2} are implanted vertically in the lower jaw. The incisal border of I_1 is horizontal whereas that of I_2 slopes distally. The lingual surfaces of both I_1 and I_2 are slightly concave and outlined by the mesial and distal marginal ridges. A lingual cingulid is present on both teeth. In his study of anterior tooth use in anthropoids, Unger (1994) found that gibbons, at least *Hylobates lar* (the species he studied), employed these teeth less than half the time while feeding. When they did, it was often for nipping and occasionally crushing.

Canines

UPPER. The upper canines in both males and females are long, trenchant teeth with mesial grooves. The teeth curve buccally and then lingually to terminate as sharp points. In general, the lingual curve is less in *H. syndactylus*, resulting in a canine that passes nearly vertically from the upper jaw. In both sexes, the canines extend well above the occlusal plane, and relative to basal crown dimensions and body mass, the canines project the most among anthropoids (Greenfield and Washburn, 1992). It is generally agreed that the lack of sexual dimorphism in canine size in gibbons is due to the similar social status of both sexes in their virtually monogamous life styles (Greenfield and Washburn, 1991).

LOWER. The lower canines have broad bases from which the crown curves buccally and slightly distally. A cingulid curves around the lingual face of the tooth and terminates as a distinct heel.

Premolars

UPPER. P^{3-4} are bicuspid. The P^3 protocone is often only a slight elevation on the expanded lingual surface of the tooth except in *H. syndactylus,* where the protocone is well developed. On P^4 the protocone is well developed in all gibbons, particularly in *H. syndactylus,* although it never attains the dimensions of the paracone. Frisch (1965) found a low frequency of lingual cingula on P^{3-4} in *H. concolor* and on P^3 in *H. moloch.* A small lingual cingulum is present on P^4 in one of the present sample of 33 *H. klossi.* Buccal cingula are rare or absent in gibbons, as reported here and by Frisch (1965).

LOWER. P_3 is sectorial whereas P_4 is bicuspid or multicuspid Frisch (1965) noted the morphological variability of P_3 in gibbons and defined two types: (1) an oblique type that possesses the common elongated crown; and (2) a triangular type that has a more developed distolingual portion to the crown. The oblique type is more common in the present sample of gibbons. The protoconid and metaconid of P_4 are set close together and connected by a protocristid. In *H. syndactylus* the metaconid approximates the protoconid in size.

In their detailed investigation of P_3 sectorial variation of male anthropoids, Greenfield and Washburn (1992, p. 186) thought it more likely 'that the canines of gibbons are derived (convergent with cercopithecoids) and that their premolars are reflecting heritage (which is the shared primitive condition seen in great apes and ceboids).' Indeed, their data indicated a very high heritage component for the anthropoid sectorial P_3.

Molars

UPPER. M^{1-3} have four cusps; the paracone and metacone are the highest, followed by the protocone and hypocone. The apex of each buccal cusp is mesial to those of the lingual cusps. A crista obliqua connects the protocone and metacone. Frisch (1965) and Kitahara-Frisch (1973) found the mode of lingual cingulum reduction to be quite variable in the different gibbon species. For example, there is almost no reduction of the lingual cingulum in *H. concolor,* moderate reduction in *H. lar* and *H. moloch,* and extreme reduction in *H. agilis* and *H. hoolock,* whereas in *H. syndactylus* the lingual cingulum is almost always absent, or only a trace may be present (Frisch, 1965). In the present sample, a lingual cingulum is present on M^1 30%, M^2 24%, and M^3 9%. Thus, the morphological field in size reduction of the lingual cingulum is from M^1 to M^3. Adloff (1908), Remane (1960), and Frisch (1965) found the cusp of Carabelli on M^1 in *H. lar*, although it is not present in the present sample. There are no buccal cingula in this sample but it has been reported in *H. concolor* (Frisch, 1965). M^3 has some reduction in all gibbons, and in rare cases, the hypocone and metacone may be absent.

LOWER. M_{1-3} generally possess five cusps with the protoconid and metaconid connected by the protocristid isolating the narrow trigonid basin from the more spacious talonid basin. The Y-5 pattern is present on M_1 100%, M_2 97%, and M_3 70% in the present sample. This is very similar to the order and frequencies reported on M_{1-3} by Frisch (1965). The position of the hypoconulid varies in gibbons from central to buccal; in the present sample it is usually buccal (80%) on all three molars. Likewise, the hypoconulid may be absent from any of the molars, but is more often absent from M_3 (10%) in the present sample. In the tribosphenic lower molar, the metaconid is distal to the protoconid, but there has been a trend to shift the metaconid mesially to face the protoconid in hominoids (Frisch, 1965). In this sample of gibbons, as in those studied by Frisch, the metaconid is still distal to the protoconid on M_1 but tends to lie opposite the protoconid on M_{2-3}, particularly on M_3. The tuberculum intermedium is present on M_1 0.07%, M_2 18%, and M_3 37% of the time in all gibbons, except that in one specimen of *H. syndactylus* it is present on M_{1-3}.

Odontometry (Appendix 1, Tables 165–178, pp. 248–54)
There are few sex differences in tooth size in extant gibbons, even in the upper $C-P_3$ complex. The mesiodistal size relations of both the upper and lower molars express a great deal of variability as mentioned by Kitahara-Frisch (1973) and supported by the data presented here.

9 Pongidae

Present distribution and habitat

This family includes three genera, *Pan*, *Gorilla* and *Pongo*; at present there are two species of chimpanzee (*Pan troglodytes* and *P. paniscus*), one species of gorilla (*Gorilla gorilla*) and one species of orangutan (*Pongo pygmaeus*). The gorilla, the largest living primate, lives in two major areas of Equatorial Africa: the western gorilla inhabits the western parts of the Congo basin, and the eastern gorilla ranges from the eastern lowlands of the Upper Congo to the mountains east of Lake Kivu. Both the western and eastern lowland gorillas occupy primary and secondary forests and marshes, whereas the mountain gorilla lives in montane and bamboo forests between 2800 and 3965 m (Jenkins, 1990).

The two species of chimpanzee, *Pan troglodytes* and *P. paniscus*, live in Africa. The former species is widely distributed from West Africa through parts of Central Africa and as far east as Lake Victoria and Lake Tanganyika. The latter, the pygmy chimpanzee or bonobo, is limited to an area bordered by the Congo and Lualaba Rivers. Chimpanzees inhabit rain forests, savannas, and montane forests up to 3000 m.

Orangutans are limited to Sumatra and Borneo, where they occupy tropical rain forests. They may be found as high as 4000 m, although they seem to prefer altitudes below 500 m (Wolfheim, 1983).

Of the three great apes, the orangutan is the most arboreal and rarely comes to the ground, whereas the gorilla is most often on the ground, particularly the large males. Chimpanzees are more or less intermediate: they may spend as much as one third of the day on the ground. When in the trees, orangutans move by a slow quadrumanous climbing technique, and on the ground they are quadrupedal, using their clenched fingers as fists for support (Tuttle, 1969). Chimpanzees and gorillas move about on the ground in a similar quadrupedal, knuckle-walking fashion, and in the trees chimpanzees employ both quadrupedal and suspensory types of locomotion. Adult mountain gorillas rarely go into the trees, whereas the lowland gorilla is much more arboreal and proceeds by arm-swinging.

153

Dietary habits

Gorillas are vegetarians and by far the most herbivorous of all great apes. Mountain gorillas eat leaves, shoots, roots, bamboo, and a little fruit, whereas the lowland gorillas eat much more fruit, leaves, roots, flowers, ferns and grasses. Chimpanzees are also vegetarians, eating fruits, leaves, seeds and grasses, but will supplement their diet with the meat of baboons, colobus monkeys, ants and termites. Orangutans eat great amounts of fruit and seem to prefer ripe, succulent fruits as well as those with hard pericarps (Unger, 1995). They also eat leaves, bark, flowers, and some animal prey.

General dental information

Permanent dentition: I^2-C^1-P^2-M^3 / I_2-C_1-P_2-M_3
Deciduous dentition: di^2-dc^1-dp^2 / di_2-dc_1-dp_2

In studies of enamel thickness among hominoids a central question has been the cause of these differences: is it rate or time? It now seems likely that the differences in enamel thickness result from variations in the duration of enamel apposition rather than in the rate (Beynon, Dean and Reid, 1991).

The study by Shellis *et al.* (1998) contained much new material regarding the ongoing problem of determining enamel thickness and the interpretation of this information for primate studies. These investigators found that, in terms of tooth size, the chimpanzee and orangutan have average or thin enamel depending on tooth type, whereas the gorilla has thin enamel. These findings suggest that the last common ancestor of great apes and humans had average enamel thickness, and therefore the thin enamel of gorillas and the thick enamel of humans are derived characters. This is contrary to Martin (1985), who believed that the last common ancestor of hominoids possessed thin enamel. Certainly much work remains in this fascinating field, and with the introduction of tomographic and micro-tomographic non-destructive methods in the last decade, the future should bring forth many new findings.

Skinner (1986) investigated enamel hypoplasia in *Pan* and *Gorilla*. He studied the permanent mandibular teeth from the right canine to the left third molar. He found a higher incidence in *Gorilla* (76%) than in *Pan* (58%) and stated that enamel hypoplasia in these primates was not a function of age, sex, or body size, but rather, a result of seasonal stress. The studies of Guatelli-Steinberg and Lukacs (1999) and Guatelli-Steinberg and Skinner (2000) have contributed much to a better understanding of enamel hypoplasias among the great apes and monkeys. Their studies indicate that both African and Asian great apes have a higher frequency of linear enamel hypoplasia (LEH) than do gibbons and monkeys; however, the dichotomy is not as extreme as some research has suggested. They suggest that the differences of LEH between the two groups may be the result of a combination of the time it takes for crown formation and exposure to environmental stress.

The occlusal surfaces of hominoid teeth possess small ridges and furrows known as wrinkles or crenulations. These are better developed on the premolars and molars. In extant hominoids, wrinkles are best developed on the low-cusped cheek teeth of *Pongo* (see Fig. 9.2). In unworn orangutan molars, the entire surface is covered with wrinkles that may obliterate the primary groove system. These wrinkles are often complex; however, they frequently pass perpendicular to the marginal ridges around the crown to end in the basins. They range from narrow, fine wrinkles to rather coarse ones, and both varieties are found in fossil and living orangutans (Hooijer, 1948).

Morphological observations

	Male	Female
Pongo pygmaeus (orangutan)	8	11
Gorilla gorilla (western lowland gorilla)	6	9
Pan troglodytes (chimpanzee)	32	61
Pan paniscus (pygmy chimpanzee or bonobo)	2	2

Incisors (Figs. 9.1–9.3)

UPPER. I^1 is a broad, spatulate tooth with straight incisal borders and the mesial and distal marginal ridges are well formed. The lingual cingulum is obvious and, with the marginal ridges, outlines a concave lingual fossa. A tuberculum dentale may be present extending as several vertical ridges from the lingual cingulum to the incisal border. In the unworn I^1 of orangutans small crenulations are also present on the lingual surface. In orangutans and gorillas I^1 is particularly hypertrophied. I^2 is always narrower than I^1, particularly in the orangutan where the incisal border is rather pointed. The lingual cingulum is present on I^2 as is a smaller tuberculum dentale. The $I^1:I^2$ ratio is 1.65 for extant orangutans; this ratio is always much larger than it is in other extant hominoids. Indeed, the

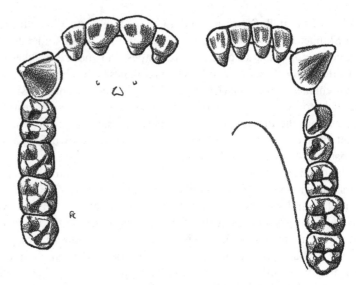

Fig. 9.1. *Pan troglodytes* male, occlusal view of permanent teeth (80 mm).

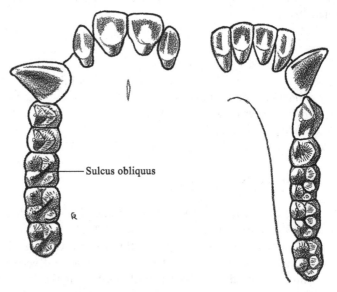

Fig. 9.2. *Pongo pygmaeus* male, occlusal view of permanent teeth (90 mm).

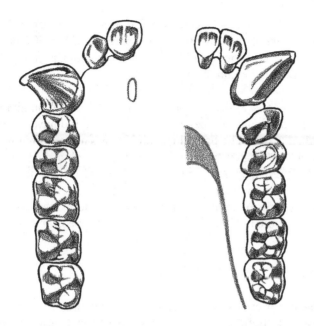

Fig. 9.3. *Gorilla gorilla* male, occlusal view of permanent teeth (112 mm).

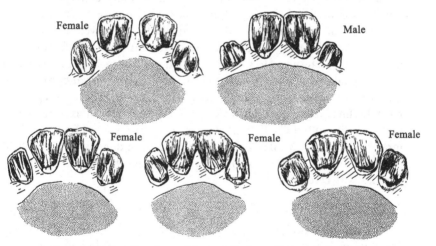

Fig. 9.4. The lingual surfaces of the upper incisors of female and male *Pan troglodytes* showing some of the variation that occurs in these teeth. Not drawn to scale. Adapted from Swindler (1968).

narrow, more pointed I^2 is a common feature of the upper incisors of orangutans. Figure 9.4 presents the lingual surfaces of I^{1-2} of *Pan troglodytes*, showing some of the variation in these teeth. Note, particularly, the variability of the tuberculum dentale.

LOWER. I_{1-2} are large in the great apes and are more or less similar in size and shape. Their lingual surfaces are subrectangular and have mesial and distal marginal ridges. The lingual cingulid is present and often large on both teeth. A true lingual tubercle is absent, although a median lingual ridge may be present in *Pan* and *Pongo*. Unger (1994) in his study of ingestive behaviors in primates, found that the orangutans used their anterior teeth more often than other primates (more than 90% of all feeding observations). They normally nipped, stripped, and incised all foods eaten.

Canines

UPPER. The upper canines are large, conical teeth and are always much larger in males. In both sexes, the canines project well below the occlusal plane of the other teeth. The mesial border is rather convex; the distal border is either straight or slightly concave from base to crown tip. A lingual cingulum is present and is particularly well developed in orangutans, and interestingly, larger in females than males, a feature also mentioned by Hooijer (1948). The lingual surface in orangutans has a mesiolingual groove that follows the curvature of the mesial border from the base to near the crown tip. Chimpanzees, but not bonobos, and gorillas have this lingual groove but it is median in position. It is interesting to note that in their study of male anthropoid upper canine variation based on measures of shape, distal edge sharpness, and linear dimensions Greenfield and Washburn (1991) placed the great apes with ceboids while including gibbons with cercopithecoids.

LOWER. The lower canines are large, more or less conical in shape, and project above the occlusal plane of the other teeth. In all three pongids they are more or less triangular when viewed occlusally, although in *Pan* the borders are rounder and less obtuse than in the other two taxa. A lingual cingulid is present in all taxa, and as in the upper canine of orangutans, it is better developed in female orangutans. A mesiolingual groove is generally present, although it is frequently absent in *Pan*.

Premolars

UPPER. P$^{3-4}$ are bicuspid with the paracone larger and higher than the protocone, although in the orangutan P^4 the protocone nearly approximates the size of the paracone. In all taxa, a mesiodistal developmental groove separates the two cusps, and in gorillas it is wide and deep. Mesial and distal marginal ridges are present outlining the occlusal surfaces. In addition, two crests are usually associated with the paracone and protocone. A mesial crest passes from the paracone, or just mesial to it, to end at the marginal ridge just mesial to the protocone; a distal crest connects the protocone to the distal aspect of the paracone. These crests separate the occlusal surface into three parts. Small crenulations are present on the occlusal surface of unworn premolars in the orangutan. P^3, especially in orangutans, has a more prominent mesiobuccal surface than P^4 so that isolated upper premolars are easily identified, a condition first described by Hooijer (1948). Buccal cingula are rare on P$^{3-4}$ in these taxa and absent in *Pan*. Hypocones and metacones have been reported on P^4 in orangutans by Hooijer (1948) and Remane (1921). These cusps are not present in the present sample of orangutans, but a small hypocone is present on P^4 in one gorilla specimen. None is present in *Pan*. A lingual cingulum is present along the mesiolingual surface of the protocone on P^3 (85%) and P^4 (90%) in gorillas. A buccal cingulum is present on P$^{3-4}$ in one male gorilla specimen. There are no cingular remnants on either premolar in the present sample of *Pan*.

LOWER. The lower premolars are heteromorphic; P$_3$ is sectorial whereas P$_4$ is multicuspid. However, P$_3$ frequently has a metaconid in the great apes that may approximate the size of the protoconid and, rarely, surpass it. The protocristid connects the protoconid to the metaconid, and if the latter cusp is absent the protocristid continues on to the lingual cingulid. The protocristid divides the occlusal surface of P$_4$ into two parts that are usually somewhat depressed and are probably homoplastic if not homologous with the trigonid and talonid basins of the lower molars. The trigonid and talonid basins vary in shape and size among pongids. In the orangutan narrow crenulations are present in these basins. P$_4$ has two cusps and a protocristid separating the occlusal surface into trigonid and talonid basins. A hypoconid and entoconid are often present on P$_4$, making it a multicuspid tooth, indeed, Remane (1921) identified a hypoconulid on P$_4$ in the gorilla, but this cusp is not present in the present sample. Buccal cingulids are present on P$_4$ 90% of the time in the gorilla, but absent in the other taxa. Cingular remnants are absent in this sample of chimpanzees.

Molars

UPPER. M^{1-3} have four cusps that vary in size in recent pongids. In general, the paracone and protocone are larger than the metacone and hypocone. This is always true regarding the hypocone; however, the protocone and metacone may be subequal. There is a progressive reduction in the size of the hypocone from M^1 to M^3, and in the orangutan this is accompanied by reduction of the metacone (Hooijer, 1948; this study). There is less hypocone reduction in gorillas; in fact hypocone reduction increases from *Gorilla* to *Pongo* to *Pan* (Frisch, 1965). In their study of the Liberian chimpanzee, Schuman and Brace (1955) found the hypocone largest on M^1 (61%), medium on M^2 (55%) and smallest on M^3 (52%). The crista obliqua connects the protocone with the metacone. The occlusal surface is separated into three fossae: a mesial fossa (fovea anterior) is delimited by a preprotocrista distally and a mesial marginal ridge mesially, a central fossa or trigon basin is between the preprotocrista and the crista obliqua, and the distal fossa (fovea posterior) is between the crista obliqua and the distal marginal ridge. Variability can be expected most often in the formation of the fovea anterior and posterior. In orangutans, the occlusal surface is covered with a series of crenulations, radiating from the center of the fovea, which begin to disappear soon after wear commences. A sulcus obliquus (distolingual groove) separates the hypocone from the other cusps but is not as distinct in *Pan* as in the other species. Protoconules and metaconules may be present and are usually more common in gorillas and orangutans; in fact a metaconule is not reported in chimpanzees, although a vestigial protoconule is recorded by Korenhof (1960). The protoconule is absent in the present sample of chimpanzees.

A lingual cingulum is rather common in *Pan*, appearing on the mesiolingual aspect of the protocone in M^1, being more extensive on M^{2-3} so as to include the protocone. On M^3 the lingual cingulum may extend distally around the hypocone to become confluent with the distal marginal ridge in *Pan*. In the present sample of *Pan troglodytes* the lingual cingulum is present on M^1 (80%), M^2 (80%), and M^3 (63%) with sexes combined, since there is no sexual dimorphism. Lingual cingula are also common in *Gorilla*, M^{1-2} (100%) and M^3 (71%) and a small lingual cingulum is present in *Pongo*, M^1 (50%), M^2 (51%), and M^3 (20%). Buccal cingula are rare in pongids, having their highest frequency in the common chimpanzee, M^1 (33%), M^2 (13%), and M^3 (33%). In most cases, buccal cingula are limited to the paracone or bridge the buccal developmental groove between the paracone and metacone.

A final comment regarding the lingual cingulum in non-human primates. There is occasionally a cusp on the lingual surface of the protocone in prosimians, New and Old World monkeys, pongids and of course, humans. In humans, it is known as Carabelli's cusp and occurs in various percentages throughout humankind (Turner and Hawkey, 1998). I believe Carabelli's cusp represents a homologous trait among primates; certainly it is homoplastic.

LOWER. M_{1-3} generally have five cusps; the incidence of the different lower molar groove and cusp numbers in extant pongids is presented in Table 9.1. It is immediately apparent that the Y-5 pattern is the dominant one in these taxa, particularly if the Y-6 pattern is included in the Y-5 category, as it is by most authors. Only in the chimpanzee is there a noticeable increase in the other lower molar configurations, and even in this taxon, it is limited mainly to the Liberian subspecies *Pan troglodytes verus*. Also in this subspecies the almost complete lack of Y-5 patterns on M_3 is difficult to explain unless it is another indication of the large interspecific variability of the dentition of extant chimpanzees. The Y-5 molar pattern can probably be traced back to *Aegyptopithecus* of the Oligocene (Skaryd, 1971). Certainly *Aegyptopithecus* had a metaconid–hypoconid contact.

The metaconid on M_1 is either opposite the protoconid or distal, whereas on the majority of M_{2-3} the metaconid is opposite or mesial to the protoconid. The hypoconulid is most often buccal on M_{1-3} (about 90% of the time), or central the remainder of the time, except on M_3 in *Pan* where it has a frequency of only 76%. The protoconid and metaconid are connected by the protocristid that forms the distal boundary of the narrow trigonid basin. Distal to the protocristid is the central portion of the crown or talonid basin, which is especially commodious in gorillas. A narrow postprotocristid connects the entoconid to the hypoconid to form the narrow, oblique transverse post-talonid basin. The occlusal separation between the metaconid and entoconid is wide and deep on M_{1-3}, a condition that is unique in gorillas among all pongids. A buccal cingulid is often present on the lower molars, most frequently in gorillas, least in orangutans and chimpanzees. In the present study, only *Gorilla* has a complete buccal cingulum. From this and several other features, some odontologists believe the dentition of gorillas represents the most primitive among extant pongids (Remane, 1921, 1960; Korenhof, 1960; Frisch, 1965).

As discussed above, the occlusal surface of hominoid molars has a

Table 9.1. *Mandibular molar groove patterns and cusp number in recent pongids (percentages and sexes combined)*

Taxon	Tooth	Number of teeth	Y-6	Y-5	Y-4	+6	+5	+4	Y	+
Pan troglodytes verus[1]	M1	157	11.5	88.5	—	—	—	—	100.0	—
	M2	143	11.9	77.6	—	2.8	7.7	—	89.5	10.5
	M3	109	2.8	44.0	0.9	1.8	33.9	16.5	47.7	52.2
Pan troglodytes schweinfurthii[2]	M1	124	—	100.0	—	—	—	—	100.0	—
	M2	81	—	100.0	—	—	—	—	100.0	—
	M3	62	—	82.3	—	—	14.5	3.2	82.3	17.7
Pan troglodytes troglodytes[2]	M1	211	—	97.6	—	—	—	—	97.6	—
	M2	175	—	98.3	0.6	—	—	—	98.6	—
	M3	129	—	71.3	14.7	—	4.7	3.1	86.0	7.8
Pan paniscus[2]	M1	103	—	100.0	—	—	—	—	100.0	—
	M2	60	—	96.6	—	—	—	—	96.6	—
	M3	32	—	75.0	6.2	—	—	—	81.2	—
Pongo pygmaeus pygmaeus[1]	M1	300	2.0	98.0	—	—	—	—	100.0	—
	M2	293	7.2	91.5	—	—	1.4	—	98.7	1.4
	M3	208	9.1	86.5	0.5	—	3.4	0.5	96.1	3.9
Gorilla gorilla gorilla[1]	M1	88	14.9	85.1	—	—	—	—	100.0	—
	M2	94	8.7	91.3	—	—	—	—	100.0	—
	M3	80	—	100.0	—	—	—	—	100.0	—
Gorilla gorilla beringei[1]	M1	71	—	100.0	—	—	—	—	100.0	—
	M2	69	43.5	56.5	—	—	—	—	100.0	—
	M3	65	33.8	66.2	—	—	—	—	100.0	—

[1]Present study.
[2]Johanson (1979).
Reprinted from Swindler, D. R. and Olshan, A. F. (1988). Comparative and evolutionary aspects of the permanent dentition, Table 20.1, p. 278. In Schwartz, J. H. (ed.), *Orang-Utan Biology*. Reprinted with Permission of Oxford University Press.

Table 9.2. *Percentage incidence of the deflecting wrinkle in pongids*

Taxon	M_1		M_2		M_3	
	n	%	n	%	n	%
Pan troglodytes verus	151	3.3	139	17.9	116	12.9
Pan troglodytes schweinfurthii	138	13.0	114	15.9	78	5.0
Pan paniscus	156	16.0	110	5.1	47	0
Gorilla gorilla gorilla	69	0	76	1.3	65	3.1
Gorilla gorilla beringei	71	0	69	2.9	65	13.9

Adapted from Swindler and Ward (1988).

variety of small wrinkles; however, one usually larger and more pronounced wrinkle is known as the deflecting wrinkle. It was first described on the lower molars of *Gigantopithecus* by Weidenreich (1945). It is a median wrinkle of the metaconid extending toward the protoconid where it forms a right angle distally to join the entoconid near the center of the occlusal surface. When present in pongids it appears more often on M_{2-3}, except for *Pan paniscus* (M_1 16%) (Table 9.2). Swindler and Ward (1988) suggested that the deflecting wrinkle represents an ancestral trait in hominoids.

Two extra cusps have variable development in hominoids, the tuberculum sextum (C6) and tuberculum intermedium (C7). Both of these cusps, as we have seen, may be present in other primates but they generally have greater expressivity among hominoids. In their study of Plio-Pleistocene hominids, Wood and Abbott (1983) utilized these two discrete dental traits along with several other morphological characters to separate the 'robust' and 'gracile' australopithecines. For example, they found C7 to be almost lacking in the robust forms, whereas in the gracile hominids it was present over 50% of the time on M_{2-3}. The frequencies of these two cusp traits are presented in Tables 9.3 and 9.4. Among extant great apes, gorillas have the highest frequency of C6 and orangutans the least. Swindler and Olshan (1988) report a decrease in the incidence of this cusp in extant compared with Pleistocene orangutans. C7, on the other hand, has its lowest expression in *Pan* whereas the mountain gorilla has the highest incidence of the trait.

The late Paul Mahler (1980) presented a comprehensive study of molar size sequences (MSS) of both the mandibular and the maxillary molars in the great apes. He found that there are more common sequences represented in both the upper and lower molars for each species; however, the

Table 9.3. *Percentage incidence of the tuberculum sextum in pongids*

Taxon	M$_1$		M$_2$		M$_3$	
	n	%	*n*	%	*n*	%
Pan troglodytes verus	157	11.5	143	11.9	109	2.8
Pongo pygmaeus	300	2.0	293	7.2	208	9.1
Gorilla gorilla gorilla	88	0	94	14.9	80	8.7
Gorilla gorilla beringei	71	0	69	43.5	65	33.8

Adapted from Swindler and Olshan (1988).

Table 9.4. *Percentage incidence of the tuberculum intermedium in pongids*

Taxon	M$_1$		M$_2$		M$_3$	
	n	%	*n*	%	*n*	%
Pan troglodytes	84	0	75	9.3	58	0
Pongo pygmaeus	126	10.3	124	36.3	96	24.0
Gorilla gorilla gorilla	39	8.3	38	31.6	33	51.5
Gorilla gorilla beringei	34	2.7	37	48.6	34	76.5

Adapted from Swindler and Olshan (1988).

following quote should be read carefully.

> While these frequencies appear most frequently, there remains a large
> amount of variation in both mandibular and maxillary MSS in all three
> species, and no sequence can be said to be typical of all great apes or
> even of any particular species.
>
> (Mahler, 1980, p. 752)

Such variation should make one suspicious of the usefulness of this trait
for comparative taxonomic purposes.

Odontometry (Appendix 1, Tables 179–194, pp. 255–62)
There are many significant sexual differences in the teeth of pongids as
depicted in these tables. The upper canine is much larger in males than in
females. The mountain gorilla is particularly notable for the very high
degree of significant sex differences in its teeth, especially the upper canine.
This high degree of sexual dimorphism has also been reported for the
lowland gorilla by Greene (1973). The chimpanzee odontometric data
reported here are similar to the data of Ashton and Zuckerman (1950).

Appendix 1: Odontometry

Measurements to create this odontometric appendix were taken on dental casts of both the permanent and deciduous dentitions following the methods and definitions presented in Chapter 1. t-Tests of tooth size sexual dimorphism were performed on the majority of species; this information appears in the male tables (*, $p < 0.05$; †, $p < 0.01$). If sexes are not designated, they are combined.

Permanent teeth

Table 1. *Varecia variegata* male

Maxillary teeth		n	m	SD	Range
I^1	M–D	3	2.1	0.12	2.0–2.2
	B–L	2	1.5	0.00	1.5–1.5
I^2	M–D	3	2.1	0.29	1.8–2.3
	B–L	2	1.5	0.07	1.4–1.5
C	M–D	5	6.3	0.33	4.2–7.7
	B–L	5	3.0	0.23	2.7–3.3
P^2	M–D	4	4.9	0.34	4.6–5.3
	B–L	4	2.8	0.38	2.2–3.0
P^3	M–D	5	6.7	0.35	6.3–7.2
	B–L	4	4.6	0.49	3.9–5.0
P^4	M–D	5	6.3	0.27	6.0–6.6
	B–L	4	7.0	0.33	6.5–7.3
M^1	M–D	6	7.3	0.46	6.5–7.8
	Ant B	6	7.6	1.53	4.5–8.4
	Post B	6	7.8	1.53	5.0–8.5
M^2	M–D	5	7.1	0.18	6.8–7.2
	Ant B	5	8.0	0.20	7.7–8.2
	Post B	5	7.8	0.36	7.4–8.3
M^3	M–D	6	5.3	0.56	4.6–6.2
	Ant B	5	5.6	0.21	5.3–5.8
	Post B	5	5.0	0.76	3.7–5.6

Table 2. *Varecia variegata* male

Mandibular teeth		n	m	SD	Range
I_1	M–D	—	—	—	—
	B–L	—	—	—	—
I_2	M–D	—	—	—	—
	B–L	—	—	—	—
C	M–D	3	2.1	0.12	2.0–2.2
	B–L	3	2.8	0.21	2.6–3.0
P_2	M–D	2	6.9	1.13	6.1–7.7
	B–L	3	2.9	0.29	2.7–3.2
P_3	M–D	4	5.4	0.35	4.9–5.7
	B–L	5	3.1	0.26	2.9–3.5
P_4	M–D	4	6.7	0.25	6.4–7.0
	B–L	4	4.1	0.25	3.9–4.4
M_1	M–D	5	7.8	0.16	7.6–8.0
	Tri B	5	4.6	0.32	4.2–5.0
	Tal B	4	5.0	0.17	4.8–5.2
M_2	M–D	4	7.0	0.14	6.8–7.1
	Tri B	4	4.7	0.34	4.3–5.0
	Tal B	3	4.7	0.06	4.6–4.7
M_3	M–D	4	5.5	0.39	5.0–5.9
	Tri B	2	3.7	0.21	3.5–3.8
	Tal B	2	3.6	0.35	3.3–3.8
Hypoconulid		—	—	—	—

Table 3. *Lepilemur mustelinus* male

Maxillary teeth		n	m	SD	Range
I^1	M–D	—	—	—	—
	B–L	—	—	—	—
I^2	M–D	—	—	—	—
	B–L	—	—	—	—
C	M–D	6	3.13	0.32	2.6–3.5
	B–L	6	1.62	0.28	1.3–2.0
P^2	M–D	6	3.00	0.17	2.8–3.2
	B–L	6	1.62	0.21	1.4–2.0
P^3	M–D	6	2.85	0.25	2.6–3.2
	B–L	6	2.17	0.19	2.0–2.4
P^4	M–D	6	2.65	0.26	2.3–3.0
	B–L	6	2.73	0.08	2.6–2.8
M^1	M–D	6	3.50	0.27	3.2–4.0
	Ant B	6	3.27	0.20	3.0–3.6*
	Post B	—	—	—	—
M^2	M–D	6	3.40	0.28	3.1–3.9
	Ant B	6	3.42	0.15	3.2–3.6*
	Post B	—	—	—	—
M^3	M–D	6	2.67	0.14	2.5–2.9
	Ant B	6	3.05	0.21	2.7–3.3
	Post B	—	—	—	—

Table 4. *Lepilemur mustelinus* male

Mandibular teeth		n	m	SD	Range
I_1	M–D	—	—	—	—
	B–L	—	—	—	—
I_2	M–D	—	—	—	—
	B–L	—	—	—	—
C	M–D	—	—	—	—
	B–L	—	—	—	—
P_2	M–D	4	2.70	0.14	2.5–2.8
	B–L	4	1.75	0.13	1.6–1.9
P_3	M–D	6	2.93	0.23	2.7–3.2
	B–L	6	1.63	0.20	1.4–1.9
P_4	M–D	6	2.75	0.24	2.4–3.0
	B–L	6	1.78	0.38	1.1.–2.2
M_1	M–D	6	3.50	0.19	3.3–3.7
	Tri B	6	2.25	0.16	2.0–2.5
	Tal B	—	—	—	—
M_2	M–D	6	3.58	0.16	3.4–3.8
	Tri B	6	2.38	0.21	2.0–2.6
	Tal B	—	—	—	—
M_3	M–D	6	3.97	0.14	3.8 4.2
	Tri B	6	2.12	0.16	1.9–2.3
	Tal B	—	—	—	—
	Hypoconulid	—	—	—	—

Table 5. *Lepilemur mustelinus* female

Maxillary teeth		n	m	SD	Range
I^1	M–D	—	—	—	—
	B–L	—	—	—	—
I^2	M–D	—	—	—	—
	B–L	—	—	—	—
C	M–D	10	3.1	0.40	2.3–3.7
	B–L	10	1.8	0.32	1.4–2.3
P^2	M–D	11	3.1	0.50	2.5–4.0
	B–L	11	1.7	0.21	1.2–2.0
P^3	M–D	11	2.9	0.11	2.7–3.0
	B–L	11	2.3	0.13	2.0–2.5
P^4	M–D	11	2.7	0.22	2.3–3.0
	B–L	11	2.8	0.22	2.3–3.1
M^1	M–D	11	3.4	0.18	3.2–3.7
	Ant B	11	3.5	0.16	3.2–3.7
	Post B	—	—	—	—
M^2	M–D	11	3.3	0.18	3.0–3.5
	Ant B	11	3.6	0.20	3.4–4.0
	Post B	—	—	—	—
M^3	M–D	6	2.8	0.20	2.4–3.0
	Ant B	6	3.1	0.18	2.7–3.4
	Post B	—	—	—	—

Table 6. *Lepilemur mustelinus* female

Mandibular teeth		n	m	SD	Range
I_1	M–D	—	—	—	—
	B–L	—	—	—	—
I_2	M–D	—	—	—	—
	B–L	—	—	—	—
C	M–D	—	—	—	—
	B–L	—	—	—	—
P_2	M–D	9	2.8	0.39	2.2–3.6
	B–L	9	1.9	0.36	1.5–2.5
P_3	M–D	11	2.7	0.30	2.3–3.2
	B–L	11	1.6	0.17	1.3–1.9
P_4	M–D	11	2.6	0.29	2.2–3.1
	B–L	11	2.0	0.16	1.6–2.2
M_1	M–D	11	3.6	0.21	3.2–3.8
	Tri B	11	2.4	0.17	2.0–2.6
	Tal B	—	—	—	—
M_2	M–D	11	3.5	0.14	3.2–3.7
	Tri B	11	2.4	0.17	2.2–2.7
	Tal B	—	—	—	—
M_3	M–D	11	4.0	0.14	3.6–4.1
	Tri B	11	2.1	0.17	2.0–2.5
	Tal B	—	—	—	—
	Hypoconulid	—	—	—	—

Table 7. *Propithecus verreauxi* male

Maxillary teeth		n	m	SD	Range
I^1	M–D	3	4.23	0.15	4.0–4.4
	B–L	3	2.87	0.06	2.8–3.0
I^2	M–D	3	3.00	0.20	2.7–3.7
	B–L	3	2.10	0.36	1.6–2.5
C	M–D	2	6.00	0.00	6.0–6.0
	B–L	2	3.80	0.42	3.3–4.3
P^3	M–D	3	6.30	0.27	5.6–6.6
	B–L	3	3.77	0.15	3.6–4.0
P^4	M–D	3	5.83	0.12	5.0–6.0*
	B–L	3	4.77	0.06	4.6–4.8*
M^1	M–D	3	7.13	0.32	6.7–7.5
	Ant B	3	6.67	0.15	6.4–6.9*
	Post B	3	6.60	0.27	6.3–6.9†
M^2	M–D	3	7.43	0.21	6.7–7.7†
	Ant B	3	7.43	0.15	7.2–7.6†
	Post B	2	7.00	0.14	6.8–7.2†
M^3	M–D	3	5.80	0.27	5.0–6.1†
	Ant B	3	6.10	0.17	5.9–6.3†
	Post B	2	4.40	0.30	3.9–4.9†

Table 8. *Propithecus verreauxi* male

Mandibular teeth		n	m	SD	Range
I_1	M–D	3	1.47	0.06	1.3–1.5
	B–L	2	2.75	0.07	2.7–2.8
I_2	M–D	3	2.77	0.25	2.5–3.6
	B–L	3	3.53	0.25	3.2–3.8
C	M–D	—	—	—	—
	B–L	—	—	—	—
P_3	M–D	3	6.03	0.71	5.2–6.8
	B–L	3	3.13	0.21	2.9–3.3
P_4	M–D	3	5.40	0.46	4.9–5.9
	B–L	3	2.87	0.21	2.5–3.2
M_1	M–D	3	7.00	0.10	6.7–7.2
	Tri B	3	4.10	0.27	3.8–4.4
	Tal B	3	4.97	0.06	4.9–5.1*
M_2	M–D	3	7.20	0.17	6.9–7.4†
	Tri B	3	4.97	0.06	4.9–5.0†
	Tal B	3	5.43	0.06	5.3–5.5
M_3	M–D	2	7.30	0.14	6.9–7.7*
	Tri B	2	5.30	0.14	5.1–5.5*
	Tal B	3	4.67	0.23	4.3–5.0
	Hypoconulid	—	—	—	—

Table 9. *Propithecus verreauxi* female

Maxillary teeth		n	m	SD	Range
I^1	M–D	5	3.66	0.43	3.0–4.1
	B–L	5	2.62	0.27	2.2–2.9
I^2	M–D	5	2.72	0.19	2.5–3.0
	B–L	5	1.86	0.35	1.5–2.5
C	M–D	7	5.77	0.51	5.2–6.5
	B–L	7	3.57	0.39	3.1–4.3
P^3	M–D	7	6.27	0.50	5.7–7.0
	B–L	7	3.56	0.27	3.2–4.0
P^4	M–D	7	5.29	0.35	4.8–5.9
	B–L	7	4.09	0.36	3.5–4.8
M^1	M–D	7	6.70	0.31	6.1–7.5
	Ant B	6	5.83	0.41	5.4–6.5
	Post B	7	5.87	0.30	5.5–6.9
M^2	M–D	7	6.43	0.36	5.9–7.5
	Ant B	7	6.33	0.48	5.6–7.6
	Post B	6	6.10	0.24	5.8–7.1
M^3	M–D	7	4.44	0.34	4.0–5.5
	Ant B	6	4.70	0.37	4.2–6.0
	Post B	5	3.14	0.34	2.8–4.9

Table 10. *Propithecus verreauxi* female

Mandibular teeth		n	m	SD	Range
I_1	M–D	5	1.38	0.13	1.2–1.6
	B–L	7	2.43	0.26	2.1–2.8
I_2	M–D	6	2.62	0.52	2.1–3.2
	B–L	8	3.21	0.33	2.7–3.8
C	M–D	—	—	—	—
	B–L	—	—	—	—
P_3	M–D	8	6.50	0.44	5.0–7.0
	B–L	8	2.81	0.23	2.6–3.3
P_4	M–D	8	5.58	0.57	4.7–6.4
	B–L	8	2.80	0.25	2.3–3.1
M_1	M–D	8	6.69	0.35	6.1–7.3
	Tri B	8	3.83	0.29	3.4–4.3
	Tal B	7	4.66	0.18	4.3–5.0
M_2	M–D	8	6.44	0.38	6.0–7.3
	Tri B	8	4.39	0.30	4.0–4.7
	Tal B	8	4.69	0.19	4.5–5.4
M_3	M–D	7	6.47	0.37	5.9–7.4
	Tri B	7	4.50	0.39	4.1–5.2
	Tal B	7	4.16	0.41	3.7–4.8
	Hypoconulid	2	1.90	0.14	1.8–2.0

Table 11. *Nycticebus coucang* male

Maxillary teeth		n	m	SD	Range
I¹	M–D	7	1.4	0.15	1.2–1.6
	B–L	9	1.4	0.14	1.2–1.6
I²	M–D	3	1.2	0.29	0.9–1.4
	B–L	2	0.9	0.14	0.8–1.0
C	M–D	18	3.6	0.34	3.2–4.3*
	B–L	18	2.7	0.26	2.4–3.2
P²	M–D	20	3.1	0.35	2.0–3.6
	B–L	21	2.4	0.23	2.0–2.9
P³	M–D	20	2.3	0.31	1.6–2.9
	B–L	18	2.6	0.38	1.6–3.0
P⁴	M–D	20	2.4	0.36	1.9–3.2
	B–L	19	3.1	0.50	1.8–4.0
M¹	M–D	21	3.5	0.40	2.9–4.5
	Ant B	17	4.3	0.48	3.6–5.0
	Post B	17	4.4	0.54	3.6–5.6
M²	M–D	21	3.2	0.32	2.6–3.8
	Ant B	19	4.2	0.44	3.5–5.0
	Post B	19	4.1	0.48	3.3–5.2
M³	M–D	18	2.5	0.33	1.8–2.9
	Ant B	18	3.5	0.35	2.9–4.3
	Post B	2	3.8	1.06	3.0–4.5

Table 12. *Nycticebus coucang* male

Mandibular teeth		n	m	SD	Range
I₁	M–D	—	—	—	—
	B–L	14	1.6	0.21	1.2–2.0
I₂	M–D	—	—	—	—
	B–L	14	1.6	0.21	1.2–2.0
C	M–D	16	1.4	0.23	1.0–1.8
	B–L	14	2.3	0.29	1.9–2.9
P₂	M–D	20	3.5	0.50	1.8–4.2
	B–L	20	3.1	0.39	2.0–3.8
P₃	M–D	17	2.2	0.29	1.8–2.8
	B–L	17	2.3	0.31	2.0–3.1
P₄	M–D	18	2.4	0.38	1.7–3.0
	B–L	17	2.5	0.31	2.0–3.0
M₁	M–D	21	3.4	0.33	3.0–4.1
	Tri B	20	2.7	0.29	2.3–3.4
	Tal B	19	2.9	0.25	2.5–3.5
M₂	M–D	20	3.4	0.28	2.9–3.9
	Tri B	19	2.8	0.30	2.4–3.5
	Tal B	18	2.8	0.27	2.3–3.3
M₃	M–D	16	3.4	0.44	2.5–4.1
	Tri B	17	2.5	0.27	1.9–2.9
	Tal B	14	2.3	0.19	1.9–2.6
	Hypoconulid	13	1.3	0.18	1.0–1.6

Table 13. *Nycticebus coucang* female

Maxillary teeth		n	m	SD	Range
I^1	M–D	6	1.5	0.21	1.2–1.8
	B–L	6	1.4	0.14	1.2–1.6
I^2	M–D	3	1.2	0.29	0.9–1.4
	B–L	2	0.9	0.14	0.8–1.0
C	M–D	7	3.3	0.14	3.2–3.5
	B L	8	2.5	0.26	2.2–3.0
P^2	M–D	9	3.0	0.42	2.3–3.6
	B–L	9	2.3	0.32	1.6–2.8
P^3	M–D	9	2.2	0.42	1.7–3.0
	B–L	9	2.6	0.33	1.9–3.0
P^4	M–D	9	2.5	0.52	2.0–3.8
	B–L	9	3.1	0.42	2.4–3.6
M^1	M–D	9	3.5	0.25	3.2–4.0
	Ant B	8	4.2	0.32	3.7–4.6
	Post B	8	4.4	0.29	4.0–4.7
M^2	M–D	9	3.2	0.25	2.9–3.6
	Ant B	9	4.1	0.23	3.7–4.4
	Post B	9	4.2	0.25	3.8–4.6
M^3	M–D	9	2.4	0.18	2.3–2.8
	Ant B	9	3.5	0.15	3.3–3.7
	Post B	2	3.0	0.07	2.9–3.0

Table 14. *Nycticebus coucang* female

Mandibular teeth		n	m	SD	Range
I_1	M–D	—	—	—	—
	B–L	8	1.6	0.15	1.4–1.9
I_2	M–D	—	—	—	—
	B–L	8	1.7	0.18	1.5–2.0
C	M–D	8	1.3	0.11	1.2–1.5
	B–L	6	2.4	0.32	2.1–3.0
P_2	M–D	8	3.2	0.49	2.2–3.6
	B–L	8	2.9	0.34	2.3–3.3
P_3	M–D	8	2.0	0.35	1.3–2.5
	B–L	9	2.3	0.41	1.7–2.9
P_4	M–D	9	2.2	0.35	1.8–2.7
	B–L	9	2.4	0.25	1.9–2.8
M_1	M–D	9	3.4	0.28	2.8–3.8
	Tri B	8	2.7	0.25	2.3–2.9
	Tal B	8	2.9	0.23	2.5–3.2
M_2	M–D	8	3.3	0.25	2.9–3.6
	Tri B	6	2.8	0.17	2.6–3.0
	Tal B	7	2.8	0.17	2.6–3.0
M_3	M–D	8	3.5	0.26	3.1–3.9
	Tri B	7	2.3	0.23	2.0–2.6
	Tal B	7	2.2	0.23	2.0–2.6
	Hypoconulid	8	1.4	0.14	1.2–1.6

Table 15. *Periodicticus potto* male

Maxillary teeth		n	m	SD	Range
I¹	M–D	—	—	—	—
	B–L	—	—	—	—
I²	M–D	—	—	—	—
	B–L	—	—	—	—
C	M–D	8	4.3	0.43	3.5–4.9
	B–L	8	3.1	0.31	2.5–3.5*
P²	M–D	11	2.8	0.17	2.6–3.2
	B–L	9	2.5	0.09	2.3–2.6
P³	M–D	8	2.5	0.30	2.2–3.0
	B–L	6	2.3	0.25	2.0–2.7
P⁴	M–D	8	2.5	0.24	2.0–2.7
	B–L	7	2.3	0.26	3.0–3.8
M¹	M–D	11	3.5	0.27	2.8–3.8
	Ant B	10	4.0	0.34	3.5–4.7
	Post B	—	—	—	—
M²	M–D	10	3.3	0.15	3.1–3.5
	Ant B	8	4.3	0.28	3.7–4.6
	Post B	—	—	—	—
M³	M–D	6	2.3	0.31	1.7–2.5
	Ant B	6	3.5	0.42	2.9–4.1
	Post B	—	—	—	—

Table 16. *Periodicticus potto* male

Mandibular teeth		n	m	SD	Range
I₁	M–D	—	—	—	—
	B–L	—	—	—	—
I₂	M–D	—	—	—	—
	B–L	—	—	—	—
C	M–D	—	—	—	—
	B–L	—	—	—	—
P₂	M–D	10	2.9	0.38	2.3–3.7
	B–L	10	2.7	0.28	2.3–3.2
P₃	M–D	7	2.3	0.29	1.7–2.6
	B–L	7	2.4	0.10	2.3–2.5
P₄	M–D	9	2.3	0.18	2.0–2.5
	B–L	8	2.5	0.19	2.2–2.7
M₁	M–D	12	3.2	0.29	2.8–3.6
	Tri B	10	2.7	0.32	2.2–3.3
	Tal B	10	2.6	0.26	2.4–3.2
M₂	M–D	12	3.3	0.29	2.9–4.0
	Tri B	11	2.9	0.31	2.4–3.7
	Tal B	11	2.8	0.32	2.5–3.7
M₃	M–D	8	3.0	0.26	2.6–3.4
	Tri B	8	2.5	0.36	2.0–3.1
	Tal B	5	2.2	0.18	2.0–2.5
Hypoconulid		—	—	—	—

Table 17. *Periodicticus potto* female

Maxillary teeth		n	m	SD	Range
I¹	M–D	—	—	—	—
	B–L	—	—	—	—
I²	M–D	—	—	—	—
	B–L	—	—	—	—
C	M–D	9	4.1	0.42	3.5–4.8
	B–L	10	2.7	0.28	2.2–3.2
P²	M–D	9	2.9	0.31	2.4–3.3
	B–L	9	2.5	0.27	2.2–3.0
P³	M–D	9	2.4	0.21	2.0–2.7
	B–L	8	2.2	0.20	1.9–2.4
P⁴	M–D	8	2.4	0.05	2.3–2.5
	B–L	9	3.3	0.33	2.9–3.7
M¹	M–D	7	3.4	0.19	3.2–3.7
	Ant B	8	3.8	0.28	3.6–4.3
	Post B	—	—	—	—
M²	M–D	7	3.4	0.17	3.1–3.6
	Ant B	8	4.4	0.33	3.9–4.9
	Post B	—	—	—	—
M³	M–D	6	2.0	0.30	1.7–2.4
	Ant B	8	3.4	0.23	2.9–4.1
	Post B	—	—	—	—

Table 18. *Periodicticus potto* female

Mandibular teeth		n	m	SD	Range
I₁	M–D	—	—	—	—
	B–L	—	—	—	—
I₂	M–D	—	—	—	—
	B–L	—	—	—	—
C	M–D	—	—	—	—
	B L				
P₂	M–D	8	2.8	0.28	2.4–3.2
	B–L	9	2.9	0.32	2.4–3.5
P₃	M–D	9	2.2	0.26	1.9–2.7
	B–L	10	2.3	0.18	2.0–2.6
P₄	M–D	8	2.4	0.17	2.0–2.5
	B–L	10	2.4	0.20	2.1–2.7
M₁	M–D	7	3.1	0.38	2.4–3.4
	Tri B	10	2.5	0.26	2.0–2.8
	Tal B	10	2.4	0.23	2.0–2.7
M₂	M–D	9	3.4	0.16	3.2–3.7
	Tri B	10	2.8	0.29	2.5–3.2
	Tal B	10	2.7	0.26	2.2–3.0
M₃	M–D	4	3.3	0.19	3.0–3.4
	Tri B	5	2.3	0.23	2.0–2.6
	Tal B	5	2.1	0.12	2.0–2.3
Hypoconulid		—	—	—	—

Table 19. *Otolemur crassicaudatus* male

Maxillary teeth		*n*	*m*	SD	Range
I¹	M–D	—	—	—	—
	B–L	—	—	—	—
I²	M–D	—	—	—	—
	B–L	—	—	—	—
C	M–D	15	4.6	0.56	3.9–5.7*
	B–L	13	3.0	0.32	2.5–3.8*
P²	M–D	18	3.9	0.55	3.2–5.7
	B–L	19	2.4	0.37	1.9–3.5
P³	M–D	20	3.1	0.37	2.5–3.7
	B–L	20	2.7	0.37	2.1–3.7
P⁴	M–D	19	3.6	0.25	3.2–4.0
	B–L	19	4.1	0.45	3.0–5.0
M¹	M–D	19	4.2	0.26	3.7–4.5
	Ant B	18	5.0	0.47	3.6–5.6
	Post B	18	5.3	0.37	4.3–5.8
M²	M–D	18	3.9	0.17	3.6–4.3
	Ant B	17	5.4	0.28	4.8–5.8
	Post B	17	5.1	0.28	4.7–5.7
M³	M–D	15	3.0	0.28	2.3–3.3
	Ant B	13	4.2	0.32	3.7–4.7
	Post B	5	3.6	0.52	2.7–4.0

Table 20. *Otolemur crassicaudatus* male

Mandibular teeth		*n*	*m*	SD	Range
I₁	M–D	—	—	—	—
	B–L	10	1.9	0.26	1.5–2.2
I₂	M–D	—	—	—	—
	B–L	10	1.9	0.26	1.5–2.2
C	M–D	10	1.2	0.14	0.9–1.3
	B–L	10	2.4	0.26	1.9–2.6
P₂	M–D	17	3.4	0.46	2.2–4.0
	B–L	17	2.7	0.31	2.3–3.3
P₃	M–D	18	3.5	0.66	1.6–4.3
	B–L	18	2.4	0.28	1.7–2.9
P₄	M–D	19	3.4	0.45	1.8–3.8
	B–L	17	2.8	0.35	2.2–3.8
M₁	M–D	18	3.9	0.18	3.5–4.2
	Tri B	18	3.3	0.17	3.0–3.7
	Tal B	17	3.3	0.21	2.9–3.6
M₂	M–D	19	3.9	0.29	3.5–5.0
	Tri B	19	3.4	0.19	2.9–3.7
	Tal B	18	3.2	0.20	2.7–3.4
M₃	M–D	18	4.3	0.31	3.7–4.8
	Tri B	18	2.8	0.23	2.4–3.3
	Tal B	18	2.4	0.19	2.1–2.8
	Hypoconulid	15	1.4	0.25	0.9–1.8

Table 21. *Otolemur crassicaudatus* female

Maxillary teeth		n	m	SD	Range
I^1	M–D	—	—	—	—
	B–L	—	—	—	—
I^2	M–D	—	—	—	—
	B–L	—	—	—	—
C	M–D	7	4.1	0.51	3.1–4.7
	B–L	6	2.9	0.36	2.1–3.2
P^2	M–D	8	4.0	0.23	3.6–4.3
	B–L	8	2.5	0.27	1.9–2.8
P^3	M–D	8	3.1	0.30	2.5–3.5
	B–L	8	2.6	0.31	2.1–3.0
P^4	M–D	10	3.7	0.23	3.3–4.2
	B–L	11	4.3	0.19	3.9–4.6
M^1	M–D	11	4.4	0.23	3.8–4.6
	Ant B	11	4.9	0.25	4.6–5.5
	Post B	11	5.3	0.20	5.0–5.6
M^2	M–D	11	4.1	0.24	3.8–4.7
	Ant B	10	5.3	0.23	4.9–5.5
	Post B	10	5.0	0.22	4.7–5.3
M^3	M–D	10	3.2	0.14	3.0–3.4
	Ant B	10	4.3	0.10	4.2–4.5
	Post B	6	2.8	0.37	2.1–3.2

Table 22. *Otolemur crassicaudatus* female

Mandibular teeth		n	m	SD	Range
I_1	M–D	—	—	—	—
	B–L	6	1.8	0.10	1.8–2.0
I_2	M–D	—	—	—	—
	B–L	6	1.9	0.10	1.8–2.0
C	M–D	6	1.3	0.15	1.0–1.4
	B–L	6	2.4	0.22	2.0–2.6
P_2	M–D	8	3.3	0.17	3.0–3.5
	B–L	8	2.7	0.41	1.7–3.0
P_3	M–D	7	3.6	0.20	3.3–3.9
	B–L	6	2.6	0.15	2.4–2.8
P_4	M–D	8	3.6	0.19	3.3–3.9
	B–L	8	2.7	0.18	2.3–2.8
M_1	M–D	10	4.0	0.21	3.7–4.3
	Tri B	10	3.2	0.18	3.0–3.5
	Tal B	11	3.2	0.18	2.9–3.5
M_2	M–D	11	3.9	0.18	3.5–4.2
	Tri B	11	3.3	0.16	3.0–3.6
	Tal B	10	3.2	0.08	3.0–3.3
M_3	M–D	11	4.3	0.15	4.2–4.7
	Tri B	10	2.9	0.12	2.7–3.0
	Tal B	10	2.5	0.07	2.4–2.6
	Hypoconulid	10	1.4	0.22	0.9–1.7

Table 23. *Galago senegalensis* male

Maxillary teeth		n	m	SD	Range
I¹	M–D	—	—	—	—
	B–L	—	—	—	—
I²	M–D	—	—	—	—
	B–L	—	—	—	—
C	M–D	16	2.3	0.19	2.0–2.6*
	B–L	17	1.6	0.32	1.4–2.6*
P²	M–D	20	2.2	0.16	1.9–2.5
	B–L	21	1.2	0.12	1.0–1.4
P³	M–D	20	1.8	0.17	1.5–2.1
	B–L	21	1.7	0.31	1.2–2.6
P⁴	M–D	22	2.2	0.17	1.9–2.5
	B–L	22	2.6	0.25	2.2–3.5
M¹	M–D	24	2.5	0.14	2.1–2.8
	Ant B	23	3.0	0.17	2.6–3.2
	Post B	23	3.3	0.21	2.7–3.7
M²	M–D	23	2.4	0.12	2.2–2.6
	Ant B	22	3.3	0.16	3.0–3.6
	Post B	19	3.4	0.16	3.1–3.7
M³	M–D	19	1.9	0.23	1.6–2.7
	Ant B	19	2.7	0.21	2.2–3.2
	Post B	—	—	—	—

Table 24. *Galago senegalensis* male

Mandibular teeth		n	m	SD	Range
I₁	M–D	—	—	—	—
	B–L	—	—	—	—
I₂	M–D	—	—	—	—
	B–L	—	—	—	—
C	M–D	—	—	—	—
	B–L	—	—	—	—
P₂	M–D	10	1.8	0.24	1.4–2.2*
	B–L	10	1.4	0.14	1.2–1.6*
P₃	M–D	21	2.1	0.30	1.5–2.6
	B–L	17	1.3	0.19	1.0–1.6
P₄	M–D	21	2.1	0.16	1.8–2.4
	B–L	21	1.6	0.20	1.4–2.3
M₁	M–D	24	2.3	0.17	2..0–2.7
	Tri B	23	1.9	0.15	1.6–2.4
	Tal B	23	1.9	0.14	1.7–2.5
M₂	M–D	23	2.3	0.13	2.1–2.7
	Tri B	23	2.0	0.14	1.8–2.5
	Tal B	22	1.9	0.26	1.0–2.5
M₃	M–D	20	2.7	0.27	2.4–3.7
	Tri B	21	1.8	0.17	1.6–2.4
	Tal B	20	1.5	0.21	1.3–2.3
	Hypoconulid	—	—	—	—

Table 25. *Galago senegalensis* female

Maxillary teeth		n	m	SD	Range
I^1	M–D	—	—	—	—
	B–L	—	—	—	—
I^2	M–D	—	—	—	—
	B–L	—	—	—	—
C	M–D	13	2.2	0.16	1.9–2.4
	B–L	13	1.4	0.12	1.2–1.5
P^2	M–D	18	2.0	0.15	1.8–2.3
	B–L	18	1.2	0.16	0.9–1.5
P^3	M–D	18	1.9	0.15	1.6–2.2
	B–L	18	1.6	0.20	1.2–2.0
P^4	M–D	18	2.2	0.20	1.7–2.6
	B–L	18	2.5	0.13	2.3–2.8
M^1	M–D	18	2.5	0.12	2.3–2.8
	Ant B	17	3.0	0.24	2.6–3.5
	Post B	18	3.3	0.22	3.0–3.7
M^2	M–D	19	2.3	0.12	2.2–2.6
	Ant B	19	3.2	0.24	2.9–3.8
	Post B	19	3.4	0.17	3.0–3.7
M^3	M–D	19	1.8	0.14	1.5–2.2
	Ant B	18	2.9	0.22	2.3–3.1
	Post B	—	—	—	—

Table 26. *Galago senegalensis* female

Mandibular teeth		n	m	SD	Range
I_1	M–D	—	—	—	—
	B–L	—	—	—	—
I_2	M–D	—	—	—	—
	B–L	—	—	—	—
C	M–D	—	—	—	—
	B–L	—	—	—	—
P_2	M–D	9	1.7	0.30	1.2–2.2
	B–L	9	1.3	0.14	1.0–1.4
P_3	M–D	15	2.0	0.18	1.7–2.4
	B–L	15	1.3	0.13	1.0–1.5
P_4	M–D	20	2.0	0.14	1.8–2.4
	B–L	20	1.6	0.29	1.2–2.7
M_1	M–D	20	2.2	0.13	2.0–2.5
	Tri B	20	1.9	0.12	1.8–2.2
	Tal B	20	2.0	0.13	1.7–2.3
M_2	M–D	20	2.3	0.12	2.0–2.5
	Tri B	20	2.0	0.11	1.8–2.2
	Tal B	20	2.0	0.12	1.7–2.2
M_3	M–D	20	2.7	0.13	2.4–2.9
	Tri B	20	1.8	0.10	1.7–2.8
	Tal B	20	1.5	0.11	1.3–1.7
Hypoconulid		—	—	—	—

Table 27. *Tarsius (spectrum, bancanus, syrichta)* male

Maxillary teeth		n	m	SD	Range
I¹	M–D	—	—	—	—
	B–L	—	—	—	—
I²	M–D	—	—	—	—
	B–L	—	—	—	—
C	M–D	5	1.7	0.15	1.5–1.9
	B–L	5	1.6	0.09	1.5–1.8
P²	M–D	6	1.4	0.11	1.3–1.6
	B–L	6	1.4	0.14	1.2–1.6
P³	M–D	5	1.7	0.15	1.5–1.9
	B–L	6	2.3	0.25	2.0–2.6
P⁴	M–D	4	2.1	0.10	2.0–2.2
	B–L	6	3.0	0.37	2.5–3.5
M¹	M–D	6	2.9	0.30	2.5–3.3
	Ant B	6	4.0	0.40	3.4–4.5
	Post B	—	—	—	—
M²	M–D	6	2.9	0.26	2.6–3.3
	Ant B	6	4.1	0.33	3.7–4.5
	Post B	—	—	—	—
M³	M–D	5	2.5	0.14	2.4–2.7
	Ant B	5	3.8	0.32	3.5–4.1
	Post B	—	—	—	—

Table 28. *Tarsius (spectrum, bancanus, syrichta)* male

Mandibular teeth		n	m	SD	Range
I₁	M–D	—	—	—	—
	B–L	—	—	—	—
I₂	M–D	—	—	—	—
	B–L	—	—	—	—
C	M–D	5	1.6	0.17	1.5–2.0
	B–L	5	1.8	0.16	1.6–2.0
P₂	M–D	4	1.5	0.17	1.3–1.7
	B–L	5	1.5	0.15	1.3–1.7
P₃	M–D	4	1.4	0.05	1.3–1.5
	B–L	5	1.6	0.16	1.4–1.8
P₄	M–D	4	1.8	0.13	1.7–2.0
	B–L	5	1.7	0.55	1.1–2.3
M₁	M–D	6	3.0	0.21	2.7–3.2
	Tri B	6	2.4	0.14	2.1–2.5
	Tal B	6	2.4	0.23	2.0–2.7
M₂	M–D	6	2.9	0.19	2.7–3.2
	Tri B	6	2.5	0.10	2.4–2.6
	Tal B	6	2.6	0.12	2.4–2.7
M₃	M–D	6	3.6	0.23	3.4–4.0
	Tri B	6	2.5	0.12	2.3–2.6*
	Tal B	6	2.4	0.23	2.2–2.7
	Hypoconulid	6	1.3	0.08	1.2–1.4

Table 29. *Tarsius (spectrum, bancanus, syrichta)* female

Table 30. *Tarsius (spectrum, bancanus, syrichta)* female

Maxillary teeth		n	m	SD	Range
I^1	M–D	—	—	—	—
	B–L	—	—	—	—
I^2	M–D	—	—	—	—
	B–L	—	—	—	—
C	M–D	8	1.6	0.17	1.3–1.8
	B–L	7	1.6	0.15	1.4–1.8
P^2	M–D	8	1.4	0.06	1.3–1.5
	B–L	8	1.4	0.12	1.3–1.6
P^3	M–D	9	1.7	0.17	1.5–2.0
	B–L	9	2.2	0.25	1.8–2.5
P^4	M–D	9	2.1	0.31	1.7–2.5
	B–L	9	2.9	0.27	2.5–3.2
M^1	M–D	9	2.8	0.24	2.5–3.1
	Ant B	9	3.9	0.36	3.4–4.5
	Post B	—	—	—	—
M^2	M–D	9	2.7	0.16	2.4–2.9
	Ant B	9	4.1	0.25	3.6–4.5
	Post B	—	—	—	—
M^3	M–D	7	2.5	0.20	2.2–2.7
	Ant B	9	3.8	0.28	3.4–4.1
	Post B	—	—	—	—

Mandibular teeth		n	m	SD	Range
I_1	M–D	—	—	—	—
	B–L	—	—	—	—
I_2	M–D	—	—	—	—
	B–L	—	—	—	—
C	M–D	4	1.8	0.19	1.5–2.0
	B–L	6	1.7	0.19	1.2–1.9
P_2	M–D	8	1.4	0.13	1.1–1.5
	B–L	8	1.4	0.09	1.2–1.5
P_3	M–D	9	1.5	0.12	1.4–1.8
	B–L	9	1.5	0.11	1.4–1.7
P_4	M–D	9	1.7	0.12	1.5–1.9
	B–L	9	1.7	0.18	1.5–2.0
M_1	M–D	9	2.8	0.19	2.6–3.2
	Tri B	9	2.2	0.27	1.7–2.6
	Tal B	9	2.4	0.24	2.0–2.8
M_2	M–D	9	2.7	0.16	2.5–2.9
	Tri B	9	2.4	0.12	2.2–2.6
	Tal B	9	2.5	0.12	2.3–2.7
M_3	M–D	9	3.4	0.26	3.0–3.8
	Tri B	9	2.3	0.14	2.0–2.4
	Tal B	9	2.3	0.09	2.1–2.4
	Hypoconulid	9	1.3	0.14	1.0–1.5

Table 31. *Saguinus geoffroyi* male

Maxillary teeth		n	m	SD	Range
I^1	M–D	10	2.4	0.15	2.2–2.7
	B–L	11	2.0	0.13	1.8–2.3
I^2	M–D	6	1.9	0.18	1.7–2.2
	B–L	6	1.7	0.06	1.6–1.8
C	M–D	10	2.8	0.13	2.6–3.1
	B–L	9	2.3	0.09	2.2–2.5
P^2	M–D	10	2.1	0.12	1.9–2.2
	B–L	10	2.7	0.13	2.6–3.0
P^3	M–D	10	1.9	0.11	1.7–2.0
	B–L	10	3.0	0.14	2.9–3.4
P^4	M–D	10	2.0	0.14	1.9–2.3
	B–L	10	3.4	0.15	3.1–3.6
M^1	M–D	11	2.8	0.11	2.6–3.0*
	Ant B	11	3.6	0.18	3.3–4.0
	Post B	—	—	—	—
M^2	M–D	10	1.6	0.07	1.4–1.7
	Ant B	10	2.5	0.13	2.3–2.7†
	Post B	—	—	—	—
M^3	M–D	—	—	—	—
	Ant B	—	—	—	—
	Post B	—	—	—	—

Table 32. *Saguinus geoffroyi* male

Mandibular teeth		n	m	SD	Range
I_1	M–D	8	1.6	0.09	1.5–1.8
	B–L	8	1.8	0.09	1.7–2.0
I_2	M–D	8	1.7	0.18	1.5–2.0
	B–L	8	1.9	0.07	1.8–2.0
C	M–D	9	2.6	0.20	2.4–3.1
	B–L	9	2.7	0.13	2.5–2.8
P_2	M–D	10	2.4	0.18	2.1–2.6
	B–L	10	2.2	0.11	2.0–2.4
P_3	M–D	10	2.2	0.18	1.9–2.5
	B–L	10	2.2	0.08	2.1–2.3
P_4	M–D	10	2.3	0.17	2.1–2.3
	B–L	10	2.3	0.10	2.2–2.5
M_1	M–D	11	2.8	0.11	2.7–3.0*
	Tri B	11	2.3	0.12	2.1–2.5
	Tal B	10	2.3	0.12	2.1–2.5
M_2	M–D	9	2.4	0.11	2.2–2.5
	Tri B	9	1.9	0.32	1.6–2.7
	Tal B	9	1.9	0.41	1.2–2.8
M_3	M–D	—	—	—	—
	Tri B	—	—	—	—
	Tal B	—	—	—	—
	Hypoconulid	—	—	—	—

Table 33. *Saguinus geoffroyi* female

Maxillary teeth		n	m	SD	Range
I^1	M–D	14	2.4	0.20	2.0–2.6
	B–L	12	1.9	0.28	1.8–2.0
I^2	M–D	12	1.9	0.13	1.7–2.2
	B–L	12	1.7	0.11	1.4–1.8
C	M–D	13	4.1	0.20	2.6–3.2
	B L	14	2.9	0.10	2.2–2.5
P^2	M–D	14	4.0	0.19	1.9–2.4
	B–L	14	2.5	0.20	2.4–3.1
P^3	M–D	14	3.1	0.15	1.8–2.3
	B–L	13	2.6	0.19	3.0–3.5
P^4	M–D	14	3.7	0.14	1.9–2.3
	B–L	14	4.3	0.26	3.0–3.9
M^1	M–D	14	4.4	0.12	2.7–3.1
	Ant B	14	4.9	0.21	3.4–4.1
	Post B	—	—	—	—
M^2	M–D	14	4.1	0.10	1.4–1.8
	Ant B	14	5.3	0.12	2.4–2.9
	Post B	—	—	—	—
M^3	M–D	—	—	—	—
	Ant B	—	—	—	—
	Post B	—	—	—	—

Table 34. *Saguinus geoffroyi* female

Mandibular teeth		n	m	SD	Range
I$_1$	M–D	12	1.7	0.11	1.5–1.8
	B–L	12	1.8	0.12	1.6–2.0
I$_2$	M–D	12	1.8	0.14	1.5–2.0
	B–L	12	1.9	0.10	1.8–2.0
C	M–D	12	2.6	0.10	2.4–2.8
	B–L	10	2.7	0.11	2.5–2.8
P$_2$	M–D	14	2.5	0.19	2.0–2.8
	B–L	14	2.3	0.13	2.0–2.4
P$_3$	M–D	14	2.3	0.18	2.0–2.7
	B–L	14	2.2	0.14	2.0–2.5
P$_4$	M–D	14	2.3	0.11	2.2–2.6
	B–L	14	2.4	0.09	2.3–2.6
M$_1$	M–D	15	2.9	0.12	2.7–3.1
	Tri B	15	2.3	0.08	2.2–2.5
	Tal B	15	2.3	0.09	2.2–2.5
M$_2$	M–D	12	2.4	0.15	2.2–2.6
	Tri B	11	2.0	0.29	1.7–2.8
	Tal B	10	2.0	0.34	1.7–2.9
M$_3$	M–D	—	—	—	—
	Tri B	—	—	—	—
	Tal B	—	—	—	—
	Hypoconulid	—	—	—	—

Table 35. *Aotus trivirgatus* male

Maxillary teeth		n	m	SD	Range
I^1	M–D	7	3.6	0.11	3.4–3.7
	B–L	7	2.7	0.05	2.6–2.7
I^2	M–D	9	2.5	0.37	2.2–3.4
	B–L	8	2.3	0.15	2.0–2.5
C	M–D	7	3.1	0.14	2.9–3.2
	B–L	7	2.9	0.14	2.7–3.1†
P^2	M–D	9	4.0	0.16	2.0–2.5
	B–L	8	2.5	0.22	3.0–3.7
P^3	M–D	9	3.1	0.11	2.0–2.3
	B–L	9	2.6	0.26	3.1–4.0
P^4	M–D	9	3.7	0.14	2.2–2.6
	B–L	9	4.3	0.22	3.6–4.3
M^1	M–D	9	4.4	0.19	3.0–3.7
	Ant B	9	4.9	0.30	3.7–4.6
	Post B	9	5.3	0.20	3.8–4.4
M^2	M–D	9	4.1	0.14	2.9–3.4
	Ant B	9	5.3	0.26	3.7–4.5*
	Post B	7	5.0	0.10	3.5–3.8*
M^3	M–D	7	3.2	0.32	2.0–2.8
	Ant B	7	4.3	0.32	2.9–3.9†
	Post B	—	—	—	—

Table 36. *Aotus trivirgatus* male

Mandibular teeth		n	m	SD	Range
I_1	M–D	6	2.3	0.12	2.2–2.5
	B–L	6	2.4	0.16	2.2–2.5
I_2	M–D	7	2.3	0.28	1.9–2.6
	B–L	7	2.6	0.19	2.3–2.7
C	M–D	4	2.8	0.13	2.7–3.0
	B–L	4	2.7	0.37	2.3–3.1
P_2	M–D	8	2.4	0.20	2.2–2.8
	B–L	8	2.4	0.23	2.1–2.8
P_3	M–D	8	2.3	0.14	2.0–2.4
	B–L	7	2.3	0.15	2.1–2.5
P_4	M–D	9	2.5	0.15	2.2–2.7
	B–L	9	2.7	0.12	2.5–2.8
M_1	M–D	9	3.3	0.20	3.0–3.6
	Tri B	9	2.9	0.15	2.6–3.0
	Tal B	8	3.0	0.17	2.7–3.3
M_2	M–D	9	3.3	0.19	3.0–3.5
	Tri B	7	3.0	0.15	2.8–3.2
	Tal B	9	2.9	0.12	2.8–3.2
M_3	M–D	8	3.3	0.38	2.8–3.9*
	Tri B	6	2.7	0.17	2.6–3.0
	Tal B	5	2.7	0.22	2.4–3.0
Hypoconulid		—	—	—	—

Table 37. *Aotus trivirgatus* female

Maxillary teeth		*n*	*m*	SD	Range
I^1	M–D	11	3.4	0.36	2.6–3.9
	B–L	11	2.7	0.1	2.5–2.8
I^2	M–D	9	2.9	0.23	1.8–2.5
	B–L	9	2.2	0.18	1.9–2.5
C	M–D	11	3.0	0.18	2.8–3.4
	B–L	12	2.6	0.11	2.4–2.7
P^2	M–D	10	2.0	0.22	1.7–2.4
	B–L	11	3.1	0.17	2.8–3.4
P^3	M–D	12	2.1	0.16	1.8–2.3
	B–L	13	3.4	0.18	3.0–3.6
P^4	M–D	13	2.3	0.23	1.9–2.6
	B–L	12	3.6	0.16	3.4–3.9
M^1	M–D	15	3.3	0.18	3.0–3.7
	Ant B	15	3.9	0.21	3.7–4.4
	Post B	15	3.9	0.22	3.6–4.3
M^2	M–D	14	3.0	0.15	2.8–3.3
	Ant B	13	3.7	0.12	3.6–4.1
	Post B	12	3.5	0.13	3.2–3.6
M^3	M–D	13	2.3	0.51	1.7–3.7
	Ant B	13	2.9	0.23	2.5–3.2
	Post B	—	—	—	—

Table 38. *Aotus trivirgatus* female

Mandibular teeth		*n*	*m*	SD	Range
I$_1$	M–D	10	2.3	0.35	1.9–3.2
	B–L	10	2.4	0.17	2.2–2.6
I$_2$	M–D	10	2.3	0.41	1.9–3.4
	B–L	10	2.4	0.16	2.2–2.7
C	M–D	10	2.7	0.16	2.4–2.9
	B–L	9	2.4	0.25	2.0–2.8
P$_2$	M–D	11	2.4	0.14	2.1–2.6
	B–L	11	2.3	0.20	2.0–2.6
P$_3$	M–D	13	2.2	0.16	1.9–2.4
	B–L	12	2.3	0.18	2.0–2.6
P$_4$	M–D	12	2.4	0.17	2.1–2.6
	B–L	13	2.7	0.15	2.4–2.9
M$_1$	M–D	15	3.3	0.14	3.0–3.5
	Tri B	15	2.9	0.14	2.7–3.2
	Tal B	15	3.0	0.13	2.8–3.2
M$_2$	M–D	14	3.2	0.20	2.9–3.5
	Tri B	13	2.9	0.09	2.8–3.1
	Tal B	13	2.9	0.10	2.7–3.0
M$_3$	M–D	11	2.9	0.17	2.7–3.2
	Tri B	11	2.6	0.16	2.3–2.9
	Tal B	10	2.5	0.17	2.3–2.8
Hypoconulid		—	—	—	—

Table 39. *Ateles geoffroyi* male

Maxillary teeth		n	m	SD	Range
I^1	M–D	10	4.7	0.36	4.2–5.3
	B–L	12	4.2	0.35	3.7–4.8
I^2	M–D	11	3.5	0.36	3.0–4.3
	B–L	11	4.1	0.26	3.7–4.5
C	M–D	3	6.1	0.64	5.7–6.8
	B–L	3	4.9	0.45	4.4–5.3
P^2	M–D	12	3.5	0.31	3.0–4.1
	B–L	12	5.3	0.23	4.2–4.9
P^3	M–D	12	3.5	0.26	3.1–4.0
	B–L	12	5.3	0.32	4.9–5.8
P^4	M–D	12	3.5	0.17	3.2–3.8
	B–L	12	5.6	0.42	5.0–6.5
M^1	M–D	14	4.9	0.22	4.5–5.2
	Ant B	14	5.5	0.30	4.8–5.9
	Post B	12	5.6	0.39	5.0–6.2
M^2	M–D	11	4.7	0.30	4.2–5.2
	Ant B	11	5.6	0.27	5.1–6.0
	Post B	8	5.8	0.31	5.2–6.2
M^3	M–D	7	4.0	0.51	3.2–4.8
	Ant B	7	5.3	0.51	4.5–5.9
	Post B	2	5.7	0.42	5.4–6.0

Table 40. *Ateles geoffroyi* male

Mandibular teeth		n	m	SD	Range
I_1	M–D	12	3.0	0.23	2.6–3.5
	B–L	11	3.6	0.27	3.0–4.0
I_2	M–D	12	3.4	0.29	2.9–4.1
	B–L	12	4.1	0.26	3.7–4.7
C	M–D	5	5.8	0.93	5.1–7.4
	B–L	6	4.8	0.21	4.5–5.1
P_2	M–D	11	4.3	0.38	3.8–5.2
	B–L	11	4.5	0.27	4.0–4.8
P_3	M–D	12	3.6	0.31	3.2–4.4
	B–L	9	4.2	0.25	3.9–4.6
P_4	M–D	12	3.6	0.24	3.2–3.9
	B–L	11	4.5	0.21	4.2–4.9
M_1	M–D	14	5.2	0.28	4.7–5.6
	Tri B	11	5.0	0.16	4.7–5.5
	Tal B	14	4.9	0.17	4.6–5.2
M_2	M–D	13	5.0	0.25	4.6–5.5
	Tri B	13	5.1	0.22	4.7–5.5
	Tal B	13	5.0	0.20	4.7–5.4
M_3	M–D	8	4.7	0.35	4.1–5.2
	Tri B	7	4.5	0.30	4.0–4.8
	Tal B	7	4.6	0.26	4.2–4.9
Hypoconulid	—	—	—	—	—

Table 41. *Ateles geoffroyi* female

Maxillary teeth		n	m	SD	Range
I^1	M–D	14	4.7	0.34	4.0–5.2
	B–L	14	4.4	0.33	3.8–5.1
I^2	M–D	12	3.5	0.36	3.0–4.2
	B–L	12	4.1	0.24	3.7–4.6
C	M–D	13	5.5	0.53	4.7–6.6
	B–L	14	5.1	0.47	4.2–5.8
P^2	M–D	13	3.6	0.21	3.3–4.0
	B–L	13	4.7	0.21	4.4–5.2
P^3	M–D	14	3.6	0.19	3.3–3.9
	B–L	13	5.5	0.23	5.2–5.9
P^4	M–D	14	3.6	0.18	3.4–3.9
	B–L	13	5.8	0.20	5.5–6.2
M^1	M–D	15	5.3	0.20	4.7–5.5
	Ant B	15	5.7	0.32	5.2–6.5
	Post B	14	5.8	0.32	5.4–6.4
M^2	M–D	13	5.0	0.27	4.5–5.5
	Ant B	14	5.8	0.38	5.2–6.6
	Post B	13	5.7	0.36	5.2–6.4
M^3	M–D	14	3.9	0.64	2.8–4.9
	Ant B	14	5.3	0.50	4.4–6.0
	Post B	3	5.6	0.15	5.4–5.7

Table 42. *Ateles geoffroyi* female

Mandibular teeth		n	m	SD	Range
I_1	M–D	14	3.0	0.13	2.8–3.2
	B–L	14	3.7	0.23	3.3–4.2
I_2	M–D	15	3.6	0.25	3.2–4.0
	B–L	15	4.2	0.25	3.9–4.8
C	M–D	13	5.0	0.59	3.5–5.7
	B–L	14	4.5	0.27	4.0–5.1
P_2	M–D	15	4.0	0.29	3.5–4.6
	B–L	15	4.4	0.27	4.0–5.0
P_3	M–D	15	3.6	0.28	3.2–4.4
	B–L	15	4.5	0.25	4.2–5.0
P_4	M–D	15	3.7	0.19	3.4–4.1
	B–L	15	4.7	0.22	4.4–5.2
M_1	M–D	15	5.3	0.20	4.8–5.6
	Tri B	13	5.1	0.25	4.8–5.6
	Tal B	12	5.0	0.22	4.7–5.4
M_2	M–D	14	5.2	0.24	4.9–5.7
	Tri B	14	5.2	0.33	4.3–5.7
	Tal B	14	5.1	0.18	4.9–5.5
M_3	M–D	14	5.0	0.32	4.3–5.5
	Tri B	14	4.8	0.28	4.2–5.2
	Tal B	14	4.7	0.21	4.2–5.0
Hypoconulid		—	—	—	—

Table 43. *Brachyteles arachnoides* (sexes pooled)

Maxillary teeth		n	m	SD	Range
I^1	M–D	5	4.1	—	3.8–4.6
	B–L	5	3.9	—	3.6–4.5
I^2	M–D	5	3.8	—	3.6–4.1
	B–L	5	3.8	—	3.5–4.2
C	M–D	5	6.5	—	5.9–7.0
	B–L	5	5.5	—	5.0–6.3
P^2	M–D	5	4.7	—	4.5–5.7
	B–L	5	5.5	—	4.6–6.0
P^3	M–D	5	5.0	—	4.5–5.6
	B–L	5	6.6	—	6.4–7.0
P^4	M–D	5	5.0	—	4.6–5.5
	B–L	5	7.3	—	7.0–7.8
M^1	M–D	5	6.9	—	6.2–7.3
	Ant B	5	7.8	—	7.3–8.2
	Post B	—	—	—	—
M^2	M–D	5	6.4	—	5.2–5.6
	Ant B	5	7.6	—	5.9–7.3
	Post B	—	—	—	—
M^3	M–D	5	5.4	—	3.0–3.4
	Ant B	5	6.4	—	4.2–4.5
	Post B	—	—	—	—

Source: Zingeser (1973).

Table 44. *Brachyteles arachnoides* (sexes pooled)

Mandibular teeth		n	m	SD	Range
I_1	M–D	5	2.8	—	2.3–3.2
	B–L	5	3.7	—	3.2–4.5
I_2	M–D	5	3.3	—	2.8–3.5
	B–L	5	4.3	—	3.9–4.6
C	M–D	5	4.5	—	4.1–4.9
	B–L	5	5.8	—	5.5–6.2
P_2	M–D	5	4.4	—	4.1–4.9
	B–L	5	5.3	—	5.0–5.8
P_3	M–D	5	4.6	—	4.0–5.0
	B–L	5	5.0	—	4.6–5.5
P_4	M–D	5	4.9	—	4.6–5.5
	B–L	5	5.0	—	5.0–5.2
M_1	M–D	5	7.2	—	6.9–7.4
	Tri B	5	5.4	—	5.0–5.7
	Tal B	5	5.6	—	5.0–6.0
M_2	M–D	5	7.1	—	6.8–7.6
	Tri B	5	5.5	—	5.0–6.0
	Tal B	5	5.7	—	5.2–6.0
M_3	M–D	5	6.2	—	5.2–6.9
	Tri B	5	5.5	—	5.1–6.3
	Tal B	5	4.8	—	4.4–5.3
	Hypoconulid	—	—	—	—

Source: Zingeser (1973).

Table 45. *Cebus apella* male

Maxillary teeth		n	m	SD	Range
I^1	M–D	34	4.6	0.23	3.9–5.0
	B–L	33	4.4	0.31	3.9–5.1*
I^2	M–D	32	3.9	0.27	3.6–4.4†
	B–L	33	4.7	0.30	4.1–5.4*
C	M–D	28	7.4	0.80	4.8–8.3†
	B–L	22	6.3	0.76	3.8–7.3†
P^2	M–D	31	3.8	0.27	3.2–4.5
	B–L	31	6.7	0.55	5.0–7.8
P^3	M–D	30	3.6	0.15	3.2–3.9
	B–L	30	7.0	0.50	5.4–7.9*
P^4	M–D	29	3.6	0.13	3.4–3.8
	B–L	30	6.9	0.30	6.3–7.7*
M^1	M–D	34	4.7	0.20	4.3–5.0
	Ant B	35	6.3	0.32	5.6–6.7†
	Post B	35	6.0	0.30	5.5–6.7†
M^2	M–D	33	4.1	0.25	3.7–4.7
	Ant B	34	5.9	0.33	5.3–6.8†
	Post B	33	5.3	0.30	4.7–5.7
M^3	M–D	21	2.8	0.45	2.3–3.3
	Ant B	22	4.5	0.51	3.5–5.8
	Post B	2	4.0	1.49	2.5–5.5

Table 46. *Cebus apella* male

Mandibular teeth		n	m	SD	Range
I_1	M–D	34	2.8	0.20	2.4–3.2
	B–L	36	4.0	0.20	3.7–4.4†
I_2	M–D	34	3.4	0.32	2.8–4.0
	B–L	35	4.5	0.27	4.2–5.3†
C	M–D	26	6.1	0.57	4.7–7.2†
	B–L	22	7.0	0.71	4.6–7.9†
P_2	M–D	29	6.0	0.67	4.6–7.0†
	B–L	29	5.1	0.56	3.6–6.0†
P_3	M–D	28	3.9	0.22	3.5–4.5†
	B–L	29	5.0	0.31	4.2–5.7†
P_4	M–D	26	3.8	0.15	3.5–4.1
	B–L	27	5.3	0.27	4.5–5.6†
M_1	M–D	35	5.0	0.24	4.6–5.8*
	Tri B	35	5.1	0.31	4.2–5.8†
	Tal B	35	4.9	0.25	4.5–5.5*
M_2	M–D	34	4.7	0.21	4.2–5.1*
	Tri B	33	5.0	0.25	4.2–5.7*
	Tal B	34	4.5	0.34	3.8–5.0
M_3	M–D	20	3.8	0.30	3.0–4.5
	Tri B	19	4.2	0.36	3.1–5.0
	Tal B	14	3.8	0.31	3.5–4.6
Hypoconulid		—	—	—	—

Table 47. *Cebus apella* female

Maxillary teeth		n	m	SD	Range
I^1	M–D	19	4.5	0.21	4.1–5.0
	B–L	18	4.3	0.32	3.4–4.6
I^2	M–D	21	3.7	0.20	3.4–4.3
	B–L	20	4.5	0.22	4.0–4.9
C	M–D	18	6.2	0.56	4.6–7.0
	B–L	18	5.5	0.55	4.0–6.1
P^2	M–D	18	3.7	0.67	3.0–4.4
	B–L	19	6.4	0.49	4.7–6.9
P^3	M–D	19	3.6	0.17	3.2–3.8
	B–L	20	6.6	0.50	5.0–7.1
P^4	M–D	19	3.5	0.18	3.0–3.7
	B–L	19	6.6	0.33	5.8–7.2
M^1	M–D	21	4.6	0.17	4.0–4.8
	Ant B	21	6.0	0.31	5.5–6.5
	Post B	21	5.7	0.28	5.3–6.2
M^2	M–D	19	4.0	0.20	3.7–4.3
	Ant B	18	5.7	0.26	5.4–6.1
	Post B	18	5.1	0.36	4.6–5.5
M^3	M–D	12	2.7	0.56	2.1–4.2
	Ant B	11	4.4	0.47	3.8–5.5
	Post B	—	—	—	—

Table 48. *Cebus apella* female

Mandibular teeth		n	m	SD	Range
I_1	M–D	20	2.7	0.17	2.5–3.0
	B–L	18	3.8	0.24	3.4–4.2
I_2	M–D	20	3.4	0.26	2.8–3.8
	B–L	20	4.3	0.27	3.9–4.7
C	M–D	19	5.5	0.29	4.8–6.0
	B–L	16	6.1	0.49	5.6–7.0
P_2	M–D	19	4.8	0.24	4.2–5.2
	B–L	18	4.7	0.43	3.6–5.6
P_3	M–D	18	3.8	0.21	3.3–4.2
	B–L	18	4.8	0.32	3.9–5.5
P_4	M–D	18	3.8	0.18	3.5–4.0
	B–L	18	5.0	0.28	4.5–5.3
M_1	M–D	20	4.9	0.17	4.5–5.2
	Tri B	20	4.8	0.37	3.7–6.3
	Tal B	20	4.8	0.22	4.3–5.2
M_2	M–D	18	4.6	0.18	4.3–5.0
	Tri B	19	4.8	0.21	4.0–5.5
	Tal B	18	4.5	0.35	4.0–5.5
M_3	M–D	17	3.7	0.36	3.0–4.3
	Tri B	17	4.1	0.32	3.5–4.8
	Tal B	14	3.8	0.32	3.0 4.2
Hypoconulid		—	—	—	—

Table 49. *Saimiri oerstedii* male

Maxillary teeth		n	m	SD	Range
I¹	M–D	10	2.8	0.16	2.5–3.0
	B–L	10	2.6	0.17	2.3–2.8
I²	M–D	10	2.3	0.22	2.0–2.6
	B–L	10	2.5	0.21	2.2–2.8
C	M–D	6	3.9	0.25	3.4–4.1
	B–L	7	3.5	0.10	3.3–3.6
P²	M–D	9	2.4	0.18	2.2–2.8
	B–L	9	3.7	0.26	3.1–4.0
P³	M–D	9	2.0	0.11	1.9–2.2
	B–L	8	3.9	0.10	3.8–4.1
P⁴	M–D	9	1.9	0.12	1.8–2.2
	B–L	9	3.9	0.11	3.7–4.0
M¹	M–D	11	2.8	0.15	2.6–3.1
	Ant B	11	3.9	0.26	3.6–4.5
	Post B	11	3.8	0.24	3.5–4.2
M²	M–D	11	2.6	0.13	2.4–2.8
	Ant B	11	3.6	0.17	3.3–3.9
	Post B	10	3.4	0.17	3.2–3.7
M³	M–D	9	1.8	0.22	1.5–2.3
	Ant B	9	2.6	0.23	2.3–3.0
	Post B	—	—	—	—

Table 50. *Saimiri oerstedii* male

Mandibular teeth		n	m	SD	Range
I₁	M–D	7	1.7	0.38	1.3–2.5
	B–L	7	2.4	0.28	1.8–2.6
I₂	M–D	7	2.1	0.25	1.7–2.4
	B–L	8	2.6	0.09	2.4–2.7
C	M–D	8	3.5	0.29	3.0–4.0
	B–L	5	3.4	0.20	3.2–3.7
P₂	M–D	10	3.4	0.44	2.5–3.8
	B–L	10	3.0	0.13	2.9–3.3
P₃	M–D	10	2.3	0.22	2.0–2.6
	B–L	10	2.7	0.15	2.5–3.0
P₄	M–D	10	2.2	0.15	2.0–2.5
	B–L	10	2.8	0.12	2.6–3.0
M₁	M–D	11	3.0	0.26	2.6–3.4
	Tri B	11	2.8	0.20	2.4–3.0
	Tal B	11	2.7	0.21	2.4–3.0
M₂	M–D	10	2.7	0.14	2.5–3.0
	Tri B	11	2.8	0.22	2.2–2.9
	Tal B	11	2.5	0.21	2.2–2.8
M₃	M–D	7	2.2	0.08	2.2–2.4
	Tri B	8	2.1	0.24	1.7–2.5
	Tal B	6	2.0	0.24	1.6–2.3
Hypoconulid		—	—	—	—

Table 51. *Saimiri oerstedii* female

Maxillary teeth		n	m	SD	Range
I¹	M–D	6	2.7	0.10	2.5–2.8
	B–L	6	2.4	0.31	2.2–3.0
I²	M–D	6	2.3	0.12	2.2–2.5
	B–L	6	2.2	0.29	2.0–2.8
C	M–D	4	2.7	0.10	2.6–2.8
	B–L	4	3.0	0.13	2.9–3.2
P²	M–D	4	2.1	0.05	2.0–2.1
	B–L	4	3.4	0.24	3.2–3.7
P³	M–D	4	2.1	0.13	1.9–2.2
	B–L	4	3.7	0.20	3.6–4.0
P⁴	M–D	4	1.9	0.14	1.8–2.1
	B–L	4	3.7	0.22	3.5–4.0
M¹	M–D	6	2.7	0.05	2.7–2.8
	Ant B	6	3.7	0.16	3.6–4.0
	Post B	6	3.7	0.19	3.5–4.0
M²	M–D	6	2.5	0.13	2.3–2.6
	Ant B	6	3.1	0.13	3.2–3.5
	Post B	6	3.3	0.22	3.2–3.7
M³	M–D	4	1.7	0.14	1.6–1.9
	Ant B	4	2.3	0.26	2.1–2.7
	Post B	—	—	—	—

Table 52. *Saimiri oerstedii* female

Mandibular teeth		n	m	SD	Range
I₁	M–D	5	1.4	0.11	1.2–1.5
	B–L	5	1.9	0.09	1.7–1.9
I₂	M–D	5	1.8	0.11	1.7–2.0
	B–L	5	2.3	0.06	2.3–2.4
C	M–D	3	3.0	0.21	2.8–3.2
	B–L	3	2.5	0.32	2.1–2.7
P₂	M–D	2	2.4	0.28	2.2–2.6
	B–L	2	2.7	0.00	2.7–2.7
P₃	M–D	3	2.2	0.00	2.2–2.2
	B–L	3	2.6	0.06	2.5–2.6
P₄	M–D	3	2.2	0.00	2.2–2.2
	B–L	3	2.6	0.00	2.6–2.6
M₁	M–D	5	2.7	0.11	2.6–2.9
	Tri B	5	2.6	0.05	2.5–2.6
	Tal B	5	2.5	0.09	2.4–2.6
M₂	M–D	5	2.6	0.09	2.5–2.7
	Tri B	5	2.4	0.08	2.3–2.5
	Tal B	5	2.3	0.06	2.3–2.4
M₃	M–D	3	2.3	0.06	2.2–2.3
	Tri B	3	1.9	0.10	1.8–2.0
	Tal B	3	1.8	0.12	1.7–1.9
Hypoconulid		—	—	—	—

Table 53. *Saimiri sciureus* male

Maxillary teeth		n	m	SD	Range
I^1	M–D	14	2.7	0.36	1.7–3.0
	B–L	12	2.7	0.28	2.3–3.2
I^2	M–D	12	2.2	0.17	2.0–2.5
	B–L	10	2.4	0.28	1.8–2.8
C	M–D	11	3.5	0.48	2.6–4.0†
	B–L	9	3.1	0.43	2.1–3.6
P^2	M–D	17	2.2	0.24	1.5–2.5
	B–L	16	3.6	0.17	3.2–3.9
P^3	M–D	17	2.0	0.17	1.7–2.0
	B–L	17	3.9	0.16	3.7–4.2*
P^4	M–D	18	1.9	0.15	1.6–2.2
	B–L	18	3.9	0.16	3.6–4.2
M^1	M–D	20	2.8	0.14	2.5–3.0
	Ant B	20	4.1	0.19	3.7–4.5
	Post B	20	4.0	0.21	3.5–4.4
M^2	M–D	20	2.5	0.14	2.1–2.7
	Ant B	19	3.6	0.35	2.8–4.4
	Post B	19	3.5	0.40	2.6–4.5
M^3	M–D	11	1.7	0.13	1.5–1.9
	Ant B	12	2.9	0.14	2.5–3.0
	Post B	—	—	—	—

Table 54. *Saimiri sciureus* male

Mandibular teeth		n	m	SD	Range
I_1	M–D	13	1.6	0.18	1.3–1.9
	B–L	14	2.3	0.20	1.9–2.5
I_2	M–D	11	2.0	0.20	1.8–2.3
	B–L	11	2.7	0.20	2.3–2.9
C	M–D	11	3.4	0.18	3.1–3.6†
	B–L	10	3.1	0.45	2.0–3.6
P_2	M–D	17	3.1	0.29	2.4–3.7†
	B–L	17	3.0	0.30	2.4–3.5†
P_3	M–D	19	2.2	0.21	1.9–2.5
	B–L	19	2.7	0.21	2.3–3.0
P_4	M–D	19	2.1	0.15	1.9–2.4
	B–L	19	2.7	0.21	2.4–3.2
M_1	M–D	21	2.9	0.19	2.5–3.3
	Tri B	20	2.8	0.12	2.6–3.0
	Tal B	21	2.8	0.15	2.6–3.2
M_2	M–D	19	2.7	0.12	2.4–2.9
	Tri B	18	2.7	0.16	2.4–3.0
	Tal B	17	2.6	0.11	2.4–2.8
M_3	M–D	12	2.2	0.14	2.0–2.4
	Tri B	13	2.2	0.15	1.9–2.5
	Tal B	9	2.0	0.21	1.7–2.4
Hypoconulid		—	—	—	—

Table 55. *Saimiri sciureus* female

Maxillary teeth		n	m	SD	Range
I^1	M–D	6	2.8	0.29	2.4–3.2
	B–L	6	2.8	0.17	2.6–3.0
I^2	M–D	4	2.2	0.08	2.1–2.3
	B–L	5	2.6	0.13	2.6–3.0
C	M–D	4	2.7	0.13	2.5–2.8
	B–L	5	2.9	0.26	2.5–3.2
P^2	M–D	6	2.0	0.14	1.8–2.2
	B–L	6	3.5	0.19	3.2–3.7
P^3	M–D	6	2.1	0.18	1.9–2.3
	B–L	6	3.8	0.11	3.6–3.9
P^4	M–D	6	1.9	0.11	1.7–2.0
	B–L	5	3.8	0.13	3.6–3.9
M^1	M–D	5	2.8	0.09	2.7–2.9
	Ant B	5	4.0	0.19	3.7–4.2
	Post B	5	3.9	0.16	3.7–4.1
M^2	M–D	5	2.5	0.09	2.4–2.6
	Ant B	5	3.7	0.15	3.5–3.8
	Post B	5	3.5	0.26	3.2–3.9
M^3	M–D	5	1.8	0.43	1.4–2.5
	Ant B	4	2.7	0.37	2.2–3.0
	Post B	—	—	—	—

Table 56. *Saimiri sciureus* female

Mandibular teeth		n	m	SD	Range
I_1	M–D	4	1.5	0.05	1.4–1.5
	B–L	4	2.3	0.19	2.1–2.5
I_2	M–D	4	2.0	0.45	1.8–2.7
	B–L	4	2.5	0.16	2.3–2.7
C	M–D	4	2.9	0.35	2.5–3.3
	B–L	4	2.7	0.27	2.5–3.1
P_2	M–D	6	2.4	0.23	2.1–2.7
	B–L	6	2.5	0.19	2.2–2.7
P_3	M–D	6	2.3	0.20	2.0–2.5
	B–L	6	2.7	0.20	2.4–2.9
P_4	M–D	6	2.3	0.10	2.1–2.4
	B–L	6	2.7	0.29	2.4–3.0
M_1	M–D	6	2.9	0.10	2.8–3.0
	Tri B	6	2.8	0.16	2.6–3.0
	Tal B	6	2.7	0.15	2.6–2.9
M_2	M–D	6	2.6	0.16	2.4–2.8
	Tri B	5	2.5	0.18	2.3–2.8
	Tal B	5	2.5	0.17	2.4–2.8
M_3	M–D	6	2.3	0.18	2.1–2.5
	Tri B	5	2.1	0.16	1.8–2.2
	Tal B	5	2.0	0.28	1.7–2.3
Hypoconulid		—	—	—	—

Table 57. *Alouatta seniculus* male

Maxillary teeth		n	m	SD	Range
I^1	M–D	17	4.2	0.41	3.3–4.7†
	B–L	16	3.7	0.31	3.1–4.4†
I^2	M–D	18	4.0	0.29	3.4–4.4†
	B–L	18	3.6	0.29	3.2–4.3*
C	M–D	14	8.1	0.61	6.9–9.2†
	B L	11	6.2	0.48	5.6–6.8|
P^2	M–D	18	5.4	0.33	4.8–6.4†
	B–L	18	6.1	0.35	5.5–6.5†
P^3	M–D	18	5.2	0.26	4.8–5.5†
	B–L	18	6.9	0.33	6.3–7.5†
P^4	M–D	18	5.1	0.17	4.8–5.4†
	B–L	18	7.4	0.48	6.6–8.4†
M^1	M–D	19	7.5	0.26	7.0–7.9†
	Ant B	18	8.0	0.32	7.5–8.7†
	Post B	18	8.2	0.39	7.7–9.0†
M^2	M–D	19	8.0	0.41	7.2–8.8†
	Ant B	18	8.5	0.34	7.8–9.1†
	Post B	18	8.6	0.36	8.1–9.4†
M^3	M–D	15	6.2	0.47	5.5–7.3
	Ant B	15	7.7	0.38	7.2–8.5†
	Post B	—	—	—	—

Table 58. *Alouatta seniculus* male

Mandibular teeth		n	m	SD	Range
I_1	M–D	16	2.9	0.26	2.4–3.3*
	B–L	16	3.4	0.26	2.9–3.7†
I_2	M–D	17	3.5	0.24	2.9–3.8†
	B–L	18	4.1	0.26	3.8–4.5†
C	M–D	13	6.9	0.63	5.6–7.8†
	B–L	14	6.5	0.42	5.8–7.4†
P_2	M–D	19	6.1	0.28	5.7–6.7†
	B–L	19	6.5	0.32	5.8–7.4†
P_3	M–D	19	5.1	0.22	4.8–5.5*
	B–L	18	5.6	0.28	5.4–6.0†
P_4	M–D	19	5.6	0.35	4.9–6.2†
	B–L	18	5.8	0.34	5.4–6.5†
M_1	M–D	20	7.6	0.32	6.9–8.0†
	Tri B	16	5.4	0.28	4.9–5.9†
	Tal B	16	5.9	0.31	5.6–6.5†
M_2	M–D	19	8.1	0.38	7.3–8.8†
	Tri B	18	5.9	0.30	5.3–6.3†
	Tal B	17	6.4	0.22	6.2–6.7†
M_3	M–D	14	8.9	0.34	8.5–9.7†
	Tri B	14	5.9	0.33	5.2–6.4†
	Tal B	12	5.9	0.35	5.4–6.5
Hypoconulid		—	—	—	—

Table 59. *Alouatta seniculus* female

Maxillary teeth		n	m	SD	Range
I^1	M–D	17	3.8	0.36	3.3–4.5
	B–L	15	3.2	0.27	2.9–3.8
I^2	M–D	14	3.7	0.37	3.2–4.3
	B–L	13	3.3	0.43	2.5–3.9
C	M–D	17	6.3	0.71	5.4–7.7
	B–L	16	4.7	0.69	3.9–6.2
P^2	M–D	19	4.8	0.22	4.3–5.0
	B–L	19	5.6	0.32	5.1–6.1
P^3	M–D	19	4.9	0.23	4.3–5.4
	B–L	19	6.5	0.42	5.9–7.1
P^4	M–D	20	4.9	0.24	4.5–5.5
	B–L	20	6.9	0.44	6.2–7.9
M^1	M–D	19	7.1	0.36	6.5–8.0
	Ant B	17	7.3	0.30	6.8–7.7
	Post B	17	7.4	0.37	6.7–8.1
M^2	M–D	20	7.5	0.49	6.5–8.6
	Ant B	19	7.7	0.40	7.0–8.5
	Post B	19	7.8	0.43	7.2–8.7
M^3	M–D	12	5.9	0.48	5.1–6.6
	Ant B	14	7.1	0.40	6.7–7.9
	Post B	—	—	—	—

Table 60. *Alouatta seniculus* female

Mandibular teeth		n	m	SD	Range
I_1	M–D	13	2.7	0.33	2.2–3.4
	B–L	12	3.1	0.36	2.5–3.8
$I_?$	M–D	14	3.1	0.19	2.8–3.5
	B–L	13	3.6	0.41	2.7–4.2
C	M–D	16	5.6	0.69	4.4–7.0
	B–L	18	5.1	0.62	4.2–6.3
P_2	M–D	18	5.0	0.28	4.5–5.5
	B–L	19	5.2	0.40	4.5–6.1
P_3	M–D	17	4.9	0.22	4.6–5.4
	B–L	18	5.1	0.36	4.7–6.1
P_4	M–D	19	5.2	0.29	4.8–5.8
	B–L	18	5.1	0.39	4.7–6.0
M_1	M–D	20	7.2	0.39	6.5–8.2
	Tri B	17	5.1	0.25	4.7–5.5
	Tal B	19	5.5	0.32	4.9–6.1
M_2	M–D	19	7.5	0.43	6.5–8.2
	Tri B	18	5.5	0.21	5.2–5.9
	Tal B	20	6.0	0.36	5.4–6.7
M_3	M–D	16	8.3	0.58	7.5–9.5
	Tri B	19	5.7	0.27	5.2–6.1
	Tal B	15	5.6	0.36	5.1–6.4
Hypoconulid		—	—	—	—

Table 61. *Alouatta palliata* male

Maxillary teeth		n	m	SD	Range
I^1	M–D	11	3.7	0.28	3.3–4.2
	B–L	11	3.7	0.24	3.4–4.1
I^2	M–D	12	3.8	0.22	3.4–4.1
	B–L	12	3.8	0.18	3.4–4.0*
C	M–D	12	7.6	0.48	6.6–8.1†
	B–L	12	5.7	0.64	4.7–6.9†
P^2	M–D	12	4.8	0.38	4.2–5.3
	B–L	12	6.5	0.29	6.0–6.9†
P^3	M–D	12	4.8	0.22	4.5–5.2
	B–L	11	7.1	0.19	6.8–7.4†
P^4	M–D	13	4.8	0.37	4.3–5.5
	B–L	12	7.6	0.34	6.9–8.0
M^1	M–D	13	7.1	0.28	6.4–7.4
	Ant B	12	8.1	0.41	7.3–8.8†
	Post B	12	8.0	0.28	7.5–8.4*
M^2	M–D	12	7.3	0.43	6.5–8.0
	Ant B	11	8.2	0.54	7.3–9.0
	Post B	11	8.0	0.41	7.4–8.8
M^3	M–D	13	5.4	0.49	4.5–6.0
	Ant B	12	7.7	0.51	6.5–8.5
	Post B	—	—	—	—

Table 62. *Alouatta palliata* male

Mandibular teeth		n	m	SD	Range
I_1	M–D	8	2.5	0.18	2.3–2.8
	B–L	8	3.2	0.18	2.9–3.4*
I_2	M–D	10	3.1	0.27	2.8–3.8
	B–L	10	3.9	0.22	3.6–4.4
C	M–D	12	5.2	0.55	4.9–6.5†
	B–L	12	6.9	0.44	6.3–7.5†
P_2	M–D	13	5.5	0.46	4.3–6.2†
	B–L	13	6.4	0.26	6.0–6.8†
P_3	M–D	13	4.8	0.39	3.9–5.3
	B–L	12	5.6	0.41	4.8–6.0†
P_4	M–D	12	5.0	0.28	4.5–5.4
	B–L	13	6.1	0.29	5.7–6.6†
M_1	M–D	15	7.1	0.22	6.7–7.4†
	Tri B	14	5.5	0.27	5.0–5.9*
	Tal B	14	6.0	0.17	5.7–6.3†
M_2	M–D	15	7.8	0.32	7.3–8.2†
	Tri B	14	5.9	0.30	5.3–6.4†
	Tal B	14	6.7	0.31	6.3–7.1†
M_3	M–D	13	7.8	0.51	6.9–8.7†
	Tri B	12	6.0	0.35	5.1–6.3*
	Tal B	12	5.4	0.41	5.0–6.0
Hypoconulid		—	—	—	—

Table 63. *Alouatta palliata* female

Maxillary teeth		n	m	SD	Range
I^1	M–D	34	3.7	0.76	2.9–4.3
	B–L	33	3.7	0.73	3.0–6.3
I^2	M–D	32	3.7	0.79	3.0–4.7
	B–L	32	3.5	0.40	3.1–4.9
C	M–D	26	6.2	0.93	3.7–8.5
	B–L	27	4.7	0.73	3.8–7.2
P^2	M–D	34	4.5	0.44	3.0–5.1
	B–L	34	6.1	0.48	5.3–6.9
P^3	M–D	35	4.7	0.33	4.0–5.4
	B–L	36	6.8	0.40	6.2–7.7
P^4	M–D	34	4.6	0.31	3.9–5.0
	B–L	36	7.3	0.59	5.6–8.1
M^1	M–D	36	6.9	0.35	6.1–7.6
	Ant B	37	7.3	0.81	5.2–8.7
	Post B	37	7.2	1.02	6.0–8.7
M^2	M–D	36	7.2	0.42	6.7–7.8
	Ant B	36	7.6	1.13	6.0–8.6
	Post B	36	7.6	0.65	6.7–8.6
M^3	M–D	29	5.4	0.65	4.2–6.4
	Ant B	27	7.3	0.78	5.8–8.6
	Post B	6	6.2	1.30	4.7–8.0

Table 64. *Alouatta palliata* female

Mandibular teeth		n	m	SD	Range
I_1	M–D	31	2.6	1.00	2.0–2.9
	B–L	29	3.0	0.26	2.6–3.6
I_2	M–D	32	2.9	0.25	2.3–3.5
	B–L	28	3.7	0.09	3.2–4.5
C	M–D	30	4.6	0.53	3.6–6.0
	B–L	28	5.5	0.66	4.3–8.0
P_2	M–D	35	4.6	0.38	3.5–5.6
	B–L	34	5.5	0.45	4.8–6.9
P_3	M–D	35	4.7	0.38	4.0–6.3
	B–L	36	5.7	0.44	4.6–6.0
P_4	M–D	37	4.9	0.47	4.3–7.2
	B–L	36	5.7	0.44	5.0–6.9
M_1	M–D	37	6.8	0.39	6.0–8.0
	Tri B	36	5.2	0.30	4.6–5.8
	Tal B	37	5.7	0.32	5.1–6.4
M_2	M–D	37	7.3	0.35	6.6–8.0
	Tri B	36	5.6	0.34	5.0–6.3
	Tal B	36	6.3	0.33	5.8–7.0
M_3	M–D	37	7.6	0.63	6.0–9.3
	Tri B	36	5.7	0.35	4.8–6.7
	Tal B	36	5.4	0.33	4.9–6.1
Hypoconulid	—	—	—	—

Table 65. *Cercopithecus cephus* male

Maxillary teeth		n	m	SD	Range
I^1	M–D	8	6.2	0.22	5.8–6.5
	B–L	8	4.9	0.26	4.4–5.1
I^2	M–D	7	3.3	0.22	3.0–3.6
	B–L	8	3.6	0.21	3.2–3.9
C	M–D	—	—	—	—
	B–L	3	4.6	0.25	4.5–4.9
P^3	M–D	7	4.0	0.24	3.7–4.3
	B–L	9	3.9	0.13	3.7–4.1
P^4	M–D	6	4.3	0.27	4.0–4.7*
	B–L	9	4.5	0.19	4.2–4.8
M^1	M–D	8	5.7	0.21	5.4–6.0*
	Ant B	9	5.3	0.19	5.0–5.6
	Post B	9	5.0	0.14	4.8–5.2
M^2	M–D	9	6.1	0.22	5.8–6.5†
	Ant B	9	6.0	0.30	5.5–6.5
	Post B	9	5.4	0.32	4.8–5.8
M^3	M–D	7	5.4	0.33	5.0–5.9†
	Ant B	7	5.2	0.25	4.9–5.5
	Post B	6	4.1	0.52	3.5–4.9

Table 66. *Cercopithecus cephus* male

Mandibular teeth		n	m	SD	Range
I_1	M–D	6	3.8	0.30	3.2–4.1
	B–L	6	4.2	0.14	4.0–4.3
I_2	M–D	4	3.5	0.31	3.1–3.8*
	B–L	6	4.0	0.09	3.9–4.1
C	M–D	4	4.2	0.13	4.0–4.3
	B–L	6	6.0	0.29	5.6–6.5†
P_3	M–D	5	8.0	0.45	7.6–8.6†
	B–L	7	3.3	0.30	2.8–3.7
P_4	M–D	8	5.0	0.47	4.4–5.9†
	B–L	8	3.5	0.19	3.1–3.7
M_1	M–D	8	5.6	0.25	5.3–6.1
	Tri B	8	4.1	0.18	3.7–4.3
	Tal B	8	4.2	0.19	3.9–4.5
M_2	M–D	8	6.1	0.15	5.9–6.4†
	Tri B	8	4.8	0.23	4.5–5.1*
	Tal B	7	4.9	0.32	4.5–5.3
M_3	M–D	6	6.0	0.20	5.6–6.1†
	Tri B	6	4.7	0.26	4.2–4.9*
	Tal B	6	4.3	0.33	3.8–4.8
	Hypoconulid	—	—	—	—

Table 67. *Cercopithecus cephus* female

Maxillary teeth		n	m	SD	Range
I^1	M–D	12	5.9	0.56	5.0–6.7
	B–L	11	4.6	0.38	3.7–5.1
I^2	M D	10	3.4	0.30	2.9–3.8
	B–L	11	3.4	0.29	2.9–3.8
C	M–D	6	5.2	0.25	5.0–5.6
	B–L	8	4.1	0.50	3.0–4.6
P^3	M–D	9	3.8	0.24	3.4–4.1
	B–L	9	3.8	0.30	3.2–4.1
P^4	M D	9	4.0	0.28	3.6–4.5
	B–L	10	4.6	0.25	4.2–5.1
M^1	M–D	11	5.4	0.36	4.7–5.9
	Ant B	11	5.1	0.34	4.4–5.5
	Post B	12	4.9	0.27	4.3–5.2
M^2	M–D	11	5.6	0.39	5.1–6.2
	Ant B	11	5.8	0.34	5.0–6.1
	Post B	12	5.2	0.38	4.4–5.7
M^3	M–D	7	4.8	0.27	4.5–5.2
	Ant B	8	5.0	0.42	4.3–5.6
	Post B	7	4.1	0.42	3.6–4.7

Table 68. *Cercopithecus cephus* female

Mandibular teeth		n	m	SD	Range
I_1	M–D	10	3.6	0.35	3.0–4.0
	B–L	9	4.0	0.28	3.5–4.3
I_2	M–D	9	3.1	0.28	2.8–3.6
	B–L	9	3.8	0.31	3.4–4.4
C	M–D	9	3.7	0.78	3.0–5.2
	B–L	9	4.5	0.63	3.2–5.3
P_3	M–D	7	6.1	0.39	5.5–6.5
	B–L	9	3.0	0.28	2.7–3.6
P_4	M–D	12	4.6	0.25	4.0–4.9
	B–L	11	3.4	0.28	3.0–3.8
M_1	M–D	12	5.4	0.32	4.8–5.9
	Tri B	10	3.8	0.32	3.1–4.1
	Tal B	9	4.0	0.32	3.3–4.4
M_2	M–D	12	5.7	0.32	5.3–6.2
	Tri B	10	4.6	0.25	4.1–4.9
	Tal B	10	4.6	0.29	4.0–5.0
M_3	M–D	8	5.5	0.29	5.1–5.8
	Tri B	7	4.5	0.33	3.9–4.9
	Tal B	6	3.9	0.26	3.5–4.2
	Hypoconulid	—	—	—	—

Table 69. *Cercopithecus nictitans* male

Maxillary teeth		n	m	SD	Range
I^1	M–D	11	6.0	0.62	4.8–6.8
	B–L	11	4.7	0.36	4.1–5.0
I^2	M–D	11	3.4	0.54	2.7–4.6
	B–L	11	3.7	0.48	3.0–4.3
C	M–D	8	7.5	1.05	5.9–8.8†
	B–L	6	4.8	0.58	3.9–5.3*
P^3	M–D	11	4.0	0.58	3.0–4.9
	B–L	11	3.9	0.32	3.2–4.5
P^4	M–D	11	4.1	0.63	3.4–5.4
	B–L	11	4.7	0.36	3.8–5.0
M^1	M–D	11	5.6	0.51	4.8–6.3
	Ant B	11	5.1	0.41	4.2–5.7
	Post B	10	4.8	0.39	4.0–5.2
M^2	M–D	11	6.2	0.60	5.3–7.2
	Ant B	11	5.9	0.49	4.9–6.4
	Post B	11	5.5	0.43	4.9–6.3
M^3	M–D	10	5.1	0.51	4.3–5.8
	Ant B	10	5.3	0.39	4.5–5.7
	Post B	9	4.3	0.58	3.3–5.2

Table 70. *Cercopithecus nictitans* male

Mandibular teeth		n	m	SD	Range
I_1	M–D	11	3.8	0.44	3.3–4.6
	B–L	11	4.2	0.39	3.7–4.9
I_2	M–D	11	3.3	0.36	2.9–4.0
	B–L	11	4.1	0.47	3.5–4.9
C	M–D	9	6.6	0.68	5.8–8.0†
	B–L	7	5.2	0.93	4.2–6.8*
P_3	M–D	8	8.2	0.77	7.2–9.2†
	B–L	10	3.7	0.36	3.2–4.3
P_4	M–D	10	4.9	0.67	4.1–5.6
	B–L	9	3.6	0.27	3.1–3.9
M_1	M–D	11	5.7	0.57	4.9–6.7
	Tri B	11	4.0	0.40	3.0–4.5
	Tal B	10	4.2	0.20	3.8–4.5
M_2	M–D	10	6.2	0.45	5.7–7.0
	Tri B	10	5.0	0.15	4.7–5.2
	Tal B	9	5.0	0.22	4.7–5.3
M_3	M–D	8	5.9	0.49	5.4–6.9
	Tri B	8	4.8	0.33	4.5–5.5
	Tal B	8	4.3	0.33	3.8–4.8
	Hypoconulid	—	—	—	—

Table 71. *Cercopithecus nictitans* female

Maxillary teeth		n	m	SD	Range
I[1]	M–D	6	6.1	0.33	5.7–6.5
	B–L	4	4.5	0.22	4.2–4.7
I[2]	M–D	5	3.3	0.14	3.1–3.4
	B–L	5	3.6	0.68	3.2–4.8
C	M–D	4	4.9	0.41	4.6–5.5
	B–L	4	4.1	0.25	3.8–4.4
P[3]	M–D	5	3.7	0.40	3.4–4.4
	B–L	5	3.9	0.31	3.6–4.4
P[4]	M–D	6	3.8	0.36	3.5–4.5
	B–L	6	4.6	0.25	4.2–4.9
M[1]	M–D	7	5.5	0.43	4.9–6.2
	Ant B	7	5.1	0.32	4.8–5.6
	Post B	6	4.9	0.34	4.5–5.4
M[2]	M–D	6	5.8	0.41	5.2–6.4
	Ant B	6	5.8	0.19	5.6–6.1
	Post B	6	5.4	0.39	5.0–6.0
M[3]	M–D	5	5.3	0.51	4.6–6.0
	Ant B	5	5.2	0.51	4.7–6.0
	Post B	5	4.2	0.73	3.2–5.0

Table 72. *Cercopithecus nictitans* female

Mandibular teeth		n	m	SD	Range
I$_1$	M–D	7	3.6	0.52	2.5–4.1
	B–L	7	4.1	0.43	3.5–4.9
I$_2$	M–D	6	3.3	0.39	2.6–3.8
	B–L	6	3.9	0.52	3.5–4.9
C	M–D	2	4.5	0.00	4.5–4.5
	B–L	2	3.3	0.28	3.1–3.5
P$_3$	M–D	5	5.8	0.51	5.5–6.7
	B–L	6	3.3	0.59	2.8–4.5
P$_4$	M–D	4	4.6	0.32	4.3–5.0
	B–L	5	3.6	0.11	3.5–3.7
M$_1$	M–D	7	5.5	0.46	4.7–6.2
	Tri B	6	3.9	0.27	3.7–4.3
	Tal B	7	4.2	0.25	3.8–4.5
M$_2$	M–D	7	5.8	0.31	5.5–6.4
	Tri B	6	4.8	0.21	4.5–5.1
	Tal B	6	4.8	0.25	4.5–5.2
M$_3$	M–D	5	6.0	0.31	5.7–6.5
	Tri B	4	4.8	0.28	4.5–5.1
	Tal B	5	4.3	0.39	3.8–4.8
	Hypoconulid	—	—	—	—

Table 73. *Cercopithecus mona* male

Maxillary teeth		n	m	SD	Range
I^1	M–D	18	5.5	0.39	5.0–6.2
	B–L	14	4.5	0.24	4.0–4.8
I^2	M–D	20	3.3	0.27	3.0–4.1
	B–L	20	3.4	0.35	2.8–4.5
C	M–D	16	7.2	1.18	6.4–9.3†
	B–L	14	5.2	0.81	4.1–7.5†
P^3	M–D	23	3.9	0.35	3.3–3.9
	B–L	23	3.8	0.35	3.3–4.9
P^4	M–D	25	4.0	0.32	3.5–4.9
	B–L	25	4.6	0.37	4.0–5.5
M^1	M–D	25	5.3	0.46	4.7–7.0
	Ant B	23	5.2	0.34	4.4–5.8
	Post B	25	5.0	0.30	4.3–5.5
M^2	M–D	26	5.8	0.43	5.2–7.4
	Ant B	26	6.1	0.41	5.3–6.7*
	Post B	26	5.6	0.40	4.7–6.6*
M^3	M–D	23	5.1	0.54	4.6–7.1
	Ant B	23	5.4	0.36	4.7–6.3†
	Post B	22	4.5	0.39	3.8–5.5

Table 74. *Cercopithecus mona* male

Mandibular teeth		n	m	SD	Range
I_1	M–D	21	3.5	0.45	2.8–4.6
	B–L	22	3.9	0.37	3.4–5.2
I_2	M–D	19	3.1	0.28	2.7–3.8†
	B–L	22	3.9	0.38	3.5–5.4
C	M–D	11	4.3	0.83	3.4–6.0*
	B–L	16	5.5	1.13	3.2–8.8†
P_3	M–D	22	7.8	0.85	6.3–10.2†
	B–L	24	3.1	0.35	2.5–3.8
P_4	M–D	24	4.5	0.37	4.0–5.8
	B–L	26	3.5	0.26	2.9–3.8
M_1	M–D	27	5.3	0.43	4.8–7.2
	Tri B	21	3.9	0.18	3.5–4.2
	Tal B	25	4.0	0.32	3.6–5.2
M_2	M–D	25	5.9	0.43	5.4–7.5
	Tri B	23	5.0	0.26	4.5–5.5†
	Tal B	25	4.9	0.29	4.5–5.8
M_3	M–D	22	5.7	0.79	5.2–9.0
	Tri B	23	4.8	0.25	4.5–5.6
	Tal B	23	4.4	0.38	3.9–5.9
Hypoconulid		—	—	—	—

Table 75. *Cercopithecus mona* female

Maxillary teeth		n	m	SD	Range
I¹	M–D	12	5.3	0.49	4.6–6.1
	B–L	11	4.4	0.37	3.8–5.1
I²	M–D	13	3.1	0.28	2.7–3.8
	B–L	13	3.3	0.40	2.8–4.3
C	M–D	8	4.9	0.46	4.3–5.8
	B–L	11	4.0	0.47	3.3–4.8
P³	M–D	11	3.7	0.37	3.1–4.2
	B–L	12	3.6	0.51	3.2–4.9
P⁴	M–D	11	4.0	0.34	3.5–4.5
	B–L	15	4.4	0.39	3.9–5.4
M¹	M–D	13	5.3	0.41	4.7–6.1
	Ant B	16	5.0	0.40	4.4–5.7
	Post B	15	4.8	0.40	4.3–5.6
M²	M–D	15	5.7	0.39	5.1–6.6
	Ant B	14	5.8	0.43	5.2–6.7
	Post B	14	5.2	0.42	4.6–6.2
M³	M–D	9	4.9	0.39	4.6–5.7
	Ant B	11	5.1	0.51	4.6–6.4
	Post B	10	4.0	0.34	3.4–4.5

Table 76. *Cercopithecus mona* female

Mandibular teeth		n	m	SD	Range
I₁	M–D	9	3.4	0.40	2.6–3.9
	B–L	7	3.8	0.54	3.1–4.7
I₂	M–D	7	2.8	0.23	2.5–3.2
	B–L	12	3.8	0.43	3.3–4.7
C	M–D	4	3.1	0.22	2.9–3.4
	B–L	9	4.3	0.54	3.6–5.5
P₃	M–D	8	6.2	1.37	4.5–8.4
	B–L	11	2.9	0.41	2.3–3.6
P₄	M–D	14	4.3	0.30	3.8–4.9
	B–L	14	3.3	0.32	2.9–3.9
M₁	M–D	15	5.2	0.36	4.6–5.8
	Tri B	11	3.7	0.36	3.3–4.4
	Tal B	13	3.9	0.36	3.5–4.6
M₂	M–D	16	5.6	0.36	5.2–6.5
	Tri B	16	4.7	0.38	4.1–5.5
	Tal B	15	4.7	0.33	4.1–5.5
M₃	M–D	10	5.4	0.34	5.0–6.1
	Tri B	11	4.5	0.41	4.0–5.5
	Tal B	10	4.1	0.31	3.8–4.8
	Hypoconulid	—	—	—	—

Table 77. *Cercopithecus mitis* male

Maxillary teeth		n	m	SD	Range
I^1	M–D	30	6.0	0.42	5.2–6.8
	B–L	25	4.9	0.92	4.2–9.0
I^2	M–D	25	3.6	0.40	2.9–4.6
	B–L	25	3.9	0.32	3.4–4.5
C	M–D	11	7.9	1.59	3.4–9.2†
	B–L	10	5.7	0.76	4.2–6.8†
P^3	M–D	27	4.3	0.43	3.5–5.0
	B–L	32	4.3	0.34	3.7–4.9
P^4	M–D	32	4.5	0.33	3.7–5.2*
	B–L	32	5.1	0.38	4.4–5.7
M^1	M–D	32	6.1	0.31	5.5–6.8
	Ant B	28	5.5	0.35	4.8–6.1
	Post B	27	5.3	0.37	4.5–6.2
M^2	M–D	31	6.6	0.34	5.8–7.4*
	Ant B	30	6.3	0.43	5.5–7.0*
	Post B	30	5.8	0.47	4.7–6.6*
M^3	M–D	27	5.8	0.43	4.9–6.7
	Ant B	26	5.7	0.45	4.6–6.6*
	Post B	24	4.8	0.41	4.0–5.9*

Table 78. *Cercopithecus mitis* male

Mandibular teeth		n	m	SD	Range
I_1	M–D	27	3.8	0.34	3.0–4.6
	B–L	28	4.4	0.39	3.6–5.5*
I_2	M–D	29	3.5	0.41	2.8–4.6
	B–L	28	4.3	0.39	3.6–5.1
C	M–D	24	7.3	0.67	5.6–8.3†
	B–L	22	5.3	0.44	4.4–5.8†
P_3	M–D	18	9.2	1.44	4.6–11.2†
	B–L	29	4.0	0.42	3.1–4.7†
P_4	M–D	29	5.2	0.46	4.5–6.3
	B–L	25	3.8	0.31	3.2–4.5
M_1	M–D	30	6.2	0.34	5.5–6.8
	Tri B	24	4.3	0.30	3.9–5.0
	Tal B	25	4.6	0.31	4.2–5.2
M_2	M–D	31	6.7	0.31	6.1–7.5*
	Tri B	26	5.3	0.44	4.6–6.3
	Tal B	27	5.4	0.40	4.6–6.1
M_3	M–D	25	6.6	0.40	5.9–7.7*
	Tri B	23	5.2	0.46	4.4–6.2
	Tal B	23	4.7	0.36	3.9–5.7
	Hypoconulid	—	—	—	—

Table 79. *Cercopithecus mitis* female

Maxillary teeth		n	m	SD	Range
I¹	M–D	18	5.9	0.42	5.3–6.6
	B–L	13	4.8	0.59	4.2–6.4
I²	M–D	18	3.6	0.28	3.1–4.0
	B–L	17	3.9	0.36	3.5–4.5
C	M–D	14	5.5	0.45	5.0–6.5
	B–L	13	4.4	0.30	3.7–4.7
P³	M–D	19	4.0	0.41	3.4–4.9
	B–L	19	4.2	0.49	3.6–5.7
P⁴	M–D	20	4.3	0.29	3.8–4.8
	B–L	20	5.0	0.32	4.6–5.9
M¹	M–D	21	5.9	0.37	5.5–6.8
	Ant B	21	5.4	0.36	4.9–6.2
	Post B	19	5.2	0.36	4.7–6.0
M²	M–D	20	6.3	0.36	5.8–7.0
	Ant B	20	6.0	0.36	5.5–7.0
	Post B	20	5.5	0.33	5.0–6.3
M³	M–D	14	5.6	0.60	4.8–7.3
	Ant B	15	5.4	0.33	5.0–5.9
	Post B	14	4.4	0.60	3.5–6.0

Table 80. *Cercopithecus mitis* female

Mandibular teeth		n	m	SD	Range
I₁	M–D	19	3.7	0.31	3.2–4.4
	B–L	20	4.2	0.36	3.2–4.7
I₂	M–D	18	3.3	0.35	2.7–4.0
	B–L	18	4.2	0.24	3.8–4.7
C	M–D	15	5.4	0.55	4.7–6.8
	B–L	13	3.7	0.56	3.0–5.3
P₃	M–D	17	6.7	0.75	5.6–8.5
	B–L	15	3.4	0.35	2.8–4.0
P₄	M–D	16	4.9	0.35	4.5–5.7
	B–L	15	3.7	0.21	3.4–4.0
M₁	M–D	21	6.0	0.28	5.7–6.6
	Tri B	18	4.3	0.26	3.9–4.9
	Tal B	17	4.5	0.21	4.2–5.0
M₂	M–D	21	6.5	0.36	5.6–7.2
	Tri B	17	5.2	0.38	4.4–5.8
	Tal B	18	5.2	0.31	4.7–5.8
M₃	M–D	18	6.3	0.31	5.7–6.8
	Tri B	16	5.4	0.41	4.2–5.6
	Tal B	15	4.5	0.29	4.0–5.2
	Hypoconulid	—	—	—	—

Table 81. *Cercopithecus neglectus* male

Maxillary teeth		n	m	SD	Range
I^1	M–D	12	5.7	0.32	5.4–6.3*
	B–L	8	4.5	0.31	4.1–4.8
I^2	M–D	9	3.5	0.20	3.2–3.8
	B–L	8	3.7	0.28	3.2–4.1
C	M–D	7	8.3	1.38	5.5–9.4†
	B–L	6	6.1	0.56	5.3–6.8†
P^3	M–D	10	4.6	0.28	4.2–5.0*
	B–L	9	4.3	0.14	4.1–4.4†
P^4	M–D	10	4.9	0.19	4.7–5.1†
	B–L	7	5.0	0.25	4.7–5.5†
M^1	M–D	13	6.3	0.19	5.9–6.6†
	Ant B	11	5.5	0.17	5.3–5.8*
	Post B	11	5.3	0.21	5.0–5.6†
M^2	M–D	12	6.9	0.23	6.5–7.2†
	Ant B	11	6.5	0.28	6.1–7.1†
	Post B	12	6.0	0.23	5.6–6.3†
M^3	M–D	11	5.9	0.30	5.5–6.5
	Ant B	10	5.8	0.26	5.3–6.1†
	Post B	9	4.8	0.34	4.3–5.3†

Table 82. *Cercopithecus neglectus* male

Mandibular teeth		n	m	SD	Range
I_1	M–D	12	3.6	0.20	3.2–3.9*
	B–L	11	4.2	0.24	3.7–4.5*
I_2	M–D	10	3.4	0.32	2.8–3.9
	B–L	11	4.1	0.27	3.6–4.3
C	M–D	6	4.5	0.39	3.9–5.1†
	B–L	8	6.9	0.67	5.5–7.5†
P_3	M–D	7	9.4	1.16	7.0–10.3†
	B–L	10	3.7	0.14	3.4–3.8†
P_4	M–D	9	5.5	0.23	5.2–5.9
	B–L	9	3.8	0.29	3.4–4.4
M_1	M–D	11	6.3	0.26	5.7–6.6†
	Tri B	9	4.4	0.14	4.1–4.6
	Tal B	10	4.6	0.17	4.2–4.8
M_2	M–D	12	6.7	0.33	6.1–7.4†
	Tri B	11	5.4	0.16	5.2–5.9*
	Tal B	11	5.4	0.13	5.2–5.6†
M_3	M–D	9	6.8	0.33	6.3–7.2
	Tri B	9	5.3	0.21	5.0–6.0
	Tal B	8	4.8	0.26	4.5–5.2
	Hypoconulid	—	—	—	—

Table 83. *Cercopithecus neglectus* female

Maxillary teeth		n	m	SD	Range
I^1	M–D	6	5.3	0.52	4.5–5.9
	B–L	3	4.3	0.17	4.2–4.5
I^2	M–D	6	3.3	0.23	3.1–3.7
	B–L	6	3.4	0.15	3.2–3.6
C	M–D	5	5.6	0.44	5.1–6.1
	B–L	6	4.0	0.17	3.8–4.3
P^3	M–D	7	4.3	0.22	4.0–4.7
	B–L	7	3.9	0.29	3.6–4.3
P^4	M–D	6	4.4	0.10	4.3–4.6
	B–L	6	4.7	0.23	4.4–4.9
M^1	M–D	7	5.9	0.28	5.6–6.3
	Ant B	6	5.2	0.25	4.9–5.6
	Post B	6	5.0	0.16	4.8–5.2
M^2	M–D	6	6.2	0.41	5.8–6.8
	Ant B	6	6.0	0.30	5.6–6.4
	Post B	7	5.4	0.30	5.0–5.9
M^3	M–D	6	5.6	0.36	5.3–6.3
	Ant B	6	5.2	0.17	5.0–5.5
	Post B	5	4.2	0.23	3.9–4.5

Table 84. *Cercopithecus neglectus* female

Mandibular teeth		n	m	SD	Range
I_1	M–D	6	3.2	0.42	2.6–3.8
	B–L	5	3.8	0.21	3.6–4.1
I_2	M–D	6	3.1	0.19	2.7–3.3
	B–L	5	3.9	0.33	3.4–4.2
C	M–D	6	3.4	0.23	3.1–3.6
	B–L	6	4.7	0.58	3.7–5.4
P_3	M–D	7	6.4	0.60	5.6–7.3
	B–L	7	3.2	0.22	2.8–3.4
P_4	M–D	7	5.2	0.40	4.7–6.0
	B–L	7	3.7	0.24	3.4–4.0
M_1	M–D	7	5.9	0.27	5.5–6.4
	Tri B	5	4.4	0.44	4.0–5.1
	Tal B	5	4.5	0.39	4.1–5.1
M_2	M–D	7	6.3	0.35	5.9–7.0
	Tri B	5	5.2	0.18	5.1–5.5
	Tal B	6	5.2	0.29	4.7–5.5
M_3	M–D	5	6.4	0.35	6.0–6.7
	Tri B	6	5.1	0.11	4.9–5.2
	Tal B	6	4.6	0.17	4.4–4.9
	Hypoconulid	—	—	—	—

Table 85. *Cercopithecus ascanius*
male

Maxillary teeth		n	m	SD	Range
I^1	M–D	17	5.4	0.45	4.6–6.2
	B–L	17	4.2	0.26	3.7–4.5
I^2	M–D	15	2.9	0.24	2.5–3.4
	B–L	17	3.1	0.23	2.7–3.4
C	M–D	10	6.1	0.60	5.3–6.8
	B–L	11	4.2	0.26	3.9–4.7
P^3	M–D	9	3.4	0.40	3.0–3.4
	B–L	18	3.6	0.27	3.0–3.9
P^4	M–D	16	3.8	0.17	3.6–4.2
	B–L	18	4.2	0.23	3.7–4.5
M^1	M–D	17	5.1	0.21	4.7–5.5
	Ant B	18	4.8	0.38	3.7–5.2
	Post B	18	4.5	0.28	4.1–4.9
M^2	M–D	18	5.5	0.24	5.2–6.0
	Ant B	18	5.5	0.29	5.0–6.1
	Post B	18	4.9	0.28	4.7–5.6
M^3	M–D	17	5.4	0.49	4.6–6.3
	Ant B	17	4.7	0.30	4.3–5.5
	Post B	10	3.7	0.55	2.9–4.5

Source: Sirianni (1974).

Table 86. *Cercopithecus ascanius*
male

Mandibular teeth		n	m	SD	Range
I_1	M–D	18	3.3	0.22	2.9–3.7
	B–L	15	3.6	0.18	3.2–3.9
I_2	M–D	15	3.0	0.27	2.5–3.5
	B–L	15	3.5	0.30	2.8–3.9
C	M–D	13	4.1	0.61	3.3–5.7
	B–L	16	5.1	0.38	4.5–5.6
P_3	M–D	16	6.8	0.56	6.0–7.6
	B–L	16	3.1	0.27	2.7–3.5
P_4	M–D	15	4.2	0.34	3.7–4.9
	B–L	16	3.3	0.20	3.1–3.8
M_1	M–D	15	5.2	0.27	4.6–5.6
	Tri B	16	3.8	0.19	3.3–4.2
	Tal B	16	4.0	0.16	3.7–4.3
M_2	M–D	19	5.5	0.22	5.1–6.0
	Tri B	17	4.6	0.22	4.3–5.0
	Tal B	17	4.6	0.22	4.3–5.0
M_3	M–D	17	5.3	0.39	4.5–5.9
	Tri B	15	4.3	0.20	4.1–4.7
	Tal B	15	3.8	0.24	3.5–4.3
	Hypoconulid	—	—	—	—

Source: Sirianni (1974).

Table 87. *Cercopithecus ascanius* female

Maxillary teeth		n	m	SD	Range
I^1	M–D	12	5.2	0.37	4.6–5.8
	B–L	12	4.1	0.27	3.6–4.6
I^2	M–D	12	2.9	0.20	2.7–3.3
	B–L	13	3.1	0.20	2.8–3.5
C	M–D	10	4.4	0.37	3.6–5.1
	B–L	9	3.6	0.28	3.1–4.0
P^3	M–D	8	3.5	0.27	2.9–3.7
	B–L	13	3.6	0.23	3.1–4.0
P^4	M–D	12	3.7	0.15	3.5–4.1
	B–L	13	4.2	0.25	3.8–4.7
M^1	M–D	12	4.9	0.31	4.6–5.8
	Ant B	13	4.8	0.30	4.3–5.4
	Post B	13	4.5	0.28	4.2–5.1
M^2	M–D	13	5.3	0.29	4.9–5.9
	Ant B	13	5.3	0.39	4.8–6.1
	Post B	13	4.9	0.33	4.5–5.5
M^3	M–D	10	5.3	0.31	3.8–4.9
	Ant B	7	4.6	0.23	4.4–5.0
	Post B	6	3.5	0.33	3.1–3.9

Source: Sirianni (1974).

Table 88. *Cercopithecus ascanius* female

Mandibular teeth		n	m	SD	Range
I_1	M–D	11	3.3	0.20	3.0–3.7
	B–L	10	3.5	0.21	3.3–3.9
I_2	M–D	8	3.0	0.29	2.6–3.4
	B–L	11	3.4	0.12	3.2–3.6
C	M–D	9	3.8	0.16	2.5–3.0
	B–L	10	4.0	0.23	3.4–4.2
P_3	M–D	5	5.4	0.59	4.6–6.2
	B–L	10	2.7	0.18	2.4–3.0
P_4	M–D	11	4.2	0.27	3.9–4.8
	B–L	13	3.1	0.22	2.8–3.5
M_1	M–D	11	5.0	0.26	4.8–5.7
	Tri B	13	3.6	0.16	3.5–3.9
	Tal B	13	3.8	0.20	3.5–4.3
M_2	M–D	11	5.3	0.24	5.0–6.0
	Tri B	13	4.4	0.22	4.1–4.6
	Tal B	12	4.4	0.25	4.0–4.8
M_3	M–D	10	5.1	0.27	4.7–5.7
	Tri B	11	4.3	0.21	3.8–4.5
	Tal B	11	3.7	0.21	3.3–4.0
	Hypoconulid	—	—	—	—

Source: Sirianni (1974).

Table 89. *Chlorocebus aethiops* male

Maxillary teeth		n	m	SD	Range
I¹	M–D	19	5.4	0.35	5.0–6.2
	B–L	20	4.6	0.33	3.9–5.1
I²	M–D	24	3.5	0.40	2.9–4.2
	B–L	23	3.9	0.33	3.3–4.6
C	M–D	17	7.4	0.73	6.0–9.2
	B–L	17	5.2	0.57	4.4–6.9
P³	M–D	21	4.0	0.35	3.2–4.7
	B–L	25	4.3	0.55	3.1–5.6
P⁴	M–D	27	4.3	0.29	3.6–4.9
	B–L	27	5.0	0.36	4.3–5.6
M¹	M–D	28	5.7	0.31	5.3–6.6
	Ant B	27	5.6	0.34	4.9–6.4
	Post B	26	5.1	0.32	4.5–5.7
M²	M–D	29	6.5	0.33	5.9–7.0
	Ant B	28	6.5	0.42	5.8–7.4
	Post B	27	5.8	0.39	5.1–6.9
M³	M–D	22	5.4	0.49	4.6–6.3
	Ant B	23	5.7	0.63	3.9–7.2
	Post B	21	4.5	0.65	3.6–6.4

Source: Sirianni (1974).

Table 90. *Chlorocebus aethiops* male

Mandibular teeth		n	m	SD	Range
I₁	M–D	26	3.6	0.31	3.1–4.3
	B–L	18	4.2	0.38	3.6–5.1
I₂	M–D	26	3.2	0.27	2.7–3.7
	B–L	23	4.3	0.54	3.6–5.7
C	M–D	19	4.3	0.58	3.3–5.3
	B–L	24	5.7	0.69	5.7–8.2
P₃	M–D	20	8.9	0.82	7.7–10.3
	B–L	24	3.6	0.34	3.0–4.5
P₄	M–D	24	4.7	0.36	4.1–5.5
	B–L	23	3.7	0.26	3.3–4.3
M₁	M–D	26	5.9	0.26	5.4–6.5
	Tri B	26	4.4	0.28	3.9–5.0
	Tal B	25	4.5	0.31	3.6–5.1
M₂	M–D	25	6.5	0.34	5.8–7.4
	Tri B	26	5.3	0.32	4.8–6.1
	Tal B	25	5.2	0.42	4.1–6.3
M₃	M–D	21	6.3	0.35	5.6–6.7
	Tri B	20	5.3	0.41	4.7–6.3
	Tal B	18	4.5	0.47	3.7–5.7
Hypoconulid	—	—	—	—

Source: Sirianni (1974).

Table 91. *Chlorocebus aethiops* female

Maxillary teeth		n	m	SD	Range
I^1	M–D	20	5.1	0.34	4.5–5.8
	B–L	20	4.4	0.36	3.7–5.0
I^2	M–D	18	3.3	0.30	2.8–3.8
	B–L	20	3.7	0.32	3.2–4.5
C	M–D	18	5.2	0.43	4.5–5.8
	B–L	18	3.9	0.40	3.5–5.0
P^3	M–D	11	3.6	0.43	3.1–4.3
	B–L	20	3.9	0.45	3.1–5.2
P^4	M–D	19	3.9	0.37	3.4–4.8
	B–L	20	4.7	0.34	4.3–5.5
M^1	M–D	20	5.4	0.26	5.1–5.9
	Ant B	19	5.2	0.36	4.7–6.0
	Post B	20	4.8	0.30	4.4–5.5
M^2	M–D	20	6.0	0.36	5.4–6.9
	Ant B	20	5.9	0.37	5.2–6.8
	Post B	20	5.4	0.24	4.9–5.8
M^3	M–D	11	5.0	0.46	4.5–5.8
	Ant B	13	5.2	0.42	4.5–6.2
	Post B	9	4.1	0.44	3.3–4.7

Source: Sirianni (1974)

Table 92. *Chlorocebus aethiops* female

Mandibular teeth		n	m	SD	Range
I_1	M–D	19	3.3	0.26	3.0–4.0
	B–L	18	3.9	0.42	3.1–4.8
I_2	M–D	14	3.1	0.22	2.7–3.3
	B–L	17	3.9	0.35	3.4–4.6
C	M–D	14	3.2	0.46	2.3–4.2
	B–L	18	4.9	0.61	4.1–6.6
P_3	M–D	16	6.7	0.91	5.5–9.5
	B–L	17	3.0	0.33	2.6–4.0
P_4	M–D	16	4.5	0.34	4.1–5.2
	B–L	18	3.4	0.23	3.0–4.0
M_1	M–D	17	5.6	0.17	5.4–5.9
	Tri B	17	4.2	0.21	3.8–4.7
	Tal B	18	4.2	0.22	3.9–4.6
M_2	M–D	19	6.0	0.26	5.7–6.8
	Tri B	18	5.0	0.26	4.6–5.5
	Tal B	18	4.9	0.21	4.6–5.3
M_3	M–D	11	5.8	0.42	5.0–6.3
	Tri B	12	4.8	0.33	4.4–5.3
	Tal B	13	4.3	0.42	6.7–5.0
	Hypoconulid	—	—	—	—

Source: Sirianni (1974).

Table 93. *Lophocebus albigena* male

Maxillary teeth		*n*	*m*	SD	Range
I^1	M–D	28	8.2	0.37	7.5–8.8†
	B–L	27	6.7	0.44	5.9–8.0†
I^2	M–D	23	5.9	0.36	5.4–6.3†
	B–L	25	5.3	0.30	4.8–6.2†
C	M–D	22	7.7	0.61	6.8–9.3†
	B–L	24	5.3	0.77	3.9 6.6†
P^3	M–D	30	5.4	0.50	4.3–6.4†
	B–L	29	5.8	0.29	5.4–6.3†
P^4	M–D	30	4.7	0.31	4.2–5.5†
	B–L	30	5.9	0.25	5.4–6.4†
M^1	M–D	31	7.0	0.25	6.5–7.4†
	Ant B	31	6.7	0.28	6.1–7.4†
	Post B	30	6.6	0.27	6.1–7.2
M^2	M–D	29	7.6	0.22	7.1–8.2†
	Ant B	29	7.4	0.26	7.0–8.0†
	Post B	29	6.8	0.34	6.2–7.4†
M^3	M–D	24	6.8	0.33	6.4–7.5†
	Ant B	26	6.9	0.29	6.5–7.5†
	Post B	18	5.6	0.48	4.5–6.2†

Table 94. *Lophocebus albigena* male

Mandibular teeth		*n*	*m*	SD	Range
I_1	M–D	27	5.8	0.51	4.5–6.6†
	B–L	21	6.3	0.40	5.6–7.3†
I_2	M–D	25	4.8	0.54	3.7–6.4†
	B–L	24	5.6	0.36	4.7–6.4†
C	M–D	14	5.1	0.79	4.3–7.3†
	B–L	20	7.4	0.29	6.7–7.9†
P_3	M–D	22	7.4	1.32	4.7–10.7†
	B–L	28	4.7	0.46	3.9–5.6†
P_4	M–D	30	5.3	0.35	4.8–6.5*
	B–L	29	4.9	0.26	4.5–5.5†
M_1	M–D	30	6.9	0.20	6.5–7.2†
	Tri B	30	5.6	0.21	5.1–6.0†
	Tal B	30	5.7	0.28	5.3–6.6†
M_2	M–D	31	7.6	0.28	6.8–8.2†
	Tri B	31	6.6	0.27	6.0–7.1†
	Tal B	31	6.2	0.28	5.7–6.8†
M_3	M–D	27	8.3	0.45	7.3–9.1†
	Tri B	27	6.4	0.31	5.9–7.1†
	Tal B	25	5.4	0.31	4.9–6.1†
	Hypoconulid	16	2.9	0.45	1.7–3.3†

Table 95. *Lophocebus albigena* female

Maxillary teeth		n	m	SD	Range
I¹	M–D	29	7.5	0.35	6.6–8.2
	B–L	27	6.3	0.42	5.3–6.9
I²	M–D	29	5.3	0.41	4.4–6.7
	B–L	28	5.0	0.34	4.4–5.7
C	M–D	28	6.1	0.40	5.2–6.7
	B–L	28	4.6	0.16	4.4–5.1
P³	M–D	28	4.9	0.41	4.1–5.6
	B–L	29	5.4	0.33	4.2–5.9
P⁴	M–D	29	4.5	0.26	3.9–5.0
	B–L	29	5.5	0.21	5.0–5.9
M¹	M–D	29	6.6	0.21	6.1–6.9
	Ant B	28	6.2	0.22	5.8–6.7
	Post B	28	6.0	0.22	5.5–6.5
M²	M–D	27	7.1	0.34	6.4–7.7
	Ant B	27	6.7	0.27	6.2–7.2
	Post B	25	6.2	0.23	5.7–6.7
M³	M–D	21	6.5	0.33	6.0–7.3
	Ant B	21	6.3	0.26	5.8–6.7
	Post B	18	5.1	0.39	4.6–6.0

Table 96. *Lophocebus albigena* female

Mandibular teeth		n	m	SD	Range
I₁	M–D	29	5.4	0.32	4.4–5.8
	B–L	26	5.9	0.35	5.4–6.8
I₂	M–D	26	4.4	0.42	3.3–5.4
	B–L	26	5.3	0.27	4.6–5.8
C	M–D	26	4.1	0.32	3.6–4.7
	B–L	25	5.5	0.49	3.6–6.4
P₃	M–D	25	5.9	1.12	4.1–9.0
	B–L	26	4.0	0.37	3.1–4.8
P₄	M–D	29	5.1	0.31	4.6–5.6
	B–L	28	4.6	0.30	4.1–5.5
M₁	M–D	29	6.5	0.15	6.2–6.8
	Tri B	29	5.2	0.21	4.8–5.6
	Tal B	29	5.3	0.20	4.7–5.6
M₂	M–D	29	7.1	0.31	6.5–7.7
	Tri B	29	6.1	0.28	5.4–6.4
	Tal B	29	5.8	0.23	5.2–6.2
M₃	M–D	22	7.7	0.53	6.7–8.7
	Tri B	24	5.9	0.30	5.2–6.5
	Tal B	20	5.1	0.24	4.6–5.7
	Hypoconulid	16	2.4	0.53	1.4–3.4

Table 97. *Cercocebus torquatus* male

Maxillary teeth		n	m	SD	Range
I¹	M–D	12	7.4	0.64	6.0–8.4
	B–L	11	6.5	0.98	4.4–8.0
I²	M–D	11	4.7	0.60	3.9–5.7
	B–L	11	5.2	0.69	4.2–6.1
C	M–D	8	8.5	1.84	6.2–10.6*
	B–L	6	6.2	1.16	5.1 7.7*
P³	M–D	8	5.7	0.35	5.0–6.0
	B–L	9	6.1	0.32	5.3–6.3
P⁴	M–D	9	5.7	0.78	5.1–7.7
	B–L	9	7.1	0.57	5.8–7.7
M¹	M–D	14	7.5	0.55	6.3–8.5
	Ant B	14	7.8	0.81	6.1–8.7
	Post B	14	6.9	0.67	6.5–7.8
M²	M–D	10	8.0	0.59	6.7–8.7
	Ant B	10	8.4	0.84	6.8–9.5
	Post B	9	7.7	0.91	6.2–8.7
M³	M–D	6	7.7	0.85	6.2–8.3
	Ant B	6	8.0	0.82	6.5–8.7
	Post B	6	6.6	0.92	4.8–7.3

Table 98. *Cercocebus torquatus* male

Mandibular teeth		n	m	SD	Range
I₁	M–D	12	5.1	0.52	3.7–5.5
	B–L	12	5.9	0.81	4.1–7.2
I₂	M–D	11	4.2	0.49	3.4–5.2
	B–L	13	5.6	0.85	4.3–7.3
C	M–D	5	5.1	0.68	4.3–5.7†
	B–L	6	7.5	1.58	5.6–9.1†
P₃	M–D	5	12.1	2.90	8.1–14.5†
	B–L	9	4.6	0.51	3.7–5.4
P₄	M–D	8	6.1	0.40	5.4–6.7
	B–L	8	5.7	1.01	3.5–6.6
M₁	M–D	13	7.3	0.57	6.1–8.2
	Tri B	14	6.1	0.65	4.6–6.9
	Tal B	13	5.8	0.52	4.6–6.4
M₂	M–D	10	8.2	0.57	7.2–8.9
	Tri B	10	7.3	0.88	5.6–8.2
	Tal B	9	6.8	0.84	5.3–7.7
M₃	M–D	6	9.7	3.43	9.1–10.2
	Tri B	6	7.2	0.75	5.7–7.7
	Tal B	6	6.3	0.70	5.0–6.8*
	Hypoconulid	4	4.2	0.31	3.7–4.4

Table 99. *Cercocebus torquatus* female

Maxillary teeth		*n*	*m*	SD	Range
I^1	M–D	10	7.4	0.48	6.4–8.1
	B–L	9	6.4	0.33	5.9–6.9
I^2	M–D	9	4.6	0.43	3.9–5.1
	B–L	10	4.9	0.34	4.3–5.5
C	M–D	7	6.5	0.11	6.4–6.7
	B–L	7	5.0	0.29	4.7–5.4
P^3	M–D	8	5.4	0.20	5.1–5.7
	B–L	8	5.9	0.21	5.6–6.3
P^4	M–D	8	5.6	0.18	5.4–5.9
	B–L	8	6.9	0.27	6.6–7.3
M^1	M–D	10	7.2	0.71	5.3–7.8
	Ant B	10	7.6	0.25	7.2–7.9
	Post B	10	6.9	0.22	6.5–7.2
M^2	M–D	8	8.1	0.26	7.7–8.5
	Ant B	8	8.4	0.23	8.0–8.7
	Post B	8	7.8	0.39	7.4–8.6
M^3	M–D	2	8.5	0.28	8.3–8.7
	Ant B	2	8.4	0.21	8.2–8.5
	Post B	2	7.2	0.21	7.0–7.3

Table 100. *Cercocebus torquatus* female

Mandibular teeth		*n*	*m*	SD	Range
I$_1$	M–D	10	5.0	0.24	4.5–5.3
	B–L	9	5.6	0.48	4.8–6.2
I$_2$	M–D	9	4.2	0.24	3.8–4.6
	B–L	10	5.3	0.47	4.2–5.8
C	M–D	8	4.1	0.28	3.5–4.5
	B–L	8	5.6	0.27	5.1–5.9
P$_3$	M–D	7	8.5	0.62	7.5–9.5
	B–L	8	4.3	0.21	4.0–4.6
P$_4$	M–D	7	5.9	0.25	5.5–6.2
	B–L	8	5.6	0.33	5.0–6.1
M$_1$	M–D	10	7.3	0.22	6.9–7.6
	Tri B	10	6.0	0.24	5.7–6.4
	Tal B	10	5.8	0.27	5.5–6.2
M$_2$	M–D	8	8.3	0.35	7.6–8.7
	Tri B	8	7.3	0.26	6.9–7.6
	Tal B	8	6.8	0.29	6.4–7.2
M$_3$	M–D	2	9.3	0.92	8.6–9.9
	Tri B	2	7.6	0.21	7.4–7.7
	Tal B	2	7.1	1.13	6.3–7.9
	Hypoconulid	2	3.0	0.35	2.7–3.2

Table 101. *Cercocebus galeritus* male

Maxillary teeth		n	m	SD	Range
I^1	M–D	7	7.9	0.60	7.0–8.5
	B–L	8	7.1	0.39	6.2–7.5
I^2	M–D	10	4.8	0.45	4.2–5.4
	B–L	8	5.2	0.29	4.9–5.7†
C	M–D	7	9.7	0.50	8.9–10.5†
	B–L	8	6.6	0.65	5.8–7.6†
P^3	M–D	10	5.5	0.55	4.7–6.2
	B–L	10	6.5	0.51	5.5–7.3†
P^4	M–D	9	5.7	0.43	4.7–6.1
	B–L	10	7.8	0.68	6.0–8.4*
M^1	M–D	10	7.3	0.29	6.8–7.8
	Ant B	10	8.1	0.72	6.6–9.2*
	Post B	10	7.4	0.43	6.6–8.1*
M^2	M–D	10	7.9	0.37	7.4–8.5
	Ant B	10	8.6	0.34	8.1–9.2*
	Post B	10	8.0	0.47	7.1–8.8
M^3	M–D	9	6.8	0.49	6.1–7.4
	Ant B	9	8.0	0.44	7.4–8.5*
	Post B	8	6.5	0.61	5.6–7.3

Table 102. *Cercocebus galeritus* male

Mandibular teeth		n	m	SD	Range
I_1	M–D	9	5.9	0.32	5.3–6.4
	B–L	9	6.5	0.31	5.8–6.8*
I_2	M–D	9	4.1	0.32	3.6–4.7
	B–L	10	5.7	0.27	5.3–6.0†
C	M–D	4	4.9	0.15	4.8–5.1†
	B–L	7	7.9	1.45	4.7–8.7†
P_3	M–D	5	11.9	0.39	11.5–12.4†
	B–L	9	4.5	0.25	4.3–5.0†
P_4	M–D	10	6.2	0.31	5.5–6.6*
	B–L	10	5.9	0.31	5.5–6.5*
M_1	M–D	10	7.4	0.39	6.9–8.2*
	Tri B	10	6.4	0.32	5.9–6.7*
	Tal B	10	6.2	0.39	5.6–6.7
M_2	M–D	10	8.1	0.39	7.5–8.9
	Tri B	8	7.5	0.27	7.2–7.9
	Tal B	9	7.2	0.34	6.7–7.8
M_3	M–D	9	8.5	0.62	7.5–9.5
	Tri B	9	7.3	0.28	6.9–7.7*
	Tal B	9	6.6	0.56	5.8–7.8*
	Hypoconulid	6	3.5	0.41	2.7–3.8

Table 103. *Cercocebus galeritus* female

Maxillary teeth		n	m	SD	Range
I^1	M–D	10	7.6	0.70	6.2–8.7
	B–L	10	6.7	0.40	6.2–7.4
I^2	M–D	9	4.6	0.32	4.1–5.1
	B–L	9	4.7	0.33	4.2–5.4
C	M–D	6	6.0	0.39	5.4–6.4
	B–L	6	4.8	0.18	4.6–5.0
P^3	M–D	10	5.1	0.26	4.5–5.4
	B–L	10	6.0	0.39	5.2–6.5
P^4	M–D	9	5.6	0.22	5.1–5.8
	B–L	10	7.2	0.47	6.3–8.0
M^1	M–D	10	7.1	0.32	6.4–7.5
	Ant B	10	7.4	0.44	6.3–8.0
	Post B	10	6.9	0.40	6.0–7.6
M^2	M–D	10	7.8	0.37	6.9–8.3
	Ant B	10	8.0	0.58	7.0–8.7
	Post B	10	7.6	0.54	6.4–8.1
M^3	M–D	7	6.8	0.26	6.5–7.2
	Ant B	8	7.4	0.52	6.6–8.3
	Post B	7	6.1	0.40	5.6–6.7

Table 104. *Cercocebus galeritus* female

Mandibular teeth		n	m	SD	Range
I_1	M–D	10	5.5	0.52	4.6–6.2
	B–L	8	5.9	0.60	4.7–6.7
I_2	M–D	8	3.8	0.37	3.1–4.2
	B–L	10	5.1	0.39	4.5–5.7
C	M–D	9	3.8	0.19	3.5–4.1
	B–L	8	5.7	0.60	4.8–6.5
P_3	M–D	8	7.5	0.41	7.0–8.3
	B–L	9	4.0	0.25	3.6–4.3
P_4	M–D	10	5.8	0.32	5.2–6.1
	B–L	9	5.5	0.58	4.5–6.5
M_1	M–D	10	7.0	0.29	6.5–7.4
	Tri B	9	5.9	0.40	5.0–6.4
	Tal B	9	5.8	0.46	4.9–6.6
M_2	M–D	10	7.8	0.34	7.0–8.3
	Tri B	10	7.2	0.57	5.9–8.1
	Tal B	10	6.8	0.68	5.3–8.0
M_3	M–D	8	8.1	0.64	7.3–8.9
	Tri B	9	6.8	0.43	5.8–7.3
	Tal B	8	6.4	0.37	5.8–6.9
	Hypoconulid	7	2.6	0.74	1.4—3.4

Table 105. *Macaca nemestrina* male

Maxillary teeth		*n*	*m*	SD	Range
I¹	M–D	15	7.6	0.56	6.4–8.5
	B–L	13	6.5	0.52	5.5–7.2
I²	M–D	15	5.5	0.52	4.4–6.4
	B–L	14	5.8	0.40	5.0–6.5†
C	M–D	11	11.7	1.03	9.4–13.0†
	B–L	9	8.0	0.57	7.2–9.1†
P³	M–D	16	5.9	0.45	5.0–6.7†
	B–L	17	6.3	0.36	5.8–7.0*
P⁴	M–D	15	5.7	0.30	5.0–6.0*
	B–L	15	6.8	0.33	6.4–7.6*
M¹	M–D	17	7.4	0.31	6.8–7.9
	Ant B	19	7.2	0.53	5.6–8.2
	Post B	15	7.9	0.37	6.3–7.5*
M²	M–D	17	8.6	0.38	7.8–9.3†
	Ant B	17	8.5	0.48	7.8–9.3†
	Post B	17	7.9	0.50	7.0–8.7†
M³	M–D	16	8.7	0.52	7.6–8.9†
	Ant B	16	8.5	0.48	7.6–9.3†
	Post B	16	7.7	0.73	6.4–9.3†

Table 106. *Macaca nemestrina* male

Mandibular teeth		*n*	*m*	SD	Range
I₁	M–D	17	5.3	0.41	4.6–5.9
	B–L	17	6.1	0.41	5.3–6.9
I₂	M–D	14	5.2	0.58	4.5–6.5
	B–L	14	5.9	0.79	3.6–6.8
C	M–D	15	10.9	0.95	9.1–12.7†
	B–L	14	7.2	0.69	6.3–8.6†
P₃	M–D	12	15.8	1.32	12.8–17.3†
	B–L	16	5.8	0.52	4.6–6.7†
P₄	M–D	16	6.6	0.50	5.8–7.6*
	B–L	15	5.5	0.38	5.0–6.2
M₁	M–D	19	7.5	0.31	6.8–8.0
	Tri B	19	5.8	0.21	5.5–6.3
	Tal B	18	5.8	0.20	5.4–6.2*
M₂	M–D	17	8.6	0.50	7.8–9.5*
	Tri B	18	7.2	0.57	6.2–8.4*
	Tal B	18	6.8	0.46	5.9–7.8†
M₃	M–D	15	10.9	0.87	9.6–12.5
	Tri B	16	7.8	0.49	7.0–8.5†
	Tal B	16	7.1	0.52	6.0–8.0†
	Hypoconulid	16	5.0	0.54	3.7–6.0†

Table 107. *Macaca nemestrina* female

Maxillary teeth		n	m	SD	Range
I^1	M–D	9	7.3	1.22	5.8–9.0
	B–L	7	6.1	0.53	5.5–6.8
I^2	M–D	7	5.0	0.77	3.9–5.8
	B–L	7	5.2	0.54	4.4–5.8
C	M–D	7	7.3	0.83	6.0–8.2
	B–L	7	5.5	0.67	4.4–6.2
P^3	M–D	7	5.2	0.57	4.2–5.8
	B–L	7	5.9	0.62	5.0–6.7
P^4	M–D	8	5.2	0.64	3.8–5.7
	B–L	8	6.3	0.61	5.4–7.0
M^1	M–D	11	7.1	0.69	5.7–7.8
	Ant B	10	7.1	0.48	6.0–7.6
	Post B	11	6.5	0.42	5.8–7.0
M^2	M–D	9	7.9	0.74	6.3–8.7
	Ant B	9	7.9	0.52	6.8–8.5
	Post B	9	7.2	0.40	6.5–7.7
M^3	M–D	8	7.8	0.77	6.6–9.0
	Ant B	7	7.7	0.65	6.7–8.7
	Post B	8	6.7	0.64	5.9–7.6

Table 108. *Macaca nemestrina* female

Mandibular teeth		n	m	SD	Range
I_1	M–D	7	5.4	0.69	4.4–6.2
	B–L	7	6.1	0.65	5.0–7.0
I_2	M–D	7	4.8	0.55	4.0–5.4
	B–L	6	5.6	0.76	4.6–6.5
C	M–D	6	7.0	1.08	4.9–8.0
	B–L	6	4.9	0.81	3.6–5.9
P_3	M–D	5	10.4	1.70	7.8–12.2
	B–L	6	5.5	2.55	3.9–10.6
P_4	M–D	9	6.0	0.85	4.4–7.0
	B–L	7	5.2	0.54	4.2–5.7
M_1	M–D	10	7.3	0.64	5.9–8.0
	Tri B	10	5.6	0.46	4.5–6.0
	Tal B	10	5.4	0.47	4.5–6.2
M_2	M–D	8	7.9	0.72	6.6–8.8
	Tri B	7	6.6	0.65	5.4–7.3
	Tal B	7	6.2	0.52	5.2–6.7
M_3	M–D	9	10.0	1.15	7.6–11.2
	Tri B	8	6.9	0.91	5.1–7.9
	Tal B	8	6.2	0.87	4.9–7.6
	Hypoconulid	6	4.7	0.77	3.6–5.8

Table 109. *Macaca mulatta* male

Maxillary teeth		n	m	SD	Range
I^1	M–D	91	6.3	0.61	5.0–8.0
	B–L	71	6.1	0.99	4.4–8.8*
I^2	M–D	32	4.9	0.60	3.5–6.4
	B–L	30	5.4	0.45	4.8–7.0†
C	M–D	27	8.9	1.93	3.9–11.3†
	B–L	25	6.7	1.01	4.8–8.6†
P^3	M–D	57	5.0	0.38	4.2–6.3*
	B–L	52	6.1	0.55	5.0–7.2
P^4	M–D	57	5.1	0.39	4.5–6.4
	B–L	54	6.5	0.56	5.5–7.6
M^1	M–D	95	7.2	0.55	5.9–8.5*
	Ant B	103	7.0	0.51	6.0–8.5†
	Post B	101	6.5	0.58	5.6–8.0
M^2	M–D	50	7.8	0.81	6.0–9.6
	Ant B	56	7.7	0.82	5.5–9.6
	Post B	53	7.1	0.88	5.0–9.0
M^3	M–D	22	7.4	0.57	6.6–8.5
	Ant B	26	7.4	0.82	6.5–9.0
	Post B	20	6.5	0.79	5.5–7.9

Table 110. *Macaca mulatta* male

Mandibular teeth		n	m	SD	Range
I_1	M–D	89	4.4	0.46	3.6–6.5†
	B–L	82	5.4	0.92	3.5–7.4†
I_2	M–D	75	4.2	0.37	3.4–5.2†
	B–L	70	5.3	0.68	4.0–7.5†
C	M–D	31	7.5	1.57	3.8–9.8†
	B–L	29	6.3	0.22	3.0–9.0†
P_3	M–D	17	11.2	1.43	6.8–14.0†
	B–L	28	4.5	2.54	3.4–5.6†
P_4	M–D	55	5.4	0.52	4.4–7.6
	B–L	47	5.0	0.50	3.9–6.5†
M_1	M–D	97	7.2	0.56	5.9–9.5*
	Tri B	95	5.6	0.47	4.6–7.0*
	Tal B	93	5.5	0.39	4.6–6.6
M_2	M–D	60	7.8	0.81	3.8–9.0
	Tri B	66	6.6	0.66	5.5–8.8
	Tal B	59	6.2	0.57	5.0–7.3
M_3	M–D	26	9.2	1.11	6.4–12.0*
	Tri B	32	6.5	0.74	4.7–7.8
	Tal B	27	5.9	0.64	4.5–7.1
	Hypoconulid	21	4.3	0.43	3.6–5.2

Table 111. *Macaca mulatta* female

Maxillary teeth		*n*	*m*	SD	Range
I^1	M–D	78	6.2	0.50	4.7–7.6
	B–L	59	5.7	1.00	3.4–9.7
I^2	M–D	33	4.7	0.62	3.7–7.2
	B–L	29	4.9	0.55	4.0–6.8
C	M–D	37	6.0	0.35	5.2–6.9
	B–L	35	5.0	0.38	4.0–5.8
P^3	M–D	55	4.9	0.41	4.0–6.0
	B–L	54	6.0	0.43	5.0–6.9
P^4	M–D	57	5.0	0.37	4.2–6.0
	B–L	56	6.3	0.46	5.4–7.2
M^1	M–D	79	7.1	0.57	5.7–8.7
	Ant B	83	6.8	0.40	5.3–8.0
	Post B	79	6.4	0.46	4.9–7.6
M^2	M–D	33	7.6	0.68	5.8–8.7
	Ant B	52	7.6	0.67	5.9–10.1
	Post B	39	7.0	0.61	5.8–8.4
M^3	M–D	16	7.3	0.71	6.0–8.7
	Ant B	15	7.2	0.84	6.4–8.9
	Post B	13	6.5	0.74	5.5–7.5

Table 112. *Macaca mulatta* female

Mandibular teeth		*n*	*m*	SD	Range
I$_1$	M–D	81	4.2	0.39	3.5–5.6
	B–L	75	5.0	0.84	3.4–7.5
I$_2$	M–D	78	4.0	0.36	3.2–5.2
	B–L	75	4.8	0.63	3.4–6.8
C	M–D	29	5.3	0.67	4.1–6.9
	B–L	23	4.2	0.52	3.5–6.0
P$_3$	M–D	20	7.5	1.34	4.8–11.8
	B–L	31	4.0	0.45	3.2–5.2
P$_4$	M–D	50	5.3	0.36	4.5–6.0
	B–L	47	4.8	0.42	4.0–5.7
M$_1$	M–D	87	7.0	0.47	6.0–8.3
	Tri B	82	5.5	0.41	4.5–6.6
	Tal B	80	5.4	0.54	4.2–8.6
M$_2$	M–D	49	7.8	0.59	5.2–8.9
	Tri B	59	6.4	0.54	5.0–7.5
	Tal B	51	6.1	0.62	3.9–7.2
M$_3$	M–D	11	8.3	1.42	5.7–11.2
	Tri B	17	6.1	0.80	4.6–8.0
	Tal B	10	5.7	0.76	4.7–7.0
	Hypoconulid	7	4.6	1.13	3.5–6.7

Table 113. *Macaca fascicularis* male

Maxillary teeth		n	m	SD	Range
I^1	M–D	51	6.6	0.67	5.0–8.2
	B–L	46	6.2	0.76	4.5–7.8
I^2	M–D	48	4.5	0.53	3.5–5.7
	B–L	49	5.0	0.52	3.7–5.9*
C	M–D	35	9.2	1.46	5.6–12.2†
	B–L	33	6.6	0.81	4.8–8.4†
P^3	M–D	53	4.8	0.38	4.0–5.6†
	B–L	55	5.5	0.50	4.4–6.5
P^4	M–D	54	4.7	0.35	4.0–5.7
	B–L	55	5.9	0.53	4.9–7.2
M^1	M–D	65	6.5	0.50	5.3–7.7*
	Ant B	65	6.4	0.55	5.5–7.6
	Post B	64	6.0	0.53	5.2–7.5
M^2	M–D	58	7.2	0.61	6.0–9.5
	Ant B	57	7.2	0.72	5.9–9.4
	Post B	56	6.7	0.69	5.5–9.2
M^3	M–D	47	7.1	0.69	6.0–9.1*
	Ant B	44	6.9	0.81	5.4–9.5*
	Post B	45	6.1	0.77	5.0–8.8*

Table 114. *Macaca fascicularis* male

Mandibular teeth		n	m	SD	Range
I_1	M–D	55	4.6	0.43	4.0–5.7
	B–L	53	5.9	0.70	4.6–7.5*
I_2	M–D	51	3.9	0.36	3.0–4.6
	B–L	49	5.2	0.66	4.0–6.6†
C	M–D	40	8.3	1.32	5.6–12.0†
	B–L	36	5.1	1.07	3.5–9.0†
P_3	M–D	40	10.9	1.72	7.1–15.5†
	B–L	51	4.5	0.54	3.5–6.4†
P_4	M–D	53	5.1	0.38	4.2–6.6
	B–L	53	4.7	0.45	3.9–6.2
M_1	M–D	66	6.5	0.52	5.6–7.7
	Tri B	58	5.1	0.44	4.5–6.2
	Tal B	64	5.0	0.46	4.3–5.8
M_2	M–D	58	7.4	0.55	6.2–9.0
	Tri B	55	6.1	0.60	4.6–7.7
	Tal B	55	5.7	0.53	4.7–6.9
M_3	M–D	45	8.8	0.94	6.8–11.2
	Tri B	45	6.0	0.85	5.5–7.8
	Tal B	43	5.5	0.60	4.3–6.8
	Hypoconulid	38	4.1	0.62	3.0–5.7

Table 115. *Macaca fascicularis* female

Maxillary teeth		n	m	SD	Range
I^1	M–D	46	6.5	0.72	4.7–7.6
	B–L	39	6.1	0.65	5.0–7.6
I^2	M–D	47	4.4	0.51	3.3–5.6
	B–L	44	4.8	0.43	3.7 5.9
C	M–D	40	5.8	0.78	4.3–7.8
	B–L	37	5.0	0.53	4.0–6.7
P^3	M–D	45	4.5	0.48	3.6–5.6
	B–L	45	5.4	0.48	4.6–6.9
P^4	M–D	45	4.7	0.44	3.8–5.8
	B–L	45	5.8	0.53	5.0–7.4
M^1	M–D	53	6.3	0.54	4.3–7.6
	Ant B	51	6.2	0.50	5.5–8.2
	Post B	49	5.9	0.57	5.0–7.9
M^2	M–D	46	7.2	0.58	5.7–8.5
	Ant B	48	7.1	0.58	6.0–8.7
	Post B	47	6.5	0.57	5.6–7.8
M^3	M–D	32	6.8	0.62	5.8–8.5
	Ant B	31	6.6	0.40	6.0–8.0
	Post B	29	5.7	0.41	5.0–6.8

Table 116. *Macaca fascicularis* female

Mandibular teeth		n	m	SD	Range
I_1	M–D	48	4.6	0.49	3.4–5.8
	B–L	48	5.6	0.71	4.0–6.9
I_2	M–D	49	3.9	0.47	3.1–5.7
	B–L	47	4.9	0.55	3.5–5.8
C	M–D	43	6.0	0.74	5.0–8.3
	B–L	40	3.8	0.70	2.8–6.0
P_3	M–D	38	7.9	1.17	6.2–11.4
	B–L	44	3.9	0.44	3.2–4.8
P_4	M–D	47	5.0	0.48	4.2–6.0
	B–L	46	4.6	0.48	3.8–5.7
M_1	M–D	56	6.4	0.49	5.0–7.5
	Tri B	52	5.1	0.39	4.0–5.9
	Tal B	52	4.9	0.39	4.0–6.2
M_2	M–D	49	7.3	0.59	6.0–8.7
	Tri B	48	6.0	0.48	5.2–7.4
	Tal B	46	5.6	0.47	4.7–6.9
M_3	M–D	34	8.5	0.82	5.9–10.5
	Tri B	35	5.9	0.51	5.0–7.6
	Tal B	33	5.2	0.53	4.2–7.6
	Hypoconulid	28	3.9	0.56	2.7–5.0

Table 117. *Macaca nigra* male

Maxillary teeth		n	m	SD	Range
I^1	M–D	5	6.3	1.30	4.8–7.5
	B–L	9	5.9	1.43	3.4–7.3
I^2	M–D	9	3.8	0.73	2.3–4.5
	B–L	9	4.9	1.22	2.7–6.0
C	M–D	5	7.6	2.56	4.8–10.0
	B–L	3	4.9	2.09	3.5–7.3
P^3	M–D	7	4.9	0.25	4.5–5.2
	B–L	6	6.0	0.19	5.8–6.3†
P^4	M–D	7	5.3	0.16	5.0–5.4
	B–L	7	6.2	0.31	5.7–6.6*
M^1	M–D	10	7.1	0.48	6.0–7.5*
	Ant B	10	6.5	0.94	4.6–7.2
	Post B	10	6.0	0.79	4.6–6.7
M^2	M–D	8	8.3	0.35	8.0–9.0*
	Ant B	8	7.8	0.46	7.2–8.5
	Post B	7	7.4	0.36	6.9–7.9
M^3	M–D	3	8.4	0.61	7.7–8.9
	Ant B	3	7.9	0.75	7.5–8.8
	Post B	3	6.8	0.69	6.4–7.6*

Table 118. *Macaca nigra* male

Mandibular teeth		n	m	SD	Range
I_1	M–D	9	4.6	0.95	2.7–5.4
	B–L	10	5.2	1.16	3.0–6.7
I_2	M–D	8	3.1	0.85	2.0–4.8
	B–L	9	4.9	1.09	3.0–5.8
C	M–D	5	5.2	0.52	4.6–5.8
	B–L	5	6.3	3.14	3.0–9.5
P_3	M–D	3	10.6	4.30	6.4–15.0
	B–L	4	4.8	0.53	4.0–5.2†
P_4	M–D	7	6.0	0.36	5.4–6.4
	B–L	7	5.1	0.09	5.0–5.2†
M_1	M–D	9	7.1	0.24	6.6–7.3*
	Tri B	8	5.2	1.02	3.5–6.1
	Tal B	9	5.3	0.91	3.9–6.5
M_2	M–D	8	7.9	0.33	7.3–8.3
	Tri B	8	7.0	0.61	5.6–7.5
	Tal B	8	6.5	0.56	5.3–7.3
M_3	M–D	3	11.0	1.14	10.2–12.3
	Tri B	4	7.1	0.61	6.6–8.0
	Tal B	3	6.4	0.85	5.8–7.4
Hypoconulid		—	—	—	—

Table 119. *Macaca nigra* female

Maxillary teeth		n	m	SD	Range
I¹	M–D	5	6.2	0.69	5.1–6.9
	B–L	5	5.2	1.02	3.4–5.9
I²	M–D	6	3.2	0.78	2.0–4.1
	B–L	6	4.1	1.13	2.0–5.0
C	M–D	5	6.3	0.74	5.0–6.8
	B–L	5	5.2	0.85	3.7–5.7
P³	M–D	4	4.5	0.47	4.0–4.9
	B–L	5	5.6	0.11	5.5–5.7
P⁴	M–D	4	4.9	0.40	4.6–5.4
	B–L	4	5.8	0.21	5.6–6.0
M¹	M–D	7	6.4	0.52	5.6–7.0
	Ant B	7	6.0	1.03	4.3–6.8
	Post B	6	5.7	0.92	3.9–6.3
M²	M–D	6	7.5	0.67	6.3–8.3
	Ant B	6	7.1	1.16	4.8–8.0
	Post B	6	6.4	0.95	4.6–7.2
M³	M–D	3	7.7	0.25	7.4–7.9
	Ant B	3	7.0	0.20	6.8–7.2
	Post B	3	5.6	0.21	5.4–5.8

Table 120. *Macaca nigra* female

Mandibular teeth		n	m	SD	Range
I₁	M–D	4	4.7	0.25	4.4–5.0
	B–L	5	5.4	0.30	5.0–5.8
I₂	M–D	5	3.2	0.95	2.0–4.3
	B–L	6	4.6	0.90	3.8–5.2
C	M–D	6	4.9	0.93	4.0–6.0
	B–L	6	4.5	1.19	2.9–5.8
P₃	M–D	2	9.9	0.71	9.4–10.4
	B–L	4	3.7	0.21	3.4–3.9
P₄	M–D	4	5.7	0.08	5.6–5.8
	B–L	4	4.7	0.21	4.5–4.9
M₁	M–D	7	6.6	0.39	5.9–7.2
	Tri B	5	4.9	1.07	3.6–5.8
	Tal B	7	4.7	0.71	3.7–5.6
M₂	M–D	6	7.4	0.64	6.2–8.1
	Tri B	5	6.3	0.97	4.6–7.0
	Tal B	6	6.0	0.62	4.8–6.5
M₃	M–D	2	9.6	0.85	9.0–10.2
	Tri B	3	6.5	0.15	6.4–6.7
	Tal B	2	5.9	0.07	5.8–5.9
	Hypoconulid	—	—	—	—

Table 121. *Papio cynocephalus* male

Maxillary teeth		n	m	SD	Range
I^1	M–D	34	10.7	0.95	9.0–13.3†
	B–L	31	9.3	0.88	6.0–10.7†
I^2	M–D	29	8.1	0.75	6.6–10.0†
	B–L	29	8.5	0.74	6.0–9.6†
C	M–D	27	14.0	2.35	10.4–18.6†
	B–L	19	11.1	1.68	7.6 13.9†
P^3	M–D	31	7.3	0.56	6.3–8.4†
	B–L	33	7.9	0.45	6.9–9.3†
P^4	M–D	34	7.7	0.43	6.7–8.4†
	B–L	34	8.7	0.42	8.0–10.0†
M^1	M–D	39	10.9	0.48	10.0–12.1†
	Ant B	39	10.1	0.56	9.1–11.6†
	Post B	38	9.5	0.46	8.3–10.5†
M^2	M–D	37	12.9	0.61	11.6–14.0†
	Ant B	38	12.2	1.80	10.1–14.3†
	Post B	36	10.8	0.65	9.0–12.0†
M^3	M–D	27	13.4	0.71	11.9–14.5†
	Ant B	27	12.6	0.83	11.0–14.5†
	Post B	24	10.8	0.67	9.3–11.8†

Table 122. *Papio cynocephalus* male

Mandibular teeth		n	m	SD	Range
I_1	M–D	37	7.9	0.65	6.0–9.0†
	B–L	33	9.0	0.62	7.7–10.2†
I_2	M–D	32	7.0	0.63	5.4–8.3†
	B–L	29	8.6	0.71	7.1–9.9†
C	M–D	17	8.6	0.97	5.3–10.1†
	B–L	22	13.7	2.67	10.7–16.3†
P_3	M–D	19	17.9	5.19	11.8–20.4†
	B–L	29	6.8	0.64	5.7–7.8†
P_4	M–D	32	8.4	0.56	7.5–9.8†
	B–L	29	7.2	0.35	6.6–7.9†
M_1	M–D	39	10.8	0.46	9.7–11.7†
	Tri B	37	8.3	0.48	7.6–9.4†
	Tal B	39	8.8	0.54	7.8–10.1†
M_2	M–D	40	12.8	0.57	11.6–13.7†
	Tri B	39	10.4	0.80	8.6–12.0†
	Tal B	39	10.1	0.77	8.7–11.9†
M_3	M–D	26	16.1	0.82	13.7–17.6†
	Tri B	26	11.3	0.84	9.7–13.0†
	Tal B	25	10.3	0.70	8.8–11.7†
	Hypoconulid	25	6.3	1.06	3.4–7.6†

Table 123. *Papio cynocephalus* female

Maxillary teeth		n	m	SD	Range
I^1	M–D	35	9.8	0.89	7.4–11.7
	B–L	32	8.4	0.61	7.1–9.6
I^2	M–D	35	7.1	0.65	6.0–8.5
	B–L	31	7.4	0.68	5.9–8.9
C	M–D	29	8.3	1.33	6.7–10.7
	B–L	29	7.3	1.04	5.1–10.1
P^3	M–D	31	6.5	0.43	5.9–8.0
	B–L	32	7.2	0.49	6.1–8.6
P^4	M–D	32	7.1	0.38	6.1–7.9
	B–L	32	8.1	0.50	7.2–8.9
M^1	M–D	38	10.3	0.33	9.4–11.1
	Ant B	36	9.3	0.37	8.6–10.0
	Post B	36	8.8	0.46	8.0–10.0
M^2	M–D	31	11.7	0.37	11.0–12.4
	Ant B	32	10.9	0.55	9.9–11.8
	Post B	29	10.1	0.70	8.5–11.9
M^3	M–D	22	11.8	0.61	10.5–13.0
	Ant B	21	11.2	1.04	9.0–13.0
	Post B	20	9.9	0.71	9.0–11.9

Table 124. *Papio cynocephalus* female

Mandibular teeth		n	m	SD	Range
I$_1$	M–D	37	7.2	0.63	4.9–8.1
	B–L	34	8.0	0.66	6.2–9.0
I$_2$	M–D	32	6.2	0.57	4.4–7.2
	B–L	31	7.3	0.63	5.5–8.4
C	M–D	28	5.9	0.79	4.6–7.9
	B–L	28	8.3	0.77	6.3–9.7
P$_3$	M–D	28	10.2	1.65	7.8–14.3
	B–L	35	5.2	0.64	4.4–7.2
P$_4$	M–D	36	7.5	0.41	6.7–8.7
	B–L	33	6.6	0.39	5.9–7.5
M$_1$	M–D	37	10.1	0.39	9.5–11.1
	Tri B	37	7.7	0.45	6.7–8.9
	Tal B	36	8.1	0.39	7.4–9.0
M$_2$	M–D	37	11.6	0.54	10.7–12.9
	Tri B	35	9.5	0.53	8.4–10.9
	Tal B	32	9.4	0.52	7.9–10.4
M$_3$	M–D	26	14.3	0.78	12.7–16.1
	Tri B	25	10.2	0.49	9.0–11.3
	Tal B	24	9.5	0.71	8.1–11.0
	Hypoconulid	23	5.5	0.69	3.7–6.6

Table 125. *Theropithecus gelada* male

Maxillary teeth		n	m	SD	Range
I^1	M–D	—	—	—	—
	B–L	16	6.7	0.5	5.9–7.9
I^2	M–D	—	—	—	—
	B–L	—	—	—	—
C	M–D	26	14.8	1.10	13.0–17.9
	B–L	26	9.0	0.60	7.7–10.6
P^3	M–D	25	6.8	0.70	6.0–8.6
	B–L	28	7.2	0.40	6.2–8.2
P^4	M–D	12	7.2	0.40	6.7–8.0
	B–L	25	8.0	0.40	7.3–8.8
M^1	M–D	—	—	—	—
	Ant B	16	9.3	0.30	8.8–10.0
	Post B	14	8.7	0.30	8.3–9.2
M^2	M–D	18	13.4	0.80	11.9–14.8
	Ant B	23	10.9	0.40	10.0–11.8
	Post B	22	10.0	0.40	9.3–11.0
M^3	M–D	37	13.7	0.70	11.6–14.9
	Ant B	37	11.3	0.70	9.3–13.0
	Post B	37	9.7	0.50	9.0–10.7

Source: G. G. Eck (unpublished data, 1975).

Table 126. *Theropithecus gelada* male

Mandibular teeth		n	m	SD	Range
I_1	M–D	—	—	—	—
	B–L	15	6.3	0.30	6.0–7.0
I_2	M–D	—	—	—	—
	B–L	17	6.0	0.50	5.3–7.1
C	M–D	27	6.4	0.50	5.4–7.8
	B–L	27	11.4	0.80	9.9–13.3
P_3	M–D	29	18.1	1.10	16.6–21.5
	B–L	30	5.1	0.20	4.7–5.6
P_4	M–D	6	8.5	0.40	8.1–9.4
	B–L	18	6.5	0.30	6.0–7.0
M_1	M–D	—	—	—	—
	Tri B	12	7.5	0.30	7.1–8.0
	Tal B	12	7.8	0.40	7.3–8.6
M_2	M–D	8	13.6	0.40	13.0–14.4
	Tri B	20	9.4	0.50	8.6–10.6
	Tal B	19	9.1	0.40	8.5–9.7
M_3	M–D	30	16.8	0.80	15.2–18.8
	Tri B	35	9.9	0.70	11.5–9.1
	Tal B	35	9.1	0.40	8.4–9.9
	Hypoconulid	—	—	—	—

Source: G. G. Eck (unpublished data, 1975).

Table 127. *Theropithecus gelada* female

Maxillary teeth		n	m	SD	Range
I¹	M–D	—	—	—	—
	B–L	16	6.0	0.40	5.3–6.5
I²	M–D	—	—		
	B–L	—	—	—	—
C	M–D	18	7.4	0.40	6.4–8.1
	B–L	18	5.4	0.30	4.9–5.8
P³	M–D	11	6.2	0.40	5.5–6.8
	B–L	24	6.7	0.40	5.7–7.2
P⁴	M–D	8	7.0	0.30	6.3–7.4
	B–L	22	7.6	0.40	6.9–8.3
M¹	M–D	3	10.7	0.40	10.1–11.0
	Ant B	15	8.7	0.20	8.3–9.0
	Post B	15	8.0	0.20	7.7–8.4
M²	M–D	9	12.9	0.70	12.1–14.3
	Ant B	20	10.5	0.40	9.7–11.2
	Post B	19	9.3	0.40	8.7–10.2
M³	M–D	18	12.8	0.50	11.9–13.5
	Ant B	20	10.7	0.60	9.5–11.8
	Post B	20	9.1	0.50	8.0–10.2

Source: G. G. Eck (unpublished data, 1975).

Table 128. *Theropithecus gelada* female

Mandibular teeth		n	m	SD	Range
I₁	M–D	—	—	—	—
	B–L	15	5.6	0.30	5.0–6.1
I₂	M–D	—	—	—	—
	B–L	17	5.3	0.30	4.7–5.8
C	M–D	21	3.4	0.40	2.1–3.9
	B–L	18	6.9	0.40	6.1–7.9
P₃	M–D	17	10.4	1.00	8.8–12.1
	B–L	19	4.5	0.40	4.1–5.7
P₄	M–D	8	8.0	0.30	7.4–8.5
	B–L	18	6.2	0.20	5.7–6.6
M₁	M–D	—	—	—	—
	Tri B	11	7.2	0.20	6.8–7.5
	Tal B	11	7.5	0.30	7.1–8.1
M₂	M–D	6	13.1	0.70	11.9–13.8
	Tri B	18	8.9	0.40	8.3–9.5
	Tal B	18	8.7	0.40	7.9–93
M₃	M–D	14	16.2	0.70	15.2–17.8
	Tri B	20	9.5	0.30	8.9–10.0
	Tal B	20	8.7	0.30	8.1–9.3
	Hypoconulid	—	—	—	—

Source: G. G. Eck (unpublished data, 1975).

Table 129. *Colobus polykomos* male

Maxillary teeth		n	m	SD	Range
I^1	M–D	43	5.1	0.47	4.0–6.1
	B–L	44	4.7	0.51	3.9–6.0
I^2	M–D	42	4.6	0.42	3.7–5.6*
	B–L	42	4.4	0.73	3.6–5.6
C	M–D	40	9.8	0.51	4.7–10.1†
	B–L	38	6.5	0.93	4.2–8.6†
P^3	M–D	49	5.3	0.56	4.4–8.0*
	B–L	48	5.6	0.45	4.6–6.4†
P^4	M–D	47	5.2	0.37	4.3–6.1
	B–L	48	6.7	0.55	5.8–8.0†
M^1	M–D	49	7.0	0.42	6.0–8.5*
	Ant B	49	6.5	0.46	5.7–8.0†
	Post B	45	6.3	0.42	5.6–7.7*
M^2	M–D	48	7.5	0.70	4.0–8.6*
	Ant B	48	7.4	0.49	6.4–8.5†
	Post B	47	6.9	0.58	5.5–8.1†
M^3	M–D	43	7.6	0.46	6.8–8.7
	Ant B	44	6.9	0.57	5.0–8.0†
	Post B	44	6.4	0.71	4.7–7.5†

Table 130. *Colobus polykomos* male

Mandibular teeth		n	m	SD	Range
I_1	M–D	38	3.7	0.40	3.0–5.4
	B–L	38	4.4	0.37	3.6–5.3
I_2	M–D	37	3.8	0.63	3.0–6.7
	B–L	38	4.8	0.44	3.7–6.1*
C	M–D	37	6.9	1.22	4.7–9.1
	B–L	36	6.9	1.39	4.4–9.4†
P_3	M–D	44	9.1	1.50	4.2–10.0
	B–L	45	5.2	0.80	3.7–6.9*
P_4	M–D	48	6.1	0.43	5.1–7.1
	B–L	46	4.9	0.49	3.9–6.1†
M_1	M–D	47	7.2	0.42	6.3–8.3†
	Tri B	45	5.5	0.40	4.5–6.2†
	Tal B	46	5.7	0.43	4.6–6.4†
M_2	M–D	47	7.7	0.65	5.0–8.8*
	Tri B	45	6.3	0.52	5.0–7.4†
	Tal B	47	6.5	0.51	5.3–7.4†
M_3	M–D	43	9.6	0.67	8.1–11.1
	Tri B	44	6.3	0.52	5.0–7.3†
	Tal B	43	6.2	0.48	5.0–7.1†
	Hypoconulid	40	4.2	0.53	3.1–5.6

Table 131. *Colobus polykomos* female

Maxillary teeth		n	m	SD	Range
I¹	M–D	28	4.9	0.34	4.1–5.7
	B–L	29	4.5	0.43	3.5–5.3
I²	M–D	26	4.4	0.35	3.9–5.3
	B–L	26	4.3	0.39	3.6–5.0
C	M–D	23	7.0	0.62	5.0–8.0
	B–L	23	5.5	0.55	4.6–7.0
P³	M–D	29	5.1	0.30	4.6–5.9
	B–L	29	5.4	0.46	4.2–6.6
P⁴	M–D	28	5.1	0.27	4.5–5.5
	B–L	29	6.2	0.54	5.0–7.5
M¹	M–D	30	6.8	0.37	6.1–7.5
	Ant B	29	6.2	0.42	5.5–6.9
	Post B	29	6.1	0.41	5.3–6.8
M²	M–D	29	7.2	0.57	5.0–8.0
	Ant B	28	6.9	0.47	6.0–8.0
	Post B	28	6.5	0.68	4.1–7.7
M³	M–D	27	7.3	0.78	4.9–9.0
	Ant B	27	6.5	0.39	5.8–7.3
	Post B	27	5.9	0.57	4.7–7.0

Table 132. *Colobus polykomos* female

Mandibular teeth		n	m	SD	Range
I_1	M–D	27	3.7	0.43	3.1–5.2
	B–L	26	4.2	0.30	3.8–4.8
I_2	M–D	23	3.7	0.33	3.2–4.5
	B–L	24	4.7	0.36	4.1–5.6
C	M–D	26	5.6	1.11	3.9–8.5
	B–L	27	5.3	0.53	3.9–7.1
P_3	M–D	28	7.1	1.01	4.7–8.8
	B–L	29	4.9	0.70	3.6–6.5
P_4	M–D	30	5.9	0.37	5.0–6.7
	B–L	30	4.7	0.38	3.9–5.4
M_1	M–D	30	7.0	0.36	6.4–7.7
	Tri B	31	5.2	0.40	4.4–5.8
	Tal B	31	5.5	0.39	4.7–6.2
M_2	M–D	30	7.4	0.40	6.6–8.1
	Tri B	30	5.9	0.48	5.0–6.9
	Tal B	30	6.2	0.50	5.2–7.6
M_3	M–D	26	9.4	0.67	8.2–10.5
	Tri B	24	6.0	0.44	5.3–7.0
	Tal B	25	5.9	0.34	5.3–6.8
	Hypoconulid	25	4.1	0.65	3.1–5.6

Table 133. *Piliocolobus badius* male

Maxillary teeth		n	m	SD	Range
I^1	M–D	23	5.5	0.42	4.8–6.3
	B–L	22	4.9	0.28	4.2–5.4
I^2	M–D	18	4.5	0.39	3.7–5.1
	B–L	19	4.7	0.32	4.1–5.4
C	M–D	17	9.8	0.94	8.3–11.3†
	B–L	14	7.2	0.60	6.2–8.0†
P^3	M–D	24	5.3	0.44	4.3–6.0†
	B–L	22	5.3	0.39	4.6–6.3
P^4	M–D	22	5.0	0.41	4.2–5.8
	B–L	22	5.8	0.46	4.9–6.6
M^1	M–D	26	7.0	0.39	6.3–7.9
	Ant B	24	6.0	0.36	5.4–6.7
	Post B	23	5.8	0.36	5.1–6.6
M^2	M–D	24	7.3	0.41	6.5–8.1
	Ant B	22	6.5	0.43	5.7–7.5
	Post B	22	6.1	0.40	5.2–7.1
M^3	M–D	22	7.0	0.40	6.5–7.8
	Ant B	20	6.1	0.43	5.5–7.2
	Post B	17	5.7	0.43	4.9–6.6

Table 134. *Piliocolobus badius* male

Mandibular teeth		n	m	SD	Range
I_1	M–D	21	3.9	0.27	3.4–4.5
	B–L	22	4.7	0.33	4.1–5.2
I_2	M–D	21	3.7	0.20	3.4–4.1
	B–L	19	5.1	0.39	4.3–5.7
C	M–D	9	6.9	1.41	5.6–9.3†
	B–L	16	8.2	0.91	6.1–9.3†
P_3	M–D	20	9.2	1.46	6.2–11.3
	B–L	24	5.2	0.77	3.6–6.6†
P_4	M–D	19	5.7	0.39	5.0–6.4
	B–L	24	4.6	0.46	3.7–5.5
M_1	M–D	23	7.2	0.39	6.5–7.9
	Tri B	19	5.0	0.35	4.5–5.7
	Tal B	20	5.2	0.38	4.7–6.0
M_2	M–D	24	7.4	0.47	6.1–8.2
	Tri B	25	5.6	0.46	4.7–6.4
	Tal B	26	5.8	0.49	5.0–6.8
M_3	M–D	22	9.1	0.51	7.9–10.0
	Tri B	23	5.7	0.39	5.0–6.4
	Tal B	21	5.7	0.46	4.9–6.9
	Hypoconulid	19	3.8	0.49	3.1–4.9

Table 135. *Piliocolobus badius* female

Maxillary teeth		n	m	SD	Range
I^1	M–D	26	5.5	0.50	4.6–6.4
	B–L	20	4.8	0.31	4.3–5.4
I^2	M–D	25	4.4	0.41	3.8 5.3
	B–L	25	4.5	0.34	3.8–5.0
C	M–D	17	6.9	1.11	5.8–10.8
	B–L	19	5.5	0.60	4.6–6.8
P^3	M–D	23	4.8	0.34	4.4–5.6
	B–L	23	5.3	0.50	4.4–6.2
P^4	M–D	21	5.0	0.25	4.6–5.5
	B–L	25	5.8	0.40	5.2–6.6
M^1	M–D	26	6.9	0.40	6.3–8.1
	Ant B	25	6.0	0.33	5.5–6.8
	Post B	26	5.8	0.33	5.3–6.4
M^2	M–D	27	7.2	0.42	6.4–8.6
	Ant B	26	6.6	0.39	5.9–7.3
	Post B	23	6.2	0.42	5.5–7.3
M^3	M–D	25	7.0	0.46	6.0–8.2
	Ant B	19	6.6	0.57	5.9–7.7
	Post B	20	5.9	0.46	5.2–6.9

Table 136. *Piliocolobus badius* female

Mandibular teeth		n	m	SD	Range
I_1	M–D	21	4.0	0.32	3.3–4.4
	B–L	20	4.7	0.39	3.9–5.4
I_2	M–D	21	3.8	0.23	3.4–4.2
	B–L	23	5.1	0.45	4.3–6.1
C	M–D	16	4.4	0.58	3.7–6.4
	B–L	21	6.2	0.56	5.0–7.5
P_3	M–D	20	8.5	0.99	6.1–10.5
	B–L	22	4.2	0.52	3.6–5.4
P_4	M–D	20	5.6	0.35	4.8–6.2
	B–L	24	4.5	0.46	4.0–5.8
M_1	M–D	26	7.1	0.43	6.4–8.2
	Tri B	21	5.0	0.32	4.6–5.8
	Tal B	22	5.2	0.30	4.7–5.8
M_2	M–D	23	7.3	0.43	6.6–8.5
	Tri B	22	5.6	0.32	5.1–6.2
	Tal B	23	5.9	0.41	5.1–6.8
M_3	M–D	20	8.9	0.61	7.8–10.1
	Tri B	19	5.8	0.37	5.1–6.6
	Tal B	17	5.7	0.39	5.0–6.5
	Hypoconulid	17	3.7	0.33	3.1–4.4

Table 137. *Nasalis larvatus* male

Maxillary teeth		n	m	SD	Range
I^1	M–D	21	5.7	0.72	4.4–8.0
	B–L	20	4.9	0.33	4.3–6.0†
I^2	M–D	20	4.3	0.31	3.8–5.0
	B–L	20	4.9	0.47	4.5–6.6†
C	M–D	18	8.2	1.18	4.1–9.4†
	B–L	18	6.4	0.50	5.4–7.1†
P^3	M–D	21	5.2	0.35	4.5–6.0†
	B–L	21	5.9	0.54	5.0–7.8†
P^4	M–D	21	5.3	0.52	4.6–6.9
	B–L	21	6.1	0.31	5.5–7.0†
M^1	M–D	20	7.0	0.32	6.0–7.4
	Ant B	21	6.3	0.18	5.9–6.7†
	Post B	21	6.3	0.27	5.4–6.6*
M^2	M–D	21	7.9	0.34	7.4–8.9*
	Ant B	21	7.3	0.46	5.7–8.0†
	Post B	21	6.9	0.45	5.9–7.9*
M^3	M–D	20	7.8	0.46	6.9–8.5
	Ant B	20	7.1	0.42	5.6–7.7*
	Post B	20	6.2	0.57	4.5–7.0

Table 138. *Nasalis larvatus* male

Mandibular teeth		n	m	SD	Range
I_1	M–D	19	4.1	0.58	3.2–4.5
	B–L	19	4.9	0.78	4.0–7.9†
I_2	M–D	21	4.1	0.56	3.4–5.7
	B–L	21	5.0	0.29	4.6–5.7†
C	M–D	19	5.6	0.31	5.1–6.1*
	B–L	20	6.9	0.75	4.5–7.9†
P_3	M–D	21	10.0	0.95	7.7–11.5†
	B–L	20	5.4	0.60	4.2–6.3†
P_4	M–D	21	5.3	0.39	4.6–6.2
	B–L	21	4.8	0.43	4.1–5.6
M_1	M–D	20	7.3	0.34	6.7–7.9†
	Tri B	21	5.3	0.39	4.9–6.5*
	Tal B	21	5.7	0.40	5.1–6.6†
M_2	M–D	21	8.0	0.51	6.6–9.1*
	Tri B	20	6.3	0.36	5.5–7.0*
	Tal B	20	6.5	0.25	6.1–7.0
M_3	M–D	18	10.1	0.49	9.1–11.4†
	Tri B	18	6.6	0.33	6.0–7.3
	Tal B	18	6.4	0.31	5.7–7.0
	Hypoconulid	18	4.7	0.38	4.0–5.0

Table 139. *Nasalis larvatus* female

Maxillary teeth		n	m	SD	Range
I^1	M–D	13	5.3	0.41	4.9–6.3
	B–L	13	4.3	0.35	3.4–4.7
I^2	M–D	13	4.2	0.45	3.6–5.0
	B–L	13	4.2	0.28	3.6–4.6
C	M–D	13	6.1	0.28	5.5–6.5
	B–L	13	5.0	0.35	4.5–5.9
P^3	M–D	13	4.8	0.27	4.2–5.0
	B–L	13	5.4	0.29	4.9–5.8
P^4	M–D	13	4.9	0.33	4.4–5.4
	B–L	13	5.8	0.28	5.1–6.0
M^1	M–D	14	6.8	0.37	6.2–7.6
	Ant B	14	6.0	0.31	5.5–6.6
	Post B	14	6.1	0.24	5.6–6.5
M^2	M–D	13	7.5	0.38	6.7–8.0
	Ant B	13	6.8	0.34	6.0–7.3
	Post B	13	6.5	0.40	5.9–7.0
M^3	M–D	12	7.5	0.37	6.8–8.0
	Ant B	12	6.7	0.32	6.0–7.0
	Post B	12	5.9	0.27	5.4–6.4

Table 140. *Nasalis larvatus* female

Mandibular teeth		n	m	SD	Range
I_1	M–D	14	3.7	0.27	3.0–4.1
	B–L	14	4.2	0.27	3.7–4.6
I_2	M–D	14	3.8	0.22	3.2–4.0
	B–L	14	4.4	0.27	4.0–4.9
C	M–D	13	5.1	0.80	3.4–5.9
	B–L	13	5.6	0.33	5.0–6.0
P_3	M–D	12	5.9	0.35	5.3–6.4
	B–L	12	4.8	0.39	4.1–5.6
P_4	M–D	13	5.1	0.26	4.6–5.6
	B–L	13	4.7	0.27	4.2–5.0
M_1	M–D	13	6.9	0.36	6.3–7.7
	Tri B	13	5.0	0.21	4.6–5.4
	Tal B	13	5.3	0.28	5.0–5.9
M_2	M–D	13	7.7	0.37	7.0–8.1
	Tri B	13	6.1	0.22	5.7–6.6
	Tal B	13	6.3	0.25	5.8–6.7
M_3	M–D	12	9.6	0.40	9.1–10.0
	Tri B	12	6.5	0.35	5.9–7.0
	Tal B	12	6.2	0.34	5.5–6.8
	Hypoconulid	12	4.6	0.33	4.1–5.0

Table 141. *Simias concolor* female

Maxillary teeth		n	m	SD	Range
I^1	M–D	5	4.6	0.34	4.2–5.1
	B–L	5	4.1	0.23	3.9–4.5
I^2	M–D	5	4.0	0.21	3.8–4.2
	B–L	5	4.1	0.19	3.8–4.3
C	M–D	4	5.2	0.15	5.1–5.4
	B–L	4	4.6	0.17	4.4–4.8
P^3	M–D	5	4.5	0.25	4.2–4.8
	B–L	5	5.3	0.23	4.9–5.5
P^4	M–D	5	4.3	0.18	4.1–4.5
	B–L	5	5.6	0.18	5.3–5.8
M^1	M–D	7	6.5	0.20	6.1–6.7
	Ant B	7	5.5	0.16	5.2–5.7
	Post B	7	5.6	0.20	5.2–5.8
M^2	M–D	5	7.0	0.15	6.8–7.2
	Ant B	5	6.0	0.24	5.6–6.2
	Post B	5	5.8	0.15	5.6–6.0
M^3	M–D	3	6.9	0.40	6.5–7.3
	Ant B	3	5.7	0.25	5.5–6.0
	Post B	3	5.0	0.31	4.7–5.3

Note: no males.

Table 142. *Simias concolor* female

Mandibular teeth		n	m	SD	Range
I_1	M–D	6	3.5	0.19	3.3–3.8
	B–L	6	3.8	0.12	3.6–3.9
I_2	M–D	5	3.4	0.19	3.2–3.6
	B–L	5	4.1	0.13	4.0–4.3
C	M–D	4	4.8	0.42	4.3–5.2
	B–L	4	4.3	0.46	3.8–4.9
P_3	M–D	4	6.4	0.29	6.0–6.7
	B–L	5	4.2	0.35	3.7–4.6
P_4	M–D	5	4.7	0.20	4.4–4.9
	B–L	5	4.4	0.22	4.1–4.6
M_1	M–D	7	6.5	0.24	6.0–6.8
	Tri B	7	4.6	0.16	4.3–4.8
	Tal B	7	4.8	0.13	4.6–5.0
M_2	M–D	5	7.0	0.27	6.7–7.4
	Tri B	5	5.5	0.23	5.2–5.7
	Tal B	5	5.4	0.27	5.1.5.8
M_3	M–D	3	8.7	0.62	8.0–9.2
	Tri B	4	5.4	0.34	5.2–5.9
	Tal B	4	5.4	0.13	5.2–5.5
	Hypoconulid	3	3.7	0.17	3.5–3.8

Note: no males.

Table 143. *Pygathrix nemaeus* male

Maxillary teeth		n	m	SD	Range
I^1	M–D	8	5.3	0.36	4.8–5.9
	B–L	7	5.4	0.19	5.2–5.6
I^2	M–D	7	4.6	0.52	4.0–5.2
	B–L	7	5.2	0.32	4.8–5.7
C	M–D	4	6.8	1.18	5.8–8.4
	B–L	4	6.1	0.48	5.5–6.5
P^3	M–D	6	4.9	0.40	4.4–5.5
	B–L	6	6.3	0.34	5.9–6.7
P^4	M–D	6	4.7	0.13	4.6–4.9
	B–L	5	6.6	0.25	6.3–6.9
M^1	M–D	7	6.9	0.36	6.4–7.4
	Ant B	7	6.6	0.21	6.3–6.8
	Post B	6	6.2	0.58	5.1–6.7
M^2	M–D	7	7.4	0.39	6.9–8.1
	Ant B	6	7.2	0.40	6.6–7.7
	Post B	6	6.9	0.49	6.2–7.5
M^3	M–D	6	7.5	0.53	6.5–8.1
	Ant B	5	6.7	0.25	6.5–7.1
	Post B	6	6.0	0.38	5.7–6.7

Table 144. *Pygathrix nemaeus* male

Mandibular teeth		n	m	SD	Range
I_1	M–D	6	3.9	0.18	3.6–4.0
	B–L	6	4.9	0.25	4.6–5.2
I_2	M–D	7	3.9	0.27	3.5–4.3
	B–L	7	5.2	0.32	4.8–5.7
C	M–D	6	6.2	0.70	5.5–7.5
	B–L	6	5.3	0.34	4.7–5.6
P_3	M–D	6	8.4	0.64	7.2–9.0
	B–L	6	5.6	0.37	5.2–6.1
P_4	M–D	6	5.3	0.16	5.0–5.4
	B–L	5	5.1	0.41	4.7–5.6
M_1	M–D	8	7.0	0.29	6.6–7.4
	Tri B	7	5.4	0.22	5.1–5.6
	Tal B	6	5.3	0.26	5.0–5.6
M_2	M–D	7	7.6	0.56	7.0–8.8
	Tri B	6	6.1	0.31	5.6–6.5
	Tal B	5	6.2	0.40	5.7–6.6
M_3	M–D	6	9.2	0.30	8.8–9.5
	Tri B	4	6.2	0.21	6.0–6.5
	Tal B	5	6.0	0.17	5.7–6.1
	Hypoconulid	5	4.4	0.48	3.8–4.9

Table 145. *Pygathrix nemaeus* female

Maxillary teeth		n	m	SD	Range
I^1	M–D	4	4.8	0.40	4.2–5.1
	B–L	4	4.7	0.33	4.3–5.1
I^2	M–D	4	4.2	0.42	3.7–4.7
	B–L	4	4.9	0.25	4.6–5.2
C	M–D	3	5.4	0.53	5.0–6.0
	B–L	3	5.1	0.50	4.6–5.6
P^3	M–D	3	4.8	0.38	4.5–5.2
	B–L	4	5.7	0.10	5.5–5.7
P^4	M–D	4	4.5	0.29	4.3–4.9
	B–L	4	6.2	0.24	5.9–6.5
M^1	M–D	5	6.6	0.35	6.1–7.0
	Ant B	5	6.3	0.22	6.0–6.5
	Post B	5	6.2	0.36	5.7–6.5
M^2	M–D	4	6.9	0.30	6.5–7.2
	Ant B	3	6.9	0.38	6.5–7.2
	Post B	—	—	—	—
M^3	M–D	4	6.7	0.26	6.4–7.0
	Ant B	4	6.5	0.36	6.2–7.0
	Post B	3	6.6	0.35	6.3–7.0

Table 146. *Pygathrix nemaeus* female

Mandibular teeth		n	m	SD	Range
I_1	M–D	5	3.8	0.51	3.0–4.4
	B–L	4	4.4	0.22	4.1–4.6
I_2	M–D	4	3.6	0.26	3.4–4.0
	B–L	4	4.8	0.31	4.3–5.0
C	M–D	4	6.0	1.06	4.7–7.3
	B–L	4	4.5	0.41	3.9–4.8
P_3	M–D	2	7.1	0.64	6.6–7.5
	B–L	4	5.0	0.13	4.8–5.1
P_4	M–D	4	5.1	0.34	4.7–5.5
	B–L	4	4.8	0.22	4.5–5.0
M_1	M–D	5	6.8	0.19	6.5–7.0
	Tri B	5	5.2	0.28	5.0–5.7
	Tal B	5	5.3	0.22	5.1–5.6
M_2	M–D	4	7.1	0.29	6.8–7.5
	Tri B	4	5.9	0.13	5.7–6.0
	Tal B	4	6.0	0.25	5.8–6.3
M_3	M–D	4	8.5	0.47	8.2–9.2
	Tri B	4	5.9	0.13	5.7–6.0
	Tal B	3	5.7	0.25	5.5–6.0
	Hypoconulid	3	4.2	0.50	3.7–4.7

Table 147. *Rhinopithecus roxellana* female

Maxillary teeth		n	m	SD	Range
I¹	M–D	12	6.1	0.61	4.5–6.8
	B–L	11	5.3	0.38	4.8–6.0
I²	M–D	12	4.9	0.27	4.5 5.3
	B–L	12	4.9	0.46	4.1–6.1
C	M–D	11	6.7	1.03	5.6–9.6
	B–L	11	5.4	0.59	4.4–6.6
P³	M–D	6	4.7	0.41	4.1–5.3
	B–L	10	6.0	0.59	5.1–6.9
P⁴	M–D	12	5.4	0.51	4.3–6.1
	B–L	13	6.9	0.43	5.7–7.4
M¹	M–D	19	8.0	0.65	6.2–9.2
	Ant B	19	7.6	0.54	6.0–8.2
	Post B	19	7.5	0.59	5.7–8.3
M²	M–D	13	8.7	0.95	6.1–10.1
	Ant B	12	9.0	0.27	8.5–9.5
	Post B	10	8.0	0.37	7.6–8.6
M³	M–D	12	8.3	1.08	5.4–9.4
	Ant B	11	8.6	0.27	8.3–9.1
	Post B	9	7.2	0.18	6.9–7.5

Note: no males.

Table 148. *Rhinopithecus roxellana* female

Mandibular teeth		n	m	SD	Range
I₁	M–D	10	4.5	0.38	3.8–5.1
	B–L	11	5.0	0.41	4.3–5.5
I₂	M–D	10	4.2	0.34	3.7–4.8
	B–L	11	5.2	0.37	4.5–5.7
C	M–D	11	4.6	0.53	3.9–5.2
	B–L	10	5.7	0.75	4.7–7.3
P₃	M–D	11	8.3	0.79	6.7–10.0
	B–L	12	4.4	0.48	3.8–5.8
P₄	M–D	11	5.4	0.49	4.6–6.2
	B–L	11	5.2	0.29	4.8–5.8
M₁	M–D	18	7.9	0.72	6.3–9.1
	Tri B	18	6.1	0.38	5.3–6.8
	Tal B	18	6.5	0.42	5.5–7.2
M₂	M–D	12	8.7	0.74	6.5–9.5
	Tri B	11	7.2	0.55	5.6–7.8
	Tal B	11	7.3	0.64	5.6–8.0
M₃	M–D	11	10.4	1.38	6.8–11.8
	Tri B	10	7.2	0.78	5.2–8.1
	Tal B	10	7.0	0.53	5.6–7.6
	Hypoconulid	10	4.6	1.15	2.0–5.8

Note: no males.

Table 149. *Trachypithecus phayrei* male

Maxillary teeth		n	m	SD	Range
I^1	M–D	4	5.1	0.21	4.9–5.3
	B–L	3	4.8	0.29	4.5–5.0
I^2	M–D	2	4.6	0.21	4.4–4.7
	B–L	3	4.5	0.50	4.0–5.0
C	M–D	1	6.6	0.00	6.6–6.6
	B–L	1	5.1	0.00	5.1–5.1
P^3	M–D	4	4.5	0.17	4.3–4.6
	B–L	3	5.2	0.29	5.0–5.5
P^4	M–D	4	4.2	0.13	4.0–4.3
	B–L	4	6.1	0.13	5.9–6.2
M^1	M–D	4	6.0	0.26	5.7–6.3
	Ant B	4	6.3	0.15	6.2–6.5
	Post B	3	5.9	0.06	5.9–6.0
M^2	M–D	4	6.4	0.36	6.1–6.9
	Ant B	4	6.7	0.25	6.4–6.9
	Post B	4	6.4	0.31	6.0–6.7
M^3	M–D	2	6.1	0.07	6.0–6.1
	Ant B	2	6.4	0.64	5.9–6.8
	Post B	2	6.0	0.14	5.9–6.1

Table 150. *Trachypithecus phayrei* male

Mandibular teeth		n	m	SD	Range
I_1	M–D	3	3.5	0.10	3.4–3.6
	B–L	3	4.5	0.31	4.2–4.8
I_2	M–D	4	3.5	0.24	3.2–3.7
	B–L	4	4.5	0.27	4.1–4.7
C	M–D	3	8.7	0.96	7.6–9.4
	B–L	1	4.6	0.00	4.6–4.6
P_3	M–D	3	4.6	0.25	4.4–4.9
	B–L	3	4.9	0.80	4.1–5.7
P_4	M–D	3	8.7	0.96	7.6–9.4
	B–L	3	4.4	0.12	4.3–4.5
M_1	M–D	4	6.4	0.19	6.2–6.6
	Tri B	4	4.9	0.25	4.6–5.2
	Tal B	4	5.1	0.39	4.7–5.5
M_2	M–D	4	6.8	0.27	6.4–7.0
	Tri B	3	5.3	0.72	4.5–5.9
	Tal B	4	5.4	0.71	4.4–6.0
M_3	M–D	1	8.0	0.00	8.0–8.0
	Tri B	2	5.5	0.07	5.4–5.5
	Tal B	2	5.0	0.21	4.8–5.1
	Hypoconulid	1	3.6	0.00	3.6–3.6

Table 151. *Trachypithecus phayrei* female

Maxillary teeth		n	m	SD	Range
I^1	M–D	3	4.9	0.27	4.6–5.1
	B–L	1	4.7	0.00	4.7–4.7
I^2	M–D	4	4.1	0.50	3.4–4.6
	B–L	4	4.6	0.48	4.1–5.1
C	M–D	3	6.9	0.64	6.4–7.6
	B–L	3	5.6	0.00	5.6–5.6
P^3	M–D	4	4.3	0.10	4.2–4.4
	B–L	4	5.3	0.34	4.9–5.6
P^4	M–D	4	4.2	0.31	3.8–4.5
	B–L	3	6.1	0.21	5.9–6.3
M^1	M–D	5	5.9	0.46	5.3–6.5
	Ant B	5	6.2	0.18	6.0–6.4
	Post B	4	5.9	0.25	5.6–6.2
M^2	M–D	4	6.2	0.42	5.7–6.7
	Ant B	4	6.6	0.31	6.3–7.0
	Post B	4	6.2	0.45	5.8–6.7
M^3	M–D	4	6.1	0.33	5.7–6.5
	Ant B	4	6.2	0.54	5.6–6.9
	Post B	3	5.3	0.95	4.3–6.2

Table 152. *Trachypithecus phayrei* female

Mandibular teeth		n	m	SD	Range
I_1	M–D	3	3.7	0.62	3.2–4.4
	B–L	4	4.3	0.13	4.2–4.5
I_2	M–D	4	3.3	0.18	3.1–3.5
	B–L	4	4.5	0.42	4.2–5.1
C	M–D	4	6.3	0.45	5.9–6.9
	B–L	4	4.8	1.05	3.7–6.0
P_3	M–D	4	7.8	1.40	7.0–9.9
	B–L	4	4.5	0.25	4.2–4.8
P_4	M–D	4	4.6	0.21	4.3–4.8
	B–L	4	4.4	0.27	4.0–4.6
M_1	M–D	5	6.1	0.39	5.8–6.8
	Tri B	4	4.9	0.10	4.8–5.0
	Tal B	5	5.0	0.24	4.7–5.3
M_2	M–D	4	6.4	0.13	6.3–6.6
	Tri B	2	5.4	0.42	5.1–5.7
	Tal B	2	5.6	0.57	5.2–6.0
M_3	M–D	4	7.9	0.57	7.4–8.7
	Tri B	2	5.7	0.00	5.7–5.7
	Tal B	3	5.5	0.45	5.0–5.9
	Hypoconulid	3	3.7	0.36	3.3–4.0

Table 153. *Kasi johnii* male

Maxillary teeth		n	m	SD	Range
I^1	M–D	7	4.8	0.41	4.4–5.4
	B–L	3	4.3	0.17	4.5–4.5
I^2	M–D	5	4.2	0.36	3.7–4.7
	B–L	5	4.3	0.38	3.8–4.8
C	M–D	4	8.3	0.35	8.0–8.8
	B–L	4	6.8	0.29	6.5–7.2
P^3	M–D	6	4.7	0.28	4.2–4.9
	B–L	5	5.1	0.76	4.2–6.0
P^4	M–D	6	4.6	0.27	4.2–4.9
	B–L	4	5.6	0.42	5.2–6.2
M^1	M–D	8	6.4	0.49	5.7–7.1
	Ant B	6	6.1	0.31	5.8–6.5
	Post B	6	5.8	0.24	5.5–6.2
M^2	M–D	5	6.7	0.40	6.3–7.3
	Ant B	5	6.5	0.10	4.9–7.4
	Post B	5	6.1	0.68	5.0–6.7
M^3	M–D	5	6.7	0.41	5.9–7.0
	Ant B	5	6.7	0.41	6.0–7.0
	Post B	5	5.8	0.75	4.6–6.6

Table 154. *Kasi johnii* male

Mandibular teeth		n	m	SD	Range
I$_1$	M–D	5	3.1	0.27	3.0–3.6
	B–L	4	4.1	0.12	4.0–4.2
I$_2$	M–D	4	3.6	0.13	3.4–3.7
	B–L	4	4.2	0.19	4.0–4.4
C	M–D	4	6.9	0.79	5.7–7.5
	B–L	4	4.9	0.57	4.0–5.2
P$_3$	M–D	5	8.9	1.45	6.5–10.2
	B–L	4	4.8	0.70	4.0–5.5
P$_4$	M–D	5	5.1	0.40	4.8–5.8
	B–L	4	4.5	0.42	4.0–5.0
M$_1$	M–D	7	6.6	0.49	5.7–7.1
	Tri B	4	5.0	0.24	4.7–5.2
	Tal B	4	5.1	0.26	4.8–5.4
M$_2$	M–D	5	6.8	0.27	6.6–7.2
	Tri B	3	5.7	0.27	5.4–5.9
	Tal B	4	5.8	0.38	5.5–6.3
M$_3$	M–D	5	8.4	1.19	6.4–9.6
	Tri B	3	5.9	0.20	5.7–6.1
	Tal B	3	5.6	0.35	5.3–6.0
	Hypoconulid	3	4.4	0.49	3.8–4.7

Table 155. *Kasi johnii* female

Maxillary teeth		n	m	SD	Range
I¹	M–D	3	4.6	0.59	3.9–5.0
	B–L	2	4.5	0.00	4.5–4.5
I²	M–D	1	3.7	0.00	3.7–3.7
	B–L	1	4.3	0.00	4.3–4.3
C	M–D	2	6.4	0.42	6.1–6.7
	B–L	2	5.3	0.42	5.0–5.6
P³	M–D	2	4.4	0.57	4.0–4.8
	B–L	1	6.0	0.00	6.0–6.0
P⁴	M–D	2	4.6	0.50	4.2–4.9
	B–L	1	6.8	0.00	6.8–6.8
M¹	M–D	3	6.1	0.59	5.7–6.8
	Ant B	2	6.4	0.00	6.4–6.4
	Post B	2	6.3	0.07	6.2–6.3
M²	M–D	2	7.0	0.28	6.8–7.2
	Ant B	1	7.6	0.00	7.6–7.6
	Post B	2	7.2	0.50	6.8–7.5
M³	M–D	2	6.6	0.85	6.0–7.2
	Ant B	2	7.0	0.99	6.3–7.7
	Post B	2	6.7	1.34	5.7–7.6

Table 156. *Kasi johnii* female

Mandibular teeth		n	m	SD	Range
I₁	M–D	2	3.0	0.00	3.0–3.0
	B–L	3	4.1	0.25	3.9–4.4
I₂	M–D	2	3.4	0.07	3.3–3.4
	B–L	2	4.4	0.28	4.2–4.6
C	M–D	2	6.1	0.35	5.8–6.3
	B–L	2	4.5	0.21	4.3–4.6
P₃	M–D	2	7.1	0.57	6.7–7.5
	B–L	2	4.6	0.64	4.1–5.0
P₄	M–D	2	4.9	0.42	4.6–5.2
	B–L	2	4.9	0.50	4.5–5.2
M₁	M–D	3	6.6	0.38	6.2–6.9
	Tri B	2	5.3	0.35	5.0–5.5
	Tal B	3	5.5	0.42	5.2–6.0
M₂	M–D	2	7.3	0.57	6.9–7.7
	Tri B	—	—	—	—
	Tal B	1	6.1	0.00	6.1–6.1
M₃	M–D	2	8.8	0.28	8.6–9.0
	Tri B	1	6.7	0.00	6.7–6.7
	Tal B	1	6.5	0.00	6.5–6.5
	Hypoconulid	1	4.8	0.00	4.8–4.8

Table 157. *Presbytis comata* male

Maxillary teeth		*n*	*m*	SD	Range
I^1	M–D	14	4.0	0.30	3.6–4.8
	B–L	14	4.5	0.39	3.9–5.2
I^2	M–D	15	4.2	0.30	3.8–4.6*
	B–L	15	4.5	0.35	3.9–5.0
C	M–D	11	6.1	0.45	5.1–6.5†
	B–L	13	4.7	0.33	4.2–5.4*
P^3	M–D	17	4.0	0.36	3.4–4.4
	B–L	17	4.8	0.37	4.2–5.5
P^4	M–D	17	3.9	0.21	3.6–4.3
	B–L	15	5.5	0.35	4.5–6.0
M^1	M–D	20	5.3	0.22	4.8–5.7
	Ant B	16	5.6	0.28	5.0–6.0*
	Post B	18	5.4	0.26	4.9–5.9*
M^2	M–D	19	5.5	0.29	5.1–6.0
	Ant B	17	6.0	0.33	5.2–6.6
	Post B	18	5.6	0.28	5.2–6.2
M^3	M–D	17	5.2	0.38	4.9–6.2
	Ant B	16	5.8	0.41	4.9–6.4
	Post B	17	5.0	0.38	4.2–5.7

Table 158. *Presbytis comata* male

Mandibular teeth		*n*	*m*	SD	Range
I_1	M–D	16	3.2	0.14	2.9–3.4
	B–L	18	4.4	0.28	3.8–4.8
I_2	M–D	16	3.4	0.25	3.1–4.1
	B–L	18	4.4	0.31	3.8–4.7
C	M–D	15	5.4	0.37	5.0–6.0†
	B–L	14	4.2	0.25	3.8–4.6†
P_3	M–D	16	6.9	0.47	5.7–7.6†
	B–L	17	4.3	0.25	3.7–4.6†
P_4	M–D	17	4.5	0.40	4.0–5.1
	B–L	9	4.1	0.40	3.7–5.0
M_1	M–D	20	5.5	0.24	5.0–5.8
	Tri B	14	4.5	0.44	4.0–5.8
	Tal B	16	5.0	0.57	4.6–6.8*
M_2	M–D	19	5.7	0.22	5.4–6.0
	Tri B	16	5.1	0.35	4.6–5.8
	Tal B	17	5.4	0.48	4.7–6.5
M_3	M–D	17	6.0	0.32	5.2–6.6
	Tri B	13	4.9	0.33	4.3–5.5
	Tal B	11	4.9	0.48	4.3–5.8
	Hypoconulid	7	4.3	1.74	2.3–6.0

Table 159. *Presbytis comata* female

Maxillary teeth		n	m	SD	Range
I¹	M–D	19	4.2	0.29	3.7–4.9
	B–L	18	4.6	0.26	4.0–5.0
I²	M–D	19	4.0	0.22	3.5–4.4
	B–L	18	4.6	0.16	4.3–4.9
C	M–D	16	5.5	0.25	5.0–6.0
	B–L	16	5.0	0.22	4.7–5.5
P³	M–D	17	4.0	0.26	3.5–4.4
	B–L	17	4.8	0.38	4.2–5.5
P⁴	M–D	16	3.9	0.22	3.4–4.2
	B–L	16	5.4	0.29	4.9–5.8
M¹	M–D	26	5.3	0.28	4.7–5.8
	Ant B	25	5.3	0.31	5.0–6.0
	Post B	24	5.2	0.25	4.8–5.8
M²	M–D	21	5.4	0.23	4.8–5.8
	Ant B	21	5.9	0.23	5.4–6.4
	Post B	18	5.4	0.23	5.0–5.8
M³	M–D	18	5.2	0.45	4.9–6.6
	Ant B	17	5.6	0.35	4.7–6.0
	Post B	17	4.8	0.30	4.4–5.4

Table 160. *Presbytis comata* female

Mandibular teeth		n	m	SD	Range
I₁	M–D	22	3.3	0.37	2.9–4.4
	B–L	22	4.3	0.18	3.9–4.6
I₂	M–D	21	3.4	0.17	3.1–3.7
	B–L	22	4.6	0.25	4.0–4.9
C	M–D	16	4.5	0.79	3.5–5.6
	B–L	15	4.8	0.78	3.5–5.7
P₃	M–D	15	5.5	1.05	4.4–5.9
	B–L	16	3.9	0.44	3.4–4.7
P₄	M–D	16	4.3	0.20	4.0–4.5
	B–L	16	4.0	0.23	3.6–4.4
M₁	M–D	24	5.4	0.29	4.8–6.0
	Tri B	21	4.3	0.24	3.9–4.7
	Tal B	23	4.7	0.28	4.2–5.4
M₂	M–D	24	5.6	0.27	4.9–6.1
	Tri B	21	5.0	0.26	4.5–5.5
	Tal B	20	5.1	0.27	4.7–5.8
M₃	M–D	20	6.1	0.61	5.0–7.5
	Tri B	17	5.0	0.28	4.6–5.4
	Tal B	16	4.7	0.67	2.7–6.0
	Hypoconulid	11	3.5	1.28	2.0–5.9

Table 161. *Trachypithecus cristata*
male

Maxillary teeth		n	m	SD	Range
I¹	M–D	20	4.7	0.49	4.0–5.9*
	B–L	19	4.1	0.44	3.6–5.3
I²	M–D	22	3.8	0.28	3.4–4.5
	B–L	22	3.9	0.35	2.7–4.3
C	M–D	20	6.2	0.87	4.0–7.8†
	B–L	20	5.2	0.91	3.1–6.7
P³	M–D	18	4.3	0.33	3.9–6.6
	B–L	18	5.0	0.49	4.0–5.7
P⁴	M–D	19	4.2	0.26	3.7–4.6
	B–L	19	5.4	0.30	5.0–6.0
M¹	M–D	22	6.0	0.45	5.2–6.6
	Ant B	22	5.9	0.30	5.3–6.5
	Post B	22	5.6	0.36	5.1–6.5
M²	M–D	21	6.3	0.39	5.5–7.0
	Ant B	21	6.6	0.33	6.0–7.2
	Post B	21	6.1	0.35	5.6–7.0*
M³	M–D	18	6.1	0.39	5.5–7.0
	Ant B	17	6.3	0.28	5.7–6.8†
	Post B	17	5.4	0.48	4.5–6.1

Table 162. *Trachypithecus cristata*
male

Mandibular teeth		n	m	SD	Range
I₁	M–D	20	3.1	0.41	2.0–3.7
	B–L	20	3.5	0.59	1.9–4.5
I₂	M–D	17	3.2	0.25	2.6–3.7
	B–L	17	3.7	0.43	2.9–4.3
C	M–D	19	4.6	0.52	3.5 5.6†
	B–L	19	5.7	0.89	2.5–6.7
P₃	M–D	19	6.2	0.36	5.5–6.9†
	B–L	19	4.0	0.26	3.4–4.7
P₄	M–D	19	4.6	0.38	4.0–5.0
	B–L	19	4.0	0.19	3.7–4.3
M₁	M–D	21	6.1	0.32	5.7–6.8
	Tri B	21	4.7	0.21	4.3–5.2
	Tal B	21	4.9	0.21	4.5–5.3
M₂	M–D	20	6.5	0.43	5.8–7.1
	Tri B	20	5.5	0.30	5.0–6.0
	Tal B	20	5.5	0.37	4.7–6.1
M₃	M–D	18	7.9	0.61	7.0–8.9
	Tri B	18	5.5	0.39	4.9–6.3
	Tal B	18	5.2	0.37	4.5–6.0
	Hypoconulid	16	3.4	0.45	2.7–4.2

Table 163. *Trachypithecus cristata* female

Maxillary teeth		n	m	SD	Range
I^1	M–D	33	4.3	0.67	2.4–5.0
	B–L	30	4.0	0.48	2.2–4.8
I^2	M–D	29	3.7	0.40	2.7–4.7
	B–L	28	3.9	0.37	2.5–4.4
C	M–D	33	5.4	0.55	3.8–6.5
	B–L	33	5.1	0.55	3.1–6.2
P^3	M–D	32	4.2	0.28	3.9–5.0
	B–L	32	4.9	0.56	3.8–6.2
P^4	M–D	32	4.1	0.26	3.5–4.6
	B–L	32	5.5	0.31	5.1–6.2
M^1	M–D	32	5.8	0.39	5.0–6.4
	Ant B	32	5.8	0.31	5.2–6.9
	Post B	32	5.6	0.35	5.0–6.5
M^2	M–D	32	6.3	0.37	5.6–7.0
	Ant B	32	6.4	0.34	5.7–7.0
	Post B	32	5.9	0.29	5.4–6.6
M^3	M–D	31	6.0	0.33	5.4–6.9
	Ant B	31	6.1	0.27	5.8–6.9
	Post B	30	5.2	0.40	4.6–6.3

Table 164. *Trachypithecus cristata* female

Mandibular teeth		n	m	SD	Range
I_1	M–D	28	3.0	0.33	2.0–3.4
	B–L	28	3.6	0.32	3.0–4.0
I_2	M–D	30	3.0	0.25	2.3–3.6
	B–L	29	3.8	0.26	3.1–4.5
C	M–D	33	4.1	0.55	3.0–5.0
	B–L	32	5.1	0.35	4.6–6.0
P_3	M–D	31	5.3	0.36	4.6–6.3
	B–L	32	3.9	0.29	3.4–4.6
P_4	M–D	32	4.5	0.37	3.7–5.3
	B–L	32	4.2	0.28	3.7–5.0
M_1	M–D	33	6.0	0.39	5.3–7.0
	Tri B	32	4.7	0.24	4.3–5.1
	Tal B	32	4.9	0.19	4.5–5.3
M_2	M–D	32	6.3	0.34	5.8–7.0
	Tri B	30	5.4	0.31	4.9–6.1
	Tal B	31	5.5	0.30	5.0–6.0
M_3	M–D	32	7.6	0.63	6.3–9.0
	Tri B	32	5.4	0.34	5.0–6.1
	Tal B	32	5.3	0.34	4.9–6.3
	Hypoconulid	28	3.5	0.43	2.7–4.7

Table 165. *Hylobates klossii* male

Maxillary teeth		n	m	SD	Range
I^1	M–D	3	4.8	0.15	4.6–4.9
	B–L	3	3.8	0.06	3.8–3.9
I^2	M–D	3	3.6	0.25	3.4–3.9
	B–L	3	3.5	0.31	3.2–3.8
C	M–D	4	7.6	0.94	6.7–8.8
	B–L	4	5.4	0.35	5.0–5.8
P^3	M–D	3	4.7	0.56	4.2–5.3
	B–L	4	4.8	0.22	4.6–5.1
P^4	M–D	4	4.1	0.25	3.7–4.3
	B–L	4	5.0	0.36	4.7–5.5
M^1	M–D	4	5.5	0.17	5.2–5.6
	Ant B	3	5.9	0.12	5.8–6.0
	Post B	4	6.1	0.18	5.9–6.3
M^2	M–D	4	5.5	0.31	5.1–5.8
	Ant B	4	6.0	0.30	5.6–6.3
	Post B	4	6.1	0.25	5.7–6.3
M^3	M–D	4	4.2	0.25	4.0–4.5
	Ant B	3	5.5	0.06	5.5–5.6
	Post B	—	—	—	—

Table 166. *Hylobates klossii* male

Mandibular teeth		n	m	SD	Range
I_1	M–D	4	2.9	0.17	2.7–3.1
	B–L	4	3.3	0.33	2.9–3.7
I_2	M–D	3	3.0	0.21	2.8–3.2
	B–L	4	3.7	0.15	3.5–3.8
C	M–D	3	5.1	0.21	4.9–5.3*
	B–L	4	6.3	0.33	6.0–6.7
P_3	M–D	4	5.9	0.56	5.2–6.5
	B–L	4	3.9	0.13	3.7–4.0
P_4	M–D	4	4.6	0.37	4.1–5.0
	B–L	4	4.1	0.13	3.9–4.2
M_1	M–D	4	6.0	0.20	5.7–6.1
	Tri B	5	4.7	0.13	4.5–5.8
	Tal B	5	4.8	0.15	4.6–5.0
M_2	M–D	4	6.1	0.41	5.7–6.6
	Tri B	4	5.0	0.15	4.8–5.1
	Tal B	4	5.1	0.10	5.0–5.2
M_3	M–D	4	5.0	0.58	4.4–5.7
	Tri B	4	4.4	0.29	4.1–4.7
	Tal B	3	4.4	0.59	3.7–4.8
Hypoconulid		—	—	—	—

Table 167. *Hylobates klossii* female

Maxillary teeth		n	m	SD	Range
I^1	M–D	4	4.5	0.35	4.3–5.0
	B–L	4	4.0	0.13	3.8–4.1
I^2	M–D	2	3.6	0.14	3.5–3.7
	B–L	3	3.7	0.12	3.6–3.8
C	M–D	2	7.0	0.14	6.9–7.1
	B–L	2	4.8	0.50	4.4–5.1
P^3	M–D	4	4.6	0.14	4.5–4.8
	B–L	4	4.7	0.26	4.4–4.9
P^4	M–D	3	4.0	0.23	3.9–4.3
	B–L	4	4.9	0.36	4.4–5.2
M^1	M–D	4	5.2	0.15	5.0–5.3
	Ant B	4	5.7	0.28	5.4–6.0
	Post B	4	5.7	0.29	5.4–5.9
M^2	M–D	4	5.3	0.25	5.0–5.5
	Ant B	4	5.7	0.32	5.4–6.1
	Post B	—	—	—	—
M^3	M–D	2	3.8	0.64	3.3–4.2
	Ant B	4	5.7	0.26	5.4–6.0
	Post B	2	5.0	0.92	4.3–5.6

Table 168. *Hylobates klossii* female

Mandibular teeth		n	m	SD	Range
I_1	M–D	3	2.8	0.06	2.8–2.9
	B–L	4	3.2	0.08	3.1–3.3
I_2	M–D	4	2.9	0.21	2.7–3.1
	B–L	4	3.6	0.21	3.4–3.9
C	M–D	3	4.6	0.15	4.4–4.7
	B–L	3	5.9	0.32	5.7–6.3
P_3	M–D	4	6.2	0.27	5.9–6.5
	B–L	4	3.8	0.10	3.7–3.9
P_4	M–D	4	4.6	0.13	4.5–4.8
	B–L	4	4.0	0.10	3.9–4.1
M_1	M–D	4	5.8	0.14	5.7–6.0
	Tri B	4	4.6	0.25	4.2–4.8
	Tal B	4	4.7	0.17	4.5–4.9
M_2	M–D	4	5.8	0.26	5.5–6.1
	Tri B	4	4.9	0.19	4.8–5.2
	Tal B	4	5.0	0.13	4.8–5.1
M_3	M–D	3	5.4	0.25	5.1–5.6
	Tri B	3	4.7	0.12	4.6–4.8
	Tal B	3	4.6	0.15	4.5–4.8
	Hypoconulid	—	—	—	—

Table 169. *Hylobates agilis* male

Maxillary teeth		n	m	SD	Range
I^1	M–D	5	5.2	0.35	4.7–5.6
	B–L	5	4.0	0.41	3.5–4.5
I^2	M–D	6	4.0	0.25	3.5–4.2
	B–L	7	4.0	0.25	3.6–4.3
C	M–D	8	7.2	0.35	6.8–7.8
	B L	8	5.3	0.38	4.7–5.8
P^3	M–D	8	4.8	0.30	4.4–5.4
	B–L	8	5.0	0.39	4.4–5.6
P^4	M–D	6	4.3	0.09	4.2–4.4
	B–L	8	5.0	0.44	4.2–5.4
M^1	M–D	8	5.7	0.20	5.5–6.0
	Ant B	6	6.0	0.17	5.7–6.2
	Post B	3	5.9	0.10	5.8–6.0
M^2	M–D	8	6.1	0.23	5.7–6.4
	Ant B	6	6.5	0.14	6.3–6.7
	Post B	6	6.1	0.15	5.8–6.2
M^3	M–D	8	5.3	0.28	5.0–5.7
	Ant B	7	6.0	0.20	5.7–6.3
	Post B	3	5.3	0.38	4.9–5.6

Table 170. *Hylobates agilis* male

Mandibular teeth		n	m	SD	Range
I_1	M–D	3	3.4	0.30	3.1–3.7
	B–L	4	3.4	0.39	3.0–3.9
I_2	M–D	5	3.6	0.34	3.4–4.2
	B–L	6	4.0	0.36	3.5–4.6
C	M–D	5	5.3	0.43	4.7–5.8
	B–L	6	6.8	0.16	6.5–7.0
P_3	M–D	7	6.0	0.34	5.6–6.5
	B–L	7	4.0	0.28	3.7–4.4
P_4	M–D	7	4.8	0.18	4.6–5.1
	B–L	7	4.2	0.19	3.9–4.4
M_1	M–D	6	6.0	0.22	5.7–6.3
	Tri B	5	4.9	0.10	4.8–5.0
	Tal B	6	5.1	0.29	4.9–5.6
M_2	M–D	7	6.4	0.28	5.9–6.7
	Tri B	8	5.5	0.16	5.1–5.6
	Tal B	7	5.6	0.30	4.9–5.8
M_3	M–D	5	6.3	0.22	6.0–6.6†
	Tri B	7	5.5	0.18	5.3–5.8†
	Tal B	6	5.3	0.28	4.9–5.7*
Hypoconulid	—	—	—	—

Table 171. *Hylobates agilis* female

Maxillary teeth		*n*	*m*	SD	Range
I¹	M–D	6	4.9	0.23	4.7–5.2
	B–L	6	4.0	0.20	3.7–4.3
I²	M–D	5	3.9	0.26	3.6–4.2
	B–L	6	4.0	0.34	3.8–4.1
C	M–D	7	6.9	0.35	6.6–7.4
	B–L	5	4.9	0.40	4.4–5.5
P³	M–D	9	4.6	0.40	4.0–5.4
	B–L	9	4.9	0.38	4.3–5.3
P⁴	M–D	6	4.2	0.21	3.8–4.4
	B–L	9	5.0	0.26	4.6–5.4
M¹	M–D	9	5.6	0.37	5.2–6.1
	Ant B	4	6.0	0.47	5.4–6.5
	Post B	4	5.9	0.44	5.4–6.4
M²	M–D	9	5.8	0.39	5.3–6.4
	Ant B	8	6.2	0.49	5.1–6.7
	Post B	7	6.0	0.28	5.6–6.5
M³	M–D	6	5.1	0.48	4.4–5.7
	Ant B	2	5.9	0.14	5.8–6.0
	Post B	8	3.9	0.28	3.7–4.4

Table 172. *Hylobates agilis* female

Mandibular teeth		*n*	*m*	SD	Range
I₁	M–D	6	3.2	0.11	3.0–3.3
	B–L	7	3.5	0.20	3.3–3.9
I₂	M–D	8	3.4	0.22	3.0–3.7
	B–L	8	4.2	0.26	3.0–4.5
C	M–D	6	4.9	0.48	4.4–5.5
	B–L	9	6.7	0.33	5.9–7.0
P₃	M–D	8	6.1	0.44	5.3–6.6
	B–L	8	4.1	0.38	3.6–4.6
P₄	M–D	8	4.7	0.29	4.2–5.2
	B–L	8	4.3	0.22	3.8–4.5
M₁	M–D	9	6.0	0.27	5.7–6.4
	Tri B	6	4.9	0.15	4.8–5.2
	Tal B	6	5.2	0.26	4.8–5.6
M₂	M–D	8	6.2	0.26	5.9–6.6
	Tri B	8	5.4	0.15	5.1–5.6
	Tal B	7	5.5	0.20	5.2–5.7
M₃	M–D	4	5.7	0.26	5.5–6.1
	Tri B	4	5.1	0.14	5.0–5.3
	Tal B	4	4.7	0.48	4.0–5.0
Hypoconulid	—	—	—	—

Table 173. *Hylobates moloch* male

Maxillary teeth		n	m	SD	Range
I^1	M–D	5	5.0	0.23	4.8–5.4
	B–L	4	4.1	0.29	3.9–4.5
I^2	M–D	4	4.0	0.36	3.5–4.3
	B–L	4	4.3	0.06	4.2–4.3
C	M–D	3	7.4	0.61	6.7–7.8
	B L	3	5.5	0.30	5.2–5.8
P^3	M–D	5	4.7	0.65	3.8–5.4
	B–L	5	5.0	0.49	4.2–5.5
P^4	M–D	5	4.3	0.26	4.1–4.7
	B–L	5	5.3	0.37	4.9–5.7
M^1	M–D	5	5.9	0.25	5.5–6.1
	Ant B	3	6.5	0.21	6.3–6.7
	Post B	3	6.0	0.31	5.7–6.3
M^2	M–D	4	6.4	0.50	5.8–7.0
	Ant B	4	6.3	0.27	5.9–6.5
	Post B	4	6.5	0.33	6.2–6.9
M^3	M–D	4	5.0	0.79	4.0–5.7
	Ant B	4	6.2	0.60	5.6–7.0
	Post B	1	5.9	0.00	5.9–5.9

Table 174. *Hylobates moloch* male

Mandibular teeth		n	m	SD	Range
I_1	M–D	4	3.3	0.14	3.2–3.5
	B–L	4	3.6	0.18	3.4–3.8
I_2	M–D	4	3.6	0.25	3.3–3.9
	B–L	4	4.4	0.25	4.2–4.7
C	M–D	3	5.5	0.12	5.4–5.6
	B–L	3	7.2	0.35	6.9–7.6
P_3	M–D	4	6.1	0.25	5.8–6.4
	B–L	5	4.4	0.11	4.3–4.6
P_4	M–D	5	4.8	0.32	4.4–5.2
	B–L	5	4.3	0.19	4.1–4.6
M_1	M–D	5	6.4	0.23	6.0–6.6
	Tri B	3	5.2	0.35	4.8–5.4
	Tal B	3	5.2	0.36	4.8–5.5
M_2	M–D	4	6.6	0.33	6.2–7.0
	Tri B	5	5.7	0.22	5.4–6.0
	Tal B	4	5.9	0.41	5.3–6.3
M_3	M–D	3	6.2	0.53	5.8–6.8
	Tri B	2	5.5	0.21	5.3–5.6
	Tal B	2	5.3	0.42	5.0–5.6
	Hypoconulid	—	—	—	—

Table 175. *Hylobates moloch* female

Maxillary teeth		n	m	SD	Range
I^1	M–D	4	5.3	0.55	4.7–5.9
	B–L	4	3.9	0.30	3.5–4.2
I^2	M–D	4	4.1	0.44	3.5–4.5
	B–L	5	4.3	0.30	4.0–4.7
C	M–D	4	7.6	0.83	7.0–8.8
	B–L	4	5.0	0.63	4.1–5.6
P^3	M–D	4	5.1	0.37	4.7–5.6
	B–L	4	5.1	0.33	4.8–5.5
P^4	M–D	5	4.5	0.32	3.9–4.7
	B–L	5	5.4	0.27	5.0–5.7
M^1	M–D	4	5.7	0.49	5.1–6.3
	Ant B	4	6.3	0.38	5.7–6.5
	Post B	4	6.3	0.39	5.8–6.7
M^2	M–D	5	5.9	0.68	5.0–6.6
	Ant B	4	6.4	0.75	5.6–7.2
	Post B	3	6.5	0.76	5.6–7.0
M^3	M–D	5	5.2	0.50	4.5–5.9
	Ant B	2	5.5	0.64	5.0–5.9
	Post B	2	5.4	1.63	4.2–6.5

Table 176. *Hylobates moloch* female

Mandibular teeth		n	m	SD	Range
I$_1$	M–D	4	3.7	0.06	3.6–3.7
	B–L	5	3.7	0.30	3.3–4.0
I$_2$	M–D	4	3.6	0.37	3.0–3.8
	B–L	5	4.0	0.30	3.6–4.3
C	M–D	4	5.4	0.37	5.0–5.8
	B–L	4	7.0	0.84	6.0–7.9
P$_3$	M–D	5	6.5	0.34	6.1–6.9
	B–L	5	4.3	0.41	3.6–4.6
P$_4$	M–D	5	5.1	0.30	4.7–5.4
	B–L	5	4.3	0.34	3.8–4.6
M$_1$	M–D	5	6.2	0.44	5.5–6.7
	Tri B	3	4.9	0.32	4.5–5.1
	Tal B	2	4.9	0.50	4.5–5.2
M$_2$	M–D	4	6.5	0.68	5.6–7.2
	Tri B	2	5.3	0.57	4.9–5.7
	Tal B	2	5.4	0.64	4.9–5.8
M$_3$	M–D	4	6.3	1.01	5.2–7.3
	Tri B	3	5.3	0.79	4.4–5.9
	Tal B	3	5.2	0.79	4.3–5.8
Hypoconulid	—	—	—	—

Table 177. *Hylobates lar* male

Maxillary teeth		n	m	SD	Range
I^1	M–D	4	5.0	0.34	4.6–5.3
	B–L	4	4.0	0.29	3.6–4.3
I^2	M–D	3	3.7	0.21	3.5–3.9
	B–L	3	3.8	0.12	3.7–3.9
C	M–D	5	7.0	0.22	6.8–7.2
	B–L	5	5.0	0.17	4.7 5.1
P^3	M–D	5	4.4	0.28	4.1–4.8
	B–L	5	4.8	0.20	4.5–5.0
P^4	M–D	3	4.0	0.25	3.8–4.3
	B–L	4	5.1	0.29	4.7–5.4
M^1	M–D	5	5.4	0.40	4.9–5.9
	Ant B	3	5.5	0.36	5.1–5.8
	Post B	4	5.9	0.61	5.1–6.6
M^2	M–D	5	5.7	0.25	5.4–6.1
	Ant B	5	6.2	0.52	5.7–6.9
	Post B	4	6.1	0.46	5.8–6.8
M^3	M–D	5	4.8	0.37	4.3–5.3
	Ant B	5	5.6	0.66	4.9–6.5
	Post B	3	5.2	0.36	4.9–5.6

Note: no females.

Table 178. *Hylobates lar* male

Mandibular teeth		n	m	SD	Range
I_1	M–D	5	3.1	0.30	2.7–3.5
	B–L	5	3.3	0.29	3.0–3.7
I_2	M–D	4	3.2	0.26	3.0–3.6
	B–L	3	3.9	0.16	3.8–4.0
C	M–D	3	4.9	0.70	4.1–5.4
	B–L	4	6.3	0.29	6.0–6.7
P_3	M–D	4	5.9	0.29	5.5–6.2
	B–L	4	4.0	0.38	3.7–4.5
P_4	M–D	5	4.6	0.16	4.4–4.8
	B–L	5	4.3	0.30	4.1–4.8
M_1	M–D	5	5.6	0.51	4.9–6.2
	Tri B	2	4.4	0.21	4.2–4.5
	Tal B	4	4.6	0.43	4.2–5.2
M_2	M–D	5	5.9	0.25	5.5–6.1
	Tri B	5	5.2	0.29	4.7–5.4
	Tal B	5	5.1	0.36	4.5–5.4
M_3	M–D	5	5.4	0.21	5.1–5.7
	Tri B	4	5.0	0.29	4.7–5.4
	Tal B	3	4.8	0.40	4.4–5.2
	Hypoconulid	—	—	—	—

Note: no females.

Table 179. *Pongo pygmaeus* male

Maxillary teeth		n	m	SD	Range
I^1	M–D	4	14.7	0.96	13.3–15.4
	B–L	6	13.4	1.76	10.4–15.2
I^2	M–D	4	8.8	0.73	8.0–9.7
	B–L	4	8.4	0.28	8.1–8.7
C	M–D	4	16.4	2.08	13.4–17.9*
	B–L	4	13.2	1.89	10.4–14.4*
P^3	M–D	6	10.3	1.03	9.0–11.4
	B–L	7	13.2	1.00	12.0–14.5*
P^4	M–D	7	10.0	0.75	9.1–11.0
	B–L	7	13.3	0.89	12.0–14.3*
M^1	M–D	7	12.8	0.81	11.7–13.5*
	Ant B	7	13.8	0.82	12.5–14.6†
	Post B	7	13.0	1.03	11.5–14.3*
M^2	M–D	7	12.7	0.55	12.0–13.7
	Ant B	7	14.1	0.89	13.4–15.6†
	Post B	7	13.1	1.29	10.6–14.4*
M^3	M–D	4	12.4	0.42	12.0–13.0*
	Ant B	3	14.1	0.50	13.6–14.6†
	Post B	—	—	—	—

Table 180. *Pongo pygmaeus* male

Mandibular teeth		n	m	SD	Range
I_1	M–D	6	9.7	1.02	9.0–11.0†
	B–L	7	10.1	1.36	8.6–12.0
I_2	M–D	6	9.7	0.66	8.7–10.6†
	B–L	7	11.0	1.37	9.2–12.8*
C	M–D	3	12.0	1.07	11.3–13.2
	B–L	3	14.8	0.59	14.4–15.5†
P_3	M–D	7	15.6	2.10	13.4–18.4
	B–L	8	11.0	1.80	8.5–14.5*
P_4	M–D	7	11.3	0.80	10.5–12.4
	B–L	8	11.9	1.21	10.4–13.3*
M_1	M–D	8	13.4	0.68	12.5–14.2
	Tri B	7	11.9	0.88	10.6–12.8*
	Tal B	7	11.9	1.24	10.3–13.3
M_2	M–D	8	14.2	0.59	13.3–15.0*
	Tri B	8	13.6	0.94	12.2–14.8†
	Tal B	8	13.0	0.99	11.5–14.5*
M_3	M–D	5	13.5	1.82	10.4–14.9
	Tri B	5	12.9	0.70	11.9–13.7*
	Tal B	5	12.0	1.37	9.6–12.8
Hypoconulid		—	—	—	—

Table 181. *Pongo pygmaeus* female

Maxillary teeth		n	m	SD	Range
I^1	M–D	4	13.8	1.04	12.9–15.3
	B–L	9	12.4	0.77	11.7–14.0
I^2	M–D	4	8.4	0.25	8.1–8.7
	B–L	5	8.3	0.74	7.5–9.3
C	M–D	4	12.5	1.56	10.2–13.7
	B–L	5	10.8	0.49	10.5 11.7
P^3	M–D	8	9.6	0.94	8.0–11.2
	B–L	6	11.8	0.72	11.0–12.9
P^4	M–D	8	9.3	0.77	8.3–10.1
	B–L	8	12.1	0.64	11.2–13.2
M^1	M–D	10	11.9	0.75	10.4–13.4
	Ant B	10	12.2	0.93	10.2–13.3
	Post B	9	11.7	0.79	10.2–12.9
M^2	M–D	9	12.1	0.79	11.2–13.4
	Ant B	9	12.9	1.13	10.3–13.8
	Post B	9	11.6	0.84	10.2–12.7
M^3	M–D	5	11.5	0.46	11.0–12.0
	Ant B	5	12.6	0.40	12.2–13.0
	Post B	4	10.3	0.66	9.5–10.9

Table 182. *Pongo pygmaeus* female

Mandibular teeth		n	m	SD	Range
I_1	M–D	9	8.5	0.38	8.0–9.1
	B–L	8	9.4	0.80	8.3–10.3
I_2	M–D	9	8.4	0.58	7.5–9.4
	B–L	9	9.7	0.78	8.5–10.5
C	M–D	3	9.7	2.55	7.8–12.6
	B–L	6	11.9	1.13	10.0–12.8
P_3	M–D	7	13.8	1.01	12.5–15.0
	B–L	7	9.3	0.55	8.4–10.1
P_4	M–D	7	10.5	0.80	9.4–11.6
	B–L	6	10.4	0.53	9.7–11.2
M_1	M–D	10	11.4	2.68	10.4–13.2
	Tri B	10	11.1	0.60	10.3–12.1
	Tal B	10	11.3	0.62	10.5–12.1
M_2	M–D	8	13.3	0.58	12.4–14.3
	Tri B	8	12.0	0.96	10.2–13.0
	Tal B	8	12.1	0.72	10.9–13.0
M_3	M–D	4	12.9	0.79	11.7–13.5
	Tri B	5	11.8	0.62	11.1–12.5
	Tal B	5	11.3	0.73	10.6–12.4
	Hypoconulid	—	—	—	—

Table 183. *Gorilla gorilla* male

Maxillary teeth		n	m	SD	Range
I¹	M–D	3	14.7	0.25	14.5–15.0†
	B–L	2	11.6	0.00	11.6–11.6*
I²	M–D	2	10.8	0.42	10.5–11.1†
	B–L	4	10.1	0.75	9.4–11.0
C	M–D	4	21.4	1.66	19.6–23.3†
	B–L	4	16.0	0.78	14.9–16.6†
P³	M–D	4	12.4	0.71	11.7–13.3
	B–L	3	15.8	0.61	15.1–16.2*
P⁴	M–D	3	11.2	0.30	10.9–11.5
	B–L	3	15.2	0.69	14.8–16.0*
M¹	M–D	5	15.7	0.79	14.5–16.4*
	Ant B	4	15.4	1.19	14.2–17.0
	Post B	4	15.0	1.25	13.2–16.0*
M²	M–D	4	17.5	0.22	17.3–17.8†
	Ant B	3	16.9	0.46	16.4–17.3*
	Post B	4	16.0	1.20	14.4–17.2*
M³	M–D	4	16.1	1.35	14.1–16.9
	Ant B	4	15.8	0.80	14.6–16.3*
	Post B	4	13.7	1.99	10.8–15.4

Table 184. *Gorilla gorilla* male

Mandibular teeth		n	m	SD	Range
I₁	M–D	4	7.9	0.30	7.6–8.3
	B–L	4	9.7	0.89	8.5–10.4*
I₂	M–D	3	10.2	0.10	10.1–10.3
	B–L	3	10.9	0.50	10.4–11.4
C	M–D	4	16.9	2.19	14.8–20.0
	B–L	5	18.1	1.08	16.2–19.0†
P₃	M–D	4	16.9	0.77	15.9–17.6
	B–L	5	11.9	0.31	11.6–12.4*
P₄	M–D	5	11.8	0.82	11.1–13.0
	B–L	4	14.6	0.26	14.3–14.9*
M₁	M–D	6	16.2	0.79	14.7–17.0
	Tri B	3	13.7	1.00	12.6–14.5
	Tal B	3	13.3	1.46	11.8–14.7
M₂	M–D	5	18.3	0.65	17.2–18.9*
	Tri B	4	16.6	0.17	16.4–16.8*
	Tal B	4	16.2	0.83	15.0–16.8*
M₃	M–D	5	18.4	1.16	16.7–19.6†
	Tri B	5	16.1	0.88	14.6–16.8†
	Tal B	3	14.8	0.55	14.2–15.2*
Hypoconulid		—	—	—	—

Table 185. *Gorilla gorilla* female

Maxillary teeth		n	m	SD	Range
I^1	M–D	8	14.0	0.39	13.4–14.6
	B–L	8	10.3	0.46	9.6–10.9
I^2	M–D	8	9.9	1.03	8.7–11.7
	B–L	8	9.4	0.83	8.1–10.4
C	M–D	7	15.0	0.56	14.3–15.6
	B–L	6	11.3	0.47	10.5–11.8
P^3	M–D	9	11.6	0.83	10.6–13.3
	B–L	9	14.8	0.38	13.8–15.1
P^4	M–D	7	11.0	0.55	10.3–11.8
	B–L	8	14.4	0.47	13.4–14.9
M^1	M–D	11	14.8	0.89	13.4–16.3
	Ant B	10	15.0	1.10	13.8–17.4
	Post B	10	14.0	0.68	13.0–15.4
M^2	M–D	10	16.0	0.89	14.0–17.3
	Ant B	9	15.8	0.75	14.3–16.7
	Post B	7	14.8	0.77	13.3–15.5
M^3	M–D	6	14.6	0.72	13.6–15.2
	Ant B	5	14.5	1.22	13.2–16.3
	Post B	3	12.9	1.61	11.6–14.7

Table 186. *Gorilla gorilla* female

Mandibular teeth		n	m	SD	Range
I_1	M–D	9	8.0	0.28	7.5–8.2
	B–L	10	8.8	0.21	8.4–9.0
I_2	M–D	8	10.2	0.97	9.0–11.9
	B–L	10	10.3	0.58	9.6–11.5
C	M–D	4	12.0	0.97	11.0–13.3
	B–L	7	13.0	0.78	11.6–13.8
P_3	M–D	6	14.8	0.72	14.0–16.1
	B–L	7	11.2	1.31	9.5–13.3
P_4	M–D	7	11.3	0.63	10.3–12.1
	B–L	8	13.3	1.06	11.5–14.7
M_1	M–D	11	15.4	0.83	14.0–16.6
	Tri B	9	13.2	0.63	12.3–14.4
	Tal B	8	13.1	0.50	12.3–13.7
M_2	M–D	8	17.0	0.88	15.7–18.2
	Tri B	8	15.1	0.79	13.6–16.3
	Tal B	8	14.8	0.90	13.1–15.7
M_3	M–D	4	15.6	0.62	14.9–16.4
	Tri B	6	14.4	0.62	13.6–15.3
	Tal B	5	12.9	0.71	12.3–14.1
Hypoconulid	—	—	—	—	—

Table 187. *Pan troglodytes* male

Maxillary teeth		n	m	SD	Range
I^1	M–D	14	12.6	0.82	10.5–13.5†
	B–L	15	10.1	0.76	9.0–11.3*
I^2	M–D	15	9.3	0.61	7.9–10.3
	B–L	14	9.2	0.58	8.4–10.8
C	M–D	14	15.0	1.57	13.0–18.0†
	B–L	14	12.0	0.93	10.5–13.8†
P^3	M–D	17	8.2	0.84	6.3–9.5
	B–L	17	10.5	0.75	9.4–12.0*
P^4	M–D	16	7.2	0.55	6.2–7.9
	B–L	17	10.5	0.76	9.3–12.2†
M^1	M–D	19	10.3	0.56	9.3–11.2
	Ant B	19	11.7	0.73	10.7–13.2†
	Post B	16	11.0	0.63	10.0–12.3
M^2	M–D	19	10.4	0.56	9.6–12.3
	Ant B	18	12.0	0.70	10.6–13.3†
	Post B	17	11.0	0.59	9.5–11.9
M^3	M–D	17	10.0	0.60	9.0–11.3†
	Ant B	15	11.6	0.80	10.0–12.8†
	Post B	14	10.4	0.63	9.6–11.6†

Table 188. *Pan troglodytes* male

Mandibular teeth		n	m	SD	Range
I$_1$	M–D	11	8.3	0.75	6.8–9.0
	B–L	13	9.7	0.57	8.8–11.0*
I$_2$	M–D	13	8.9	0.80	7.6–10.2*
	B–L	17	10.0	0.75	8.6–12.0
C	M–D	13	12.7	1.58	10.9–15.9†
	B–L	15	13.0	1.19	10.4–14.7†
P$_3$	M–D	19	10.2	1.17	8.1–12.4
	B–L	18	9.1	0.99	7.5–11.0†
P$_4$	M–D	17	7.8	0.72	6.6–9.2
	B–L	16	9.1	0.87	7.4–10.5*
M$_1$	M–D	18	11.0	0.59	10.0–11.9
	Tri B	15	10.0	0.68	9.0–11.2*
	Tal B	17	10.2	0.60	8.7–11.0
M$_2$	M–D	19	11.4	0.89	9.5–12.9*
	Tri B	17	10.9	0.75	9.9–12.5†
	Tal B	19	10.7	0.66	9.2–11.8*
M$_3$	M–D	17	10.8	0.84	9.4–12.3
	Tri B	15	10.6	0.60	9.9–11.5†
	Tal B	13	9.8	0.50	8.9–10.6
	Hypoconulid	—	—	—	—

Table 189. *Pan troglodytes* female

Maxillary teeth		n	m	SD	Range
I^1	M–D	51	11.9	0.88	10.0–13.4
	B–L	50	9.6	0.82	8.3–11.7
I^2	M–D	43	8.8	0.87	6.9–10.9
	B–L	45	8.9	0.80	7.6–11.0
C	M–D	32	11.7	1.68	6.5–15.0
	B–L	30	9.5	1.53	7.4–13.9
P^3	M–D	46	8.1	0.94	6.5–12.3
	B–L	45	9.9	0.98	7.1–11.8
P^4	M–D	46	7.2	0.96	5.8–11.4
	B–L	46	9.8	0.84	6.4–11.4
M^1	M–D	51	10.1	0.72	9.0–11.9
	Ant B	50	10.9	1.01	7.0–12.8
	Post B	49	10.5	1.02	6.1–12.2
M^2	M–D	50	10.1	0.74	9.0–11.9
	Ant B	45	11.1	0.92	7.2–13.3
	Post B	48	10.3	1.05	5.5–12.2
M^3	M–D	37	9.5	0.83	8.0–11.1
	Ant B	36	10.6	0.78	8.8–12.5
	Post B	33	9.4	0.83	7.5–11.6

Table 190. *Pan troglodytes* female

Mandibular teeth		n	m	SD	Range
I_1	M–D	42	8.0	0.78	5.5–9.6
	B–L	45	9.1	0.89	7.2–11.0
I_2	M–D	41	8.4	0.63	7.0–9.6
	B–L	41	9.5	0.91	7.9–11.5
C	M–D	30	10.4	1.47	8.2–14.0
	B–L	31	10.2	1.44	8.0–14.1
P_3	M–D	45	9.9	0.96	8.4–12.4
	B–L	45	8.1	0.99	6.6–10.8
P_4	M–D	43	7.5	0.67	6.0–9.1
	B–L	43	8.5	0.76	7.2–10.0
M_1	M–D	45	10.8	0.63	9.8–12.7
	Tri B	42	9.4	0.79	7.4–11.0
	Tal B	45	9.6	0.77	8.1–11.4
M_2	M–D	47	11.0	0.65	9.5–12.2
	Tri B	46	10.2	0.86	8.3–12.3
	Tal B	46	10.2	0.86	8.5–13.1
M_3	M–D	40	10.4	0.82	9.0–12.2
	Tri B	36	9.7	0.74	8.0–10.7
	Tal B	37	9.4	0.78	8.0–10.8
	Hypoconulid	—	—	—	—

Table 191. *Pan paniscus* male

Maxillary teeth		n	m	SD	Range
I^1	M–D	15	10.3	0.9	8.9–11.9
	B–L	15	7.9	0.6	7.2–9.2
I^2	M–D	13	7.9	0.7	6.9–9.2
	B–L	13	7.3	0.6	6.7–8.5
C	M–D	15	11.1	0.9	9.7–13.3
	B–L	15	8.8	0.8	7.6–10.7
P^3	M–D	17	7.4	0.6	6.6–8.4
	B–L	17	9.3	0.6	8.4–10.3
P^4	M–D	16	6.3	0.5	5.7–7.6
	B–L	16	9.0	0.6	8.0–10.3
M^1	M–D	7	8.5	0.6	7.9–9.4
	Ant B	6	9.5	0.5	9.2–10.4
	Post B	—	—	—	—
M^2	M–D	7	8.4	0.7	7.6–9.6
	Ant B	7	10.2	0.4	9.6–10.6
	Post B	—	—	—	—
M^3	M–D	6	7.8	0.3	7.5–8.2
	Ant B	6	9.8	0.4	9.4–10.3
	Post B	—	—	—	—

Source: Johanson (1979).

Table 192. *Pan paniscus* male

Mandibular teeth		n	m	SD	Range
I_1	M–D	16	7.4	0.7	6.1–8.7
	B–L	15	7.0	0.5	6.3–8.1
I_2	M–D	17	7.5	0.6	6.3–8.3
	B–L	17	7.1	0.4	6.7–8.4
C	M–D	16	10.0	0.7	8.7–11.4
	B–L	16	7.6	0.4	6.8–8.5
P_3	M–D	17	8.1	0.5	7.2–8.9
	B–L	17	7.4	1.2	5.8–9.7
P_4	M–D	18	7.1	0.4	6.1–7.7
	B–L	18	7.8	0.7	6.5–9.2
M_1	M–D	5	9.1	0.7	8.2–9.9
	Tri B	5	8.8	0.5	8.1–9.4
	Tal B	5	9.0	0.7	7.9–9.8
M_2	M–D	7	9.9	0.4	9.4–10.6
	Tri B	7	9.1	0.6	8.4–10.0
	Tal B	7	8.8	0.3	8.3–9.2
M_3	M–D	7	9.2	0.5	8.1–9.4
	Tri B	7	8.4	0.5	7.8–9.2
	Tal B	7	7.9	0.2	7.5–8.6
	Hypoconulid	—	—	—	—

Source: Johanson (1979).

Table 193. *Pan paniscus* female

Maxillary teeth		n	m	SD	Range
I^1	M–D	20	10.4	0.7	9.0–11.5
	B–L	21	7.6	0.4	6.8–8.5
I^2	M–D	20	7.9	0.7	7.1–10.1
	B–L	20	7.1	0.4	6.4–7.7
C	M–D	18	9.0	0.5	8.2–9.9
	B–L	18	6.9	0.4	6.3–7.6
P^3	M–D	26	7.2	0.4	6.2–7.8
	B–L	27	9.2	0.4	8.3–10.2
P^4	M–D	23	6.1	0.5	5.0–6.6
	B–L	24	8.8	0.4	7.7–9.8
M^1	M–D	6	8.3	0.4	7.6–8.8
	Ant B	6	9.7	0.4	9.3–10.4
	Post B	—	—	—	—
M^2	M–D	7	8.4	0.7	7.6–9.6
	Ant B	7	10.2	0.4	9.6–10.6
	Post B	—	—	—	—
M^3	M–D	4	8.2	0.2	8.0–8.4
	Ant B	3	10.2	0.3	9.9–10.5
	Post B	—	—	—	—

Source: Johanson (1979).

Table 194. *Pan paniscus* female

Mandibular teeth		n	m	SD	Range
I_1	M–D	22	7.2	0.7	5.6–8.5
	B–L	22	6.8	0.3	6.2–7.5
I_2	M–D	24	7.3	0.8	5.2–9.0
	B–L	24	6.9	0.3	6.4–7.5
C	M–D	20	8.8	0.7	7.5–10.9
	B–L	20	6.5	0.7	5.8–8.9
P_3	M–D	24	8.2	0.4	7.5–9.1
	B–L	24	7.0	0.9	5.2–8.4
P_4	M–D	24	7.0	0.8	5.4–9.1
	B–L	24	7.6	0.6	5.6–8.5
M_1	M–D	5	9.5	0.4	9.0–10.1
	Tri B	4	8.9	0.2	8.7–9.2
	Tal B	3	8.8	0.5	8.4–9.3
M_2	M–D	5	9.7	0.3	9.3–10.0
	Tri B	5	9.2	0.4	8.8–9.7
	Tal B	5	8.9	0.3	8.7–9.3
M_3	M–D	5	9.2	0.5	8.4–9.8
	Tri B	5	8.7	0.4	8.2–9.1
	Tal B	5	8.2	0.3	7.8–8.6
Hypoconulid		—	—	—	—

Source: Johanson (1979).

Deciduous teeth

Table 195. *Macaca nemestrina* male

Maxillary teeth		n	m	SD	Range
di¹	M–D	20	5.7	0.19	5.4–6.2
	B–L	20	2.7	0.25	2.0–3.3
di²	M–D	20	4.6	0.37	4.2–5.6
	B–L	20	2.2	0.34	1.7–2.9
dc	M–D	20	4.9	0.28	4.4–5.4
	B–L	20	3.1	0.29	2.7–3.6
dp³	M–D	20	6.1	0.38	5.3–6.7
	B–L	20	4.1	0.34	3.5–4.9
dp⁴	M–D	20	6.6	0.31	6.1–7.4
	B–L	20	4.8	0.42	4.2–5.6

Table 196. *Macaca nemestrina* female

Maxillary teeth		n	m	SD	Range
di¹	M–D	20	5.4	0.78	2.5–6.4
	B–L	20	2.8	0.33	2.3–3.5
di²	M–D	20	4.6	0.39	3.9–5.4
	B–L	20	2.2	0.22	1.9–2.7
dc	M–D	20	4.8	0.28	4.1–5.2
	B–L	20	3.2	0.42	2.7–4.7
dp³	M–D	20	6.2	0.40	5.6–7.1
	B–L	20	4.3	0.51	3.6–6.0
dp⁴	M–D	20	6.6	0.30	6.2–7.2
	B–L	20	4.9	0.48	4.3–6.3

Table 197. *Macaca nemestrina* male

Mandibular teeth		n	m	SD	Range
di₁	M–D	20	3.4	0.21	3.0–3.8
	B–L	20	1.8	0.26	1.5–2.3
di₂	M–D	20	3.9	0.25	3.5–4.6
	B–L	20	1.9	0.24	1.6–2.7
dc	M–D	20	4.6	0.51	2.9–5.4
	B–L	20	2.5	0.15	2.3–2.8
dp₃	M–D	20	6.7	0.39	5.7–7.5
	B–L	20	3.3	0.30	3.0–3.9
dp₄	M–D	20	6.8	0.35	6.4–7.6
	B–L	20	4.2	0.25	3.6–4.6

Table 198. *Macaca nemestrina* female

Mandibular teeth		n	m	SD	Range
di₁	M–D	20	3.4	0.31	2.9–3.9
	B–L	20	1.9	0.24	1.5–2.5
di₂	M–D	20	3.8	0.30	3.2–4.4
	B–L	20	1.8	0.23	1.5–2.3
dc	M–D	20	4.5	0.32	3.7–5.2
	B–L	20	2.5	0.16	2.1–2.9
dp₃	M–D	20	6.8	0.34	6.3–7.5
	B–L	20	3.5	0.29	3.0–4.0
dp₄	M–D	20	6.7	0.32	6.3–7.4
	B–L	20	4.2	0.27	3.7–4.8

Table 199. *Alouatta palliata* male

Maxillary teeth		n	m	SD	Range
di^1	M–D	—	—	—	—
	B–L	—	—	—	—
di^2	M–D	—	—	—	—
	B–L	—	—	—	—
dc	M–D	6	4.2	0.36	3.8–4.8
	B–L	6	2.9	0.29	2.4–3.3
dp^2	M–D	6	3.8	0.10	3.7–4.0
	B–L	6	3.3	0.21	2.9–3.5
dp^3	M–D	6	4.0	0.16	3.8–4.3
	B–L	6	4.1	0.23	3.9–4.5
dp^4	M–D	6	4.1	0.37	5.6–6.7
	B–L	6	5.6	0.33	5.2–6.2

Table 200. *Alouatta palliata* female

Maxillary teeth		n	m	SD	Range
di^1	M–D	—	—	—	—
	B–L	—	—	—	—
di^2	M–D	—	—	—	—
	B–L	—	—	—	—
dc	M–D	7	4.1	0.31	3.7–4.6
	B–L	7	2.7	0.22	2.5–3.2
dp^2	M–D	7	3.7	0.10	3.5–4.1
	B–L	7	3.2	0.22	3.0–3.5
dp^3	M–D	7	4.0	0.23	3.8–4.3
	B–L	7	3.2	0.18	3.0–3.5
dp^4	M–D	7	5.9	0.29	5.4–6.3
	B–L	7	5.4	0.36	4.9–6.0

Table 201. *Alouatta palliata* male

Mandibular teeth		n	m	SD	Range
di$_1$	M–D	—	—	—	—
	B–L	—	—	—	—
di$_2$	M–D	—	—	—	—
	B–L	—	—	—	—
dc	M–D	5	3.7	0.45	3.3–4.5
	B–L	5	3.0	0.38	2.6–3.5
dp$_2$	M–D	6	3.8	0.17	3.6–4.1
	B–L	6	2.9	0.19	2.7–3.2
dp$_3$	M–D	6	4.1	0.18	3.9–4.4
	B–L	6	3.5	0.27	3.2–4.0
dp$_4$	M–D	6	6.2	0.28	5.9–6.7*
	B–L	6	4.2	0.32	3.9–4.8

Table 202. *Alouatta palliata* female

Mandibular teeth		n	m	SD	Range
di$_1$	M–D	—	—	—	—
	B–L	—	—	—	—
di$_2$	M–D	—	—	—	—
	B–L	—	—	—	—
dc	M–D	7	3.3	0.19	3.1–3.6
	B–L	7	2.7	0.22	2.5–3.1
dp$_2$	M–D	7	3.5	0.18	3.4–3.9
	B–L	7	2.8	0.19	2.5–3.1
dp$_3$	M–D	8	3.9	0.20	3.6–4.2
	B–L	7	3.3	0.18	3.0–3.5
dp$_4$	M–D	7	5.9	0.18	5.7–6.3
	B–L	7	4.0	0.19	3.8–4.3

Table 203. *Trachypithecus cristata* male

Maxillary teeth		n	m	SD	Range
di^1	M–D	5	3.4	0.21	3.2–3.7
	B–L	4	2.1	0.30	1.8–2.4
di^2	M–D	5	3.0	0.34	2.6–3.5
	B–L	5	2.5	0.31	2.0–2.8
dc	M–D	8	4.3	0.63	3.5–5.1
	B–L	8	2.9	0.36	2.5–3.4
dp^3	M–D	9	5.1	0.61	4.1–6.1
	B–L	9	3.6	0.47	2.7–4.1
dp^4	M–D	9	6.0	0.63	5.1–5.8
	B–L	9	4.7	0.59	4.0–5.8

Table 204. *Trachypithecus cristata* female

Maxillary teeth		n	m	SD	Range
di^1	M–D	12	3.4	0.35	2.9–4.1
	B–L	12	2.2	0.31	1.9–3.1
di^2	M–D	12	3.1	0.31	2.5–3.7
	B–L	11	2.4	0.37	1.8–3.2
dc	M–D	18	4.1	0.49	3.3–5.0
	B–L	18	2.9	0.40	1.8–3.5
dp^3	M–D	18	4.7	0.82	3.1–6.2
	B–L	18	3.7	0.71	2.2–4.8
dp^4	M–D	18	5.7	0.88	3.8–6.9
	B–L	18	4.8	0.64	3.4–5.8

Table 205. *Trachypithecus cristata* male

Mandibular teeth		n	m	SD	Range
di$_1$	M–D	—	—	—	—
	B–L	—	—	—	—
di$_2$	M–D	—	—	—	—
	B–L	—	—	—	—
dc	M–D	8	3.4	0.58	2.4–4.5
	B–L	8	2.6	0.32	2.2–3.0
dp$_3$	M–D	9	5.4	0.58	4.3–6.0
	B–L	9	2.8	0.36	2.3–3.2
dp$_4$	M–D	9	6.2	0.72	5.1–7.3
	B–L	9	3.9	0.52	3.1–4.8

Table 206. *Trachypithecus cristata* female

Mandibular teeth		n	m	SD	Range
di$_1$	M–D	—	—	—	—
	B–L	—	—	—	—
di$_2$	M–D	—	—	—	—
	B–L	—	—	—	—
dc	M–D	15	3.5	0.57	2.5–4.6
	B–L	15	2.5	0.42	1.4–3.1
dp$_3$	M–D	17	5.3	0.83	4.0–6.8
	B–L	17	2.7	0.31	2.1–3.3
dp$_4$	M–D	18	5.9	0.82	4.6–7.5
	B–L	18	3.9	0.64	3.2–5.6

Table 207. *Papio cynocephalus* male

Maxillary teeth		n	m	SD	Range
di^1	M–D	12	7.4	0.36	6.6–7.8*
	B–L	15	4.4	0.45	3.7–5.0
di^2	M–D	12	6.7	0.32	6.3–7.2*
	B–L	16	4.3	0.42	3.6–4.9
dc	M–D	13	6.6	0.41	5.8–7.2
	B–L	13	4.6	0.28	4.1–5.0
dp^3	M–D	16	8.2	0.43	7.2–8.8
	B–L	16	6.1	0.46	5.2–7.1
dp^4	M–D	16	9.5	0.44	8.9–10.6*
	B–L	16	7.4	0.45	6.8–8.3†

Table 208. *Papio cynocephalus* female

Maxillary teeth		n	m	SD	Range
di^1	M–D	15	7.0	0.45	6.3–7.8
	B–L	12	4.2	0.28	3.9–4.8
di^2	M–D	16	6.3	0.39	5.5–7.0
	B–L	12	3.8	0.39	3.1–4.3
dc	M–D	17	6.3	0.34	5.8–7.0
	B–L	17	4.4	0.26	3.8–4.8
dp^3	M–D	16	8.2	0.36	7.6–8.7
	B–L	16	5.9	0.57	5.1–7.0
dp^4	M–D	17	9.3	0.57	8.3–10.0
	B–L	17	7.0	0.62	6.0–8.2

Table 209. *Papio cynocephalus* male

Mandibular teeth		n	m	SD	Range
di$_1$	M–D	12	4.5	0.32	4.0–5.1
	B–L	12	3.3	0.25	2.7–3.6
di$_2$	M–D	12	5.3	0.31	4.7–5.8
	B–L	12	3.3	0.47	2.4–4.4
dc	M–D	12	6.2	0.59	5.1–7.1
	B–L	12	3.8	0.39	3.0–4.4
dp$_3$	M–D	17	9.1	0.39	8.5–9.7
	B–L	17	4.9	0.52	4.0–5.9
dp$_4$	M–D	17	9.6	0.52	8.8–10.3*
	B–L	17	6.2	0.56	5.4–7.4

Table 210. *Papio cynocephalus* female

Mandibular teeth		n	m	SD	Range
di$_1$	M–D	15	4.4	0.29	4.1–5.3
	B–L	15	3.3	0.63	2.2–4.4
di$_2$	M–D	15	5.1	0.52	4.0–5.7
	B–L	15	3.3	0.65	2.3–4.1
dc	M–D	16	5.9	0.46	5.1–6.8
	B–L	16	3.7	0.21	3.2–4.0
dp$_3$	M–D	17	8.7	0.59	7.9–9.4
	B–L	17	4.8	0.48	4.1–5.6
dp$_4$	M–D	18	5.4	0.46	8.5–10.0
	B–L	18	6.1	0.44	5.3–6.8

Table 211. *Chlorocebus aethiops*

Maxillary teeth		n	m	SD	Range
di^1	M–D	2	4.7	0.28	4.5–4.9
	B–L	2	2.7	0.35	2.5–3.0
di^2	M–D	2	2.7	0.35	2.5–3.0
	B–L	2	2.4	0.10	2.3–2.5
dc	M–D	4	4.3	0.38	4.0–4.9
	B–L	4	2.6	0.87	2.5–2.9
dp^3	M–D	5	5.2	0.30	4.8–5.6
	B–L	5	3.8	0.53	3.1–4.6
dp^4	M–D	5	5.6	0.31	5.3–6.1
	B–L	5	4.5	0.13	4.1–4.9

Table 212. *Chlorocebus aethiops*

Mandibular teeth		n	m	SD	Range
di_1	M–D	2	2.5	0.07	2.5–2.6
	B–L	2	2.3	0.14	2.2–2.4
di_2	M–D	2	2.6	0.35	2.4–2.9
	B–L	2	2.3	0.28	2.1–2.5
dc	M–D	4	3.6	0.14	3.4–3.7
	B–L	4	2.3	0.09	2.2–2.4
dp_3	M–D	5	5.5	0.18	5.3–5.8
	B–L	5	3.0	0.41	2.7–3.7
dp_4	M–D	5	5.6	0.13	5.5–5.8
	B–L	5	3.8	0.20	3.7–4.2

Table 213. *Rhinopithecus roxellana*

Maxillary teeth		n	m	SD	Range
di^1	M–D	4	4.5	0.36	4.1–5.0
	B–L	4	2.9	0.21	2.7–3.2
di^2	M–D	4	3.6	0.15	3.5–3.8
	B–L	3	2.9	0.11	2.8–3.0
dc	M–D	4	4.5	0.09	4.5–4.7
	B–L	4	3.2	0.17	3.0–3.4
dp^3	M–D	4	5.8	0.17	5.7–6.1
	B–L	4	4.5	0.36	4.1–4.9
dp^4	M–D	4	6.7	0.46	6.1–7.2
	B–L	4	5.8	0.20	5.7–6.1

Table 214. *Rhinopithecus roxellana*

Mandibular teeth		n	m	SD	Range
di_1	M–D	4	2.7	0.37	2.4–3.1
	B–L	4	2.4	0.18	2.3–2.7
di_2	M–D	4	2.9	0.35	2.6–3.3
	B–L	4	2.6	0.12	2.5–2.8
dc	M–D	4	3.7	0.42	3.3–4.3
	B–L	4	2.6	0.05	2.6–2.7
dp_3	M–D	4	6.2	0.31	5.8–6.5
	B–L	4	3.7	0.15	3.5–3.8
dp_4	M–D	4	6.8	0.12	6.7–7.0
	B–L	4	5.0	0.21	4.7–5.2

Table 215. *Piliocolobus badius*

Maxillary teeth		n	m	SD	Range
di^1	M–D	7	3.8	0.37	3.1–4.1
	B–L	7	2.3	0.27	1.8–2.6
di^2	M–D	5	2.7	0.20	2.5–3.0
	B–L	5	2.1	0.10	2.0–2.2
dc	M–D	10	4.0	0.32	3.6–4.5
	B–L	10	2.4	0.21	2.0–2.8
dp^3	M–D	10	5.0	0.33	9.4–5.6
	B–L	10	3.5	0.25	3.2–4.1
dp^4	M–D	10	5.9	0.23	5.5–6.2
	B–L	10	4.6	0.40	4.2–5.5

Table 216. *Piliocolobus badius*

Mandibular teeth		n	m	SD	Range
di_1	M–D	5	2.1	0.30	1.8–2.6
	B–L	5	2.0	0.24	1.8–2.4
di_2	M–D	5	2.2	0.30	1.8–2.5
	B–L	5	2.2	0.29	2.0–2.7
dc	M–D	8	3.2	0.72	2.0–4.1
	B L	8	2.1	0.16	1.9–2.4
dp_3	M–D	10	5.0	1.17	2.7–5.8
	B–L	10	2.7	0.23	2.5–3.3
dp_4	M–D	10	5.8	0.77	3.9–6.7
	B–L	10	3.9	0.41	3.3–4.7

Table 217. *Lophocebus albigena*

Maxillary teeth		n	m	SD	Range
di^1	M–D	1	5.0	—	5.0–5.0
	B–L	4	3.6	0.17	3.4–3.8
di^2	M–D	4	5.1	0.20	5.0–5.4
	B–L	4	2.9	0.26	2.7–3.3
dc	M–D	4	4.3	0.30	4.0–4.6
	B–L	5	3.3	0.25	3.0–3.6
dp^3	M–D	5	4.8	0.17	4.7–5.1
	B–L	6	5.1	0.34	4.7–5.6
dp^4	M–D	6	6.0	0.40	5.5–6.5
	B–L	7	6.1	0.39	5.5–6.7

Table 218. *Lophocebus albigena*

Mandibular teeth		n	m	SD	Range
di_1	M–D	4	3.2	0.36	2.7–3.5
	B–L	4	2.9	0.15	2.7–3.0
di_2	M–D	4	3.5	0.16	3.1–3.8
	B–L	4	2.4	0.15	2.3–2.6
dc	M–D	5	4.2	0.28	3.9–4.5
	B–L	5	2.7	0.23	2.6–3.2
dp_3	M–D	6	6.3	0.42	5.9–7.0
	B–L	6	3.8	0.29	3.4–4.2
dp_4	M–D	6	7.0	0.24	6.8–7.5
	B–L	6	4.7	0.36	4.4–5.4

Table 219. *Cercopithecus mitis*

Maxillary teeth		n	m	SD	Range
di^1	M–D	11	4.3	0.34	3.8–4.9
	B–L	11	2.7	0.27	2.3–3.1
di^2	M–D	11	2.5	0.20	2.1–2.8
	B–L	11	2.1	0.28	1.8–2.7
dc	M–D	12	3.9	0.46	3.2–4.8
	B–L	12	2.5	0.22	2.2–3.0
dp^3	M–D	13	4.9	0.45	4.0–4.1
	B–L	13	3.5	0.37	2.9–4.1
dp^4	M–D	12	5.3	0.45	4.3–5.9
	B–L	12	4.1	0.38	3.6–4.9

Table 220. *Cercopithecus mitis*

Mandibular teeth		n	m	SD	Range
di$_1$	M–D	9	2.3	0.12	2.1–2.5
	B–L	9	2.1	0.32	1.6–2.6
di$_2$	M–D	9	2.4	0.25	2.1–2.8
	B–L	9	2.2	0.30	1.8–2.7
dc	M–D	10	3.3	0.33	3.0–4.0
	B–L	10	2.1	0.22	1.8–2.6
dp$_3$	M–D	13	5.3	0.42	4.6–6.0
	B–L	13	2.5	0.43	2.2–3.8
dp$_4$	M–D	12	5.5	0.52	4.5–6.4
	B–L	12	3.4	0.28	3.0–3.9

Appendix 2: Dental eruption sequences

Dental eruption sequences are important indicators of life history patterns, are relatively robust to environmental disturbances, are useful for aging individuals, and have proved valuable in paleontological studies. In fact, Aristotle himself knew that tooth eruption could be used to estimate the age of an animal (Smith *et al.*, 1994). Smith *et al.* (1994) said it most succinctly when they wrote 'that teeth are a linchpin between the present and the past.' The paper by Smith *et al.* (1994) is the most thorough compendium so far available on the ages of dental eruption in extant primates.

The dental eruption sequences of permanent teeth presented here include data from prosimians to the great apes and represents the work of many individuals for many years. With respect to tooth eruption, the first permanent tooth to erupt in all primates is the first molar; however, after that, the sequences differ among the taxa in various ways as seen in the data presented here. For example, the last tooth to erupt is extremely variable; it may be the third molar, a canine or even a premolar.

Sources: Harvati, 2000 (colobines); Schultz, 1935 (*Hylobates* and *Pongo*); Smith *et al.* 1994 (prosimians, cercopithecines, platyrrhines, *Pan* and *Gorilla*). Definitions, as used by Harvati (2000): brackets indicate any sequence polymorphism; the equal sign (=) between two teeth indicates that the two alternative sequences can occur an equal number of times; and the question mark indicates no data for that position. Where sexes are not designated the sexes are combined.

Prosimians

The upper and lower incisors were not recorded, except for *L. catta.*

Lemur catta
$$\frac{M^1 \; M^2 \qquad I^2 \; I^1 \; P^4 P^3 \quad C \; P^2 \; M^3}{M_1 \; M_2 \;\; I_1 \; I_2 \; C \qquad P_3 \; P_4 \;\; P_2 \; M_3}$$

Eulemur macaco
$$\frac{M^1 \; M^2 \;\; P^2 \quad P^3 \; M^3}{M_1 \; M^2 \; P_2 \; P_3 \quad M_3}$$

Eulemur fulvus (male)
$$\frac{M^1 \quad M^2 \quad C \; P^2 \; P^3 \; M^3}{M_1 \quad M_2 \; P_2 \; M_3 \qquad\quad P_3}$$

Varecia variegata
$$\frac{M^1 \; M^2 \quad P^2 \quad M^3 \quad C}{M_1 \quad M_2 \quad P_2 \quad M_3 \quad P_3}$$

New World monkeys

Callithrix jacchus
$$\frac{M^1 \; I^1 \qquad M^2 \; P^4 \; P^3 \qquad I^2 \; P^2 \quad C}{M_1 \qquad I_1 \; M_2 \; P_4 \qquad I_2 \; P_3 \quad P_2 \quad C}$$

Saguinus fuscicollis
$$\frac{M^1 \; I^1 \quad I^2 \; P^4 \; M^2 \quad P^2 \quad P^3 \; C}{M_1 \; I_1 \quad I_2 \; M_2 \; P_4 \qquad P_2 P_3 \quad C}$$

Saguinus nigricollis
$$\frac{M^1 \; I^1 \quad I^2 \; M^2 \; P^4 \; P^3 \quad P^2 \quad C}{M_1 \; I_1 \; I_2 \; M_2 \quad P_4 \qquad P_2 P_3 \quad C}$$

Cebus apella
$$\frac{M^1 \; I^1 \quad I^2 \quad M^2}{M_1 \; I_1 \qquad I_2 \quad M_2}$$

Saimiri sciureus
$$\frac{M^1 \quad M^2 \; I^1 \; I^2 \qquad P^4 \; P^3 \; P^2 \qquad\quad M^3 \quad C}{M_1 \; M_2 \quad I_1 \qquad I_2 \; P_4 \qquad\quad P_3 \; P_2 \; M_3 \quad C}$$

Aotus trivirgatus
$$\frac{M^1 \quad M^2 \; I^1 \qquad I^2 \quad M^3 \; P^4 \; P^3 \; P^2 \qquad\qquad C}{M_1 \quad M_2 \qquad I_1 \; M_3 \; I_2 \; P_4 \qquad\qquad\quad P_2 \; P_3 \; C}$$

Old World monkeys

Macaca fascicularis (male)
$$\frac{M^1 \; I^1 \; I^2 \quad C \; P^3 \; M^2 \quad P^4 \qquad\qquad M^3}{M_1 \quad I_1 \;\; I_2 \quad M_2 \qquad\qquad C \; P_3 \; P_4 \quad M_3}$$

Macaca fascicularis (female)
$$\frac{M^1 \; I^1 \; I^2 \quad C \; P^3 \; P^4 \; M^2 \qquad\qquad M^3}{M_1 \quad I_1 \; I_2 \quad C \; M_2 \qquad\qquad P_3 \; P_4 \quad M_3}$$

272   *Appendix 2: Dental eruption sequences*

Macaca fuscata (male)

$$\frac{M^1\ I^1\ I^2\ M^2\ P^3\ P^4\qquad C\ M^3}{M_1\ I_1\ I_2\ M_2\quad P_3\ P_4\qquad C\quad M_3}$$

Macaca fuscata (female)

$$\frac{M^1\ I^1\ I^2\ M^2\ C\ P^3\ P^4\qquad M^3}{M_1\ I_1\ I_2\ M_2\ C\ P_3\qquad P_4\qquad M_3}$$

Macaca mulatta (male)

$$\frac{M^1\ I^1\ I^2\ M^2\ P^3\ P^4\qquad C\qquad M^3}{M_1\ I_1\ I_2\ M_2\qquad P_4\ P_3\ C\qquad M_3}$$

Macaca mulatta (female)

$$\frac{M^1\ I^1\ I^2\ M^2\ P^3\ C\ P^4\qquad M^3}{M_1\ I_1\ I_2\ M_2\ C\ P_3\qquad P_4\qquad M_3}$$

Macaca nemestrina (male)

$$\frac{M^1\ I^1\ I^2\ M^2\ P^4\ P^3\qquad C\ M^3}{M_1\ I_1\ I_2\ M_2\quad P_4\quad P_3\ C\ M_3}$$

Macaca nemestrina (female)

$$\frac{M^1\ I^1\ I^2\qquad M^2\ P^3\ P^4\ C\qquad\qquad M^3}{M_1\ I_1\qquad I_2\ M_2\ C\qquad\qquad P_3\ P_4\ M_3}$$

Papio anubis (male)

$$\frac{M^1\ I^1\ I^2\ M^2\ P^3\ C\ P^4\qquad M^3}{M_1\ I_1\ I_2\ M_2\qquad C\ P_3\ P_4\qquad M_3}$$

Papio anubis (female)

$$\frac{M^1\ I^1\ I^2\ M^2\qquad C\ P^3\ P^4\qquad M^3}{M_1\ I_1\ I_2\qquad C\ M_2\ P_3\ P_4\qquad M_3}$$

Papio cynocephalus (male)

$$\frac{M^1\ I^1\ I^2\ M^2\ P^3\ P^4\qquad C\qquad M^3}{M_1\ I_1\ I_2\ M_2\qquad P_4\ C\ P_3\ M_3}$$

Papio cynocephalus (female)

$$\frac{M^1\ I^1\ I^2\qquad M^2\ C\ P^3\ P^4\qquad M^3}{M_1\ I_1\qquad I_2\ M_2\ C\ P_3\qquad P_4\ M_3}$$

Cercopithecus aethiops (male)

$$\frac{M^1\ I^1\ I^2\ M^2\qquad P^4\ P^3\ C\ M^3}{M_1\ I_1\ I_2\ M_2\quad P_4\ P_3\ C\ M_3}$$

Colobus angolensis

$$\frac{M^1\ [M^2\ I^1]\ [I^2\ P^3{=}P^4]\ C\qquad M^3}{M_1\quad M_2\ I_1\ I_2\ P_4\ [P_3\ M_3\quad C]}$$

Colobus guereza

$$\frac{M^1\ I^1\ M^2\ I^2\ [P^3{=}P^4]\ [M^3\ C]}{M_1\ I_1\ I_2\ M_2\ P_4\ [P_3\ M_3\ C]}$$

Nasalis

$$\frac{M^1?\ I^1?\ I^2\ M^2\ P^3?\ P^4\ C\ M^3}{M_1?\ I_1?\ I_2\ M_2\ [P_3?\ P_4\ C]\ M_3}$$

Procolobus

$$\frac{M^1\ [I^1\ M^2]\ I^2\ [P^3\ P^4]\ [C\qquad M^3]}{M_1\ [I_1\ I_2\ M_2]\ P_4\ [P_3\ C\qquad M_3]}$$

Trachypithecus

$$\frac{M^1 \quad I^1 \quad M^2 \quad I^2 \quad [P^3? \quad P^4 \quad C] \quad M^3}{M_1 \quad I^1 \quad [I_2 \quad M_2] \quad P_4 \quad P_3 \quad [C \quad M_3]}$$

Presbytis

$$\frac{M^1 \quad M^2 \quad I^1 \quad I^2 \quad [M^3 \quad P^4 \quad P^3 \quad C]}{M_1 \quad M_2 \quad I_1 \quad I_2 \quad [M_3 \quad P_4 \quad P_3] \quad C}$$

Pygathrix

$$\frac{M^1 \quad [I^1 \quad M^2] \quad I^2 \quad P^3? \quad P^4 \quad C \quad M_3}{M_1 \quad I_1 \quad M_2 \quad I_2 \quad C \quad P_3? \quad P_4 \quad M_3}$$

Pongids

Hylobates

$$\frac{M^1 \quad I^1 \quad I^2 \quad M^2 \quad P^3 \quad P^4 \quad C \quad M^3}{M_1 \quad I_1 \quad I_2 \quad M_2 \quad P_3 \quad P_4 \quad C \quad M_3}$$

Pongo pygmaeus

$$\frac{M^1 \quad I^1 \quad I^2 \quad M^2 \quad P^3 \quad P^4 \quad C \quad M^3}{M_1 \quad I_1 \quad I_2 \quad M_2 \quad P_4 \quad P_3 \quad C \quad M_3}$$

Pan troglodytes (male)

$$\frac{M^1 \quad I^1 \quad I^2 \quad M^2 \quad P^3 \quad P^4 \quad C \quad M^3}{M_1 \quad I_1 \quad I_2 \quad M_2 \quad P_3 \quad P_4 \quad C \quad M_3}$$

Pan troglodytes (female)

$$\frac{M^1 \quad I^1 \quad M^2 \quad I^2 \quad P^3 \quad P^4 \quad C \quad M^3}{M_1 \quad I_1 \quad I_2 \quad M_2 \quad P_3 \quad P_4 \quad C \quad M_3}$$

Gorilla gorilla

$$\frac{M^1 \quad I^1 \quad I^2 \quad M^2 \quad P^4 \quad P^3 \quad C \quad M^3}{M_1 \quad I_1 \quad I_2 \quad M_2 \quad P_4 \quad P_3 \quad C \quad M_3}$$

Glossary

amelogenesis the formation and maturation of enamel

antimeres the corresponding teeth opposite each other, e.g. the upper right and left first molars

apomorphic a derived character state resulting from recent adaptations

bilophodont molar construction in which the mesial and distal pairs of cusps in both upper and lower molars are connected by parallel transverse ridges (lophs); characteristic of Old World monkeys

brachydont a low-crowned tooth usually with well-formed roots that become closed when fully erupted

buccal toward the lateral (cheek) side of a tooth

bunodont a tooth with low, rounded, cone-shaped cusps

caniniform in the shape of a canine tooth

catarrhine primates members of the infraorder Catarrhini of Old World anthropoid primates that include Old World monkeys, apes, and humans

cingulum (pl. cingula) a ridge or shelf of enamel passing around the crown of a tooth

Cretaceous the last geologic period of the Mesozoic Era that ended approximately 65 million years ago

cusp an elevation on the occlusal surface of a tooth ending as a conical, rounded, or flat surface

deciduous shed; teeth that are replaced by other teeth during the course of a mammal's life, e.g. the baby or milk teeth

dental alveolus (pl. alveoli) the socket in the jaw that receives the tooth roots

dental comb the procumbent lower incisors and canines of prosimians used for scraping and scooping resin from trees as well as for grooming

dental formula the number and types of teeth in each quadrant of the upper and lower jaws; for example, I^2-C-P^2-M^3 / I_2-C-P_2-M_3 is the dental formula of the permanent dentition of Old World monkeys, apes and humans and indicates that there are two incisors, one canine, two premolars, and three molars in each quadrant, or a total of 32 permanent teeth

dental lamina the thickening of the oral epithelium along each dental arch where tooth formation is initiated

dentinoenamel junction (DEJ) the line marking the junction of the dentine with the enamel

diastema a cleft or space; any space between teeth, e.g. the diastema between the upper lateral incisor and canine

Eocene the second epoch of the Cenozoic Era, 55 35 million years ago

epharmosis (underbite) protrusion of the lower incisors relative to the upper incisors with normal molar relationship

faunivory (adj. faunivorous) feeding primarily on animal matter

folivory (adj. folivorous) feeding primarily on leaves

frugivory (adj. frugivorous) feeding primarily on fruits

gomphosis the fibrous joint formed by the union of the roots of the teeth with the walls of the dental alveoli (sockets)

Gondwanaland one of the two supercontinents that broke away from Pangaea when it split apart about 200 million years ago

gummivory (adj. gummivorous) feeding primarily on gums

hetermorphic of different sizes and shapes

heterochrony a change in the timing of events during development

heterodont teeth differentiated for different functions, e.g. incisors, canines, premolars and molars

homodont teeth that are uniformly of the same shape, e.g. the teeth of dolphins

homomorphic of similar shape and size

homoplasy (adj. homoplastic) similar features that are present in different animal species that are not the result of common ancestry

hypoplasia a type of enamel defect caused by disruption of enamel matrix formation

hypsodont a high-crowned tooth with deeply folded enamel, e.g. those of the elephant

incisiform shaped like an incisor

insectivory (adj. insectivorous) feeding primarily on insects

labial toward the lips

Laurasia one of the two supercontinents that broke away from Pangaea when it split apart about 200 million years ago (see Gondwanaland)

lingual toward the tongue

lophodont a tooth with the cusps connected by a series of lophs (lophids) that may run transversely or obliquely; when worn, the enamel pattern resembles a series of crescents, the selenodont pattern of many ungulates

mammelon a rounded or conical elevation on the incisal ridge of newly erupted incisors

meristic series duplicated or repeated structures, e.g. vertebrae

Mohs scale a scale of hardness from talc (1) to diamond (10)

molariform molar-like in form and function

occlusal the surface of a premolar or molar tooth that meets the opposing teeth in the closure of the jaws

occlusion when the teeth of the upper and lower arches come into contact in any functional relationship, occlusion is achieved

perikymata a series of regular ridges on the outer surface of teeth that represent the striae of Retzius within the enamel

periodontal ligament the gingival attachment to the tooth, also called the suspensory ligament

platyrrhine primates members of the infraorder Platyrrhini or New World primates

plesiomorphic an ancestral or primitive character

sectorial an elongated crown, usually the lower third premolar (P_3), that has a honing facet for shearing against the distolingual surface of the upper canine in Old World monkeys and apes

strepsirhine member of the suborder Strepsirhini that includes the Lemurifomes

Striae of Retzius regular long-period incremental lines in enamel, representing successive layers of enamel formation. Aetiology unknown

stylar shelf (external cingulum) remnants of the stylar shelf are from mesial to distal the parastyle, mesostyle, and metastyle; these are common in prosimians and platyrrhine monkeys

sulcus obliquus the oblique sulcus separating the hypocone from the trigon, particularly well developed in pongids

tribosphenic molar a mammalian molar possessing a protocone lingual to the paracone and metacone and a talonid basin on the lower molar that receives the protocone

Tritubercular theory a theory explaining the evolution of the mammalian molar pattern; often referred to as the Cope–Osborn theory

tuberculum dentale an enamel enlargement on the middle of the lingual cingulum of maxillary incisors that may have none, one, or several extensions passing from it toward the incisal border

tuberculum intermedium a tubercle on the distal surface of the metaconid or in the groove between the metaconid and entoconid on lower molars, also called the postmetaconulid

tuberculum sextum a tubercle between the entoconid and the hypoconulid, also known as the postentoconulid

References

Aas, I. H. (1983). Variability of a dental morphological trait. *Acta Odontologica Scandinavica*, **41**, 257–63.

Adloff, P. (1908). *Das Gebiss des Menschen und der Anthropomorphen.* Berlin: Julius Springer.

Aiello, L. and Dean, C. (1990). *An Introduction to Human Evolutionary Anatomy.* London: Academic Press.

Alexandersen, V. (1963). Double-rooted lower human canine teeth. In *Dental Anthropology,* ed. D. R . Brothwell, pp. 235 244, Oxford: Pergamon Press.

Alvesalo, L. (1971). The influence of sex chromosome genes on tooth size in man. *Proceedings of the Finnish Dental Society* **67**, 3–54.

Alvesalo, L. and Tigerstedt, P. M. A. (1974). Heritabilities of human tooth dimensions. *Hereditas* **77**, 311–18.

Anapol, F. and Lee, S. (1994). Morphological adaptation to diet in platyrrhine primates. *American Journal of Physical Anthropology* **94**, 239–61.

Anemone, R. L., Watts, E. S. and Swindler, D. R. (1991). Dental development of known age chimpanzees, *Pan troglodytes*, (Primates, Pongidae). *American Journal of Physical Anthropology* **86**, 229–42.

Ankel-Simons, F. (1996). Deciduous dentition of the Aye-Aye, *Daubentonia madagascariensis. American Journal of Primatology* **39**, 87–97.

Ashton, E. H. and Zuckerman, S. (1950). Some quantitative dental characteristics of the chimpanzee, gorilla and orang-utan. *Philosophical Transactions of the Royal Society of London* **B234**, 471–84.

Ayers, J. M. (1989). Comparative feeding ecology of the uakari and bearded saki, *Cacajao* and *Chiropotes. Journal of Human Evolution* **18**, 697–716.

Baume, L. J. and Becks, H. (1950). The development of the dentition of *Macaca mulatta. American Journal of Orthodontics* **36**, 723–48.

Benefit, B. R. (1994). Phylogenetic, paleodemographic, and taphonomic implications of *Victoriapithecus* deciduous teeth from Maboko, Kenya. *American Journal of Physical Anthropology* **95**, 277–332.

Bennejeant, C. (1936). *Anomalies et Variations Dentaires chez les Primates.* Paris, Clermont-Ferrand: Imprimeries Paul Vallier.

Beynon, A. D., Dean, M. C. and Reid, D. J. (1991). On thick and thin enamel in hominoids. *American Journal of Physical Anthropology* **86**, 295–309.

Biggerstaff, R. H. (1970). Morphological variations for the permanent mandibular first molars in human monozygotic and dizygotic twins. *Archives of Oral Biology* **22**, 721–30.

Bock, W. L. and von Wahlert, T. (1965). Adaptation and the form-function complex. *Evolution* **19**, 269–99.

Boyde, A. (1969). Correlation of ameloblast size with enamel prism pattern; use of scanning electron microscope to make surface area measurements. *Zeitschrift für Zellforschurg* **93**, 583–93.

Boyde, A. (1976). Amelogenesis and the development of teeth. In *Scientific Foundations of Dentistry*, ed. B. Cohn and I. R. H. Kramer, pp. 335–52. London: Heinemann.

Boyde, A. and Martin, L. B. (1984). The microstructure of primate dental enamel. In *Food Acquisition and Processing in Primates*, ed. D. J. Chivers, B. A. Wood and A. Billsborough, pp. 341–67. New York: Plenum Press.

Brook, A. H. (1998). A unifying model for the aetiology of enamel defects. In *Dental Morphology, 1998*, ed. J. T. Mayhall and T. Heikkinen, pp. 128–32. Oulu: Oulu University Press.

Butler, P. M. (1939). Studies of the mammalian dentition. Differentiation of the post-canine teeth. *Proceedings of the Zoological Society of London* **B109**, 1–36.

Butler, P. M. (1956). The ontogeny of molar patterns. *Biological Reviews* **31**, 30–70.

Butler, P. M. (1967). Dental merism and tooth development. *Journal of Dental Research* **46**, suppl. 5, 845–50.

Butler, P. M. (1990). Early trends in the evolution of tribosphenic molars. *Biological Reviews* **65**, 329–32.

Butler, P. M. (1992). Tribosphenic molars in the Cretaceous. In *Structure, Function and Evolution of Teeth*, ed. P. Smith and E. Tchernov, pp. 125–38. London: Freund Publishing House Ltd.

Buttner-Janusch, J. and Andrew, R. J. (1962). The use of the incisors by primates in grooming. *American Journal of Physical Anthropology* **20**, 127–9.

Campbell, T. D. (1925). *The Dentition and Palate of the Australian Aboriginal.* Adelaide: Hassell Press.

Cartmill, M. (1972). *Daubentonia*, woodpeckers and klinorhynchy. *American Journal of Physical Anthropology* **37**, 432. (Abstract.)

Cartmill, M. (1974). *Daubentonia, Dactylopsila*, woodpeckers and klinorhynchy. In *Prosimian Biology*, ed. R. D. Martin, G. A. Doyle and A. C. Walker, pp. 655–72. London: Duckworth.

Clements, E. M. B. and Zuckerman, S. (1953). The order of eruption of the permanent teeth in the Hominoidea. *American Journal of Physical Anthropology* **11**, 313–22.

Colyer, F. (1936). *Variations and Diseases of the Teeth of Animals.* London: John Bale Sons and Danielsson.

Cope, D. A. (1993). Measures of dental variation as indicators of multiple taxa in samples of sympatric *Cercopithecus* species. In *Species, Species Concepts, and Primate Evolution*, ed. W. H. Kimbel and L. B. Martin, pp. 211–37. New York: Plenum Press.

Cope, E. D. (1888). On the tritubercular molar in the human dentition. *Journal of Morphology* **2**, 7–23.

Crompton, R. H. and Andau, P. M. (1986). Locomotion and habitat utilization in free-ranging *Tarsius bancanus*: A preliminary report. *Primates* **27**, 337–55.

Curtin, S. H. and Chivers, D. J. (1978). Leaf-eating primates of peninsular Malaysia: the siamang and the dusky leaf-eating monkey. In *The Ecology of Arboreal Folivores*, ed. G. G. Montgomery, pp. 441–64. Washington, D.C.: Smithsonian Institution Press.

Dean, M. C. (1989). The developing dentition and tooth structure in primates. *Folia Primatologica* 53, 160–77.

Dean, M. C. (1993). Daily rates of dentine formation in macaque tooth roots. *International Journal of Osteoarchaeology* 3, 199–206.

Dean, M. C. (2000). Incremental markings in enamel and dentine: what they can tell us about the way teeth grow. In *Development, Function and Evolution of Teeth*, ed. M. F. Teaford, M. M. Smith and M. W. J. Ferguson, pp. 119–130. Cambridge University Press.

Dean, M. C. and Wood, B. A. (1981). Developing pongid dentition and its use for aging individual crania in comparative cross-sectional growth studies. *Folia Primatologica* 36, 111–32.

Delson, E. (1973). Fossil colobine monkeys of the Circum-Mediterranean region, Parts 1 and 2. Ann Arbor: University of Michigan Microfilms.

Delson, E. (1975). Evolutionary history of the Cercopithecidae. In *Approaches to Primate Paleobiology: Contributions to Primatology*, ed. F. Szalay, vol. 5, pp. 167–217. Basel: S. Karger.

De Terra, M. (1905). *Beitrage zu Einer Odontographie der Menschenrassen*. Berlin: Berlinische Verlagsanstalt.

Disotell, T. R. (1994). Generic level relationships of the Papionini (Cercopithecoidea). *American Journal of Physical Anthropology* 94, 47–57.

Disotell, T. R. (1996). The phylogeny of Old World monkeys. *Evolutionary Anthropology* 5, 18–24.

Drusini, A., Calliari, I. and Volpe, A. (1991). Root dentine transparency: Age determination of human teeth using computerized densitometric analysis. *American Journal of Physical Anthropology* 85, 25–30.

Dumont, E. R. (1995). Enamel thickness and dietary adaptation among extant primates and chiropterans. *Journal of Mammalogy* 76, 1127–36.

Eaglen, R. H. (1980). Toothcomb homology and toothcomb function in extant strepsirhines. *International Journal of Primatology* 1, 275–86.

Eaglen, R. H. (1984). Incisor size and diet revisited: The view from a platyrrhine perspective. *American Journal of Physical Anthropology* 64, 263–76.

Eaglen, R. H. (1986). Morphometrics of the anterior dentition in strepsirhine primates. *American Journal of Physical Anthropology* 71, 185–201.

Emel, L. M. and Swindler, D. R. (1992). Underbite and the scaling of facial dimensions in colobine monkeys. *Folia Primatologica* 58, 177–89.

Erdbrink, D. P. (1965). A quantification of the *Dryopithecus* and other lower molar patterns in man and some of the apes. *Zeitschrift für Morphologie und Anthropologie* 57, 70–108.

Falk, D. (2000). *Primate Diversity*. New York: W. W. Norton & Company.

Fanning, E. A. (1961). A longitudinal study of tooth formation and root absorption. *New Zealand Dental Journal* 57, 202–17.

Fleagle, J. G. (1999). *Primate Adaptation and Evolution* 2nd edn. London: Academic Press.

Ford, S. M. and Bargielski, M. M. (1985). Dental allometry in the Platyrrhini and changes in a dwarfing lineage. *American Journal of Physical Anthropology* **66**, 169.

Frisch, J. E. (1965). Trends in the evolution of the hominoid dentition. *Bibliotheca Primatologica* **3**, 1–130.

Gantt, D. G. (1982). Hominid evolution — a tooth's inside view. In *Teeth: Form, Function and Evolution*, ed. B. Kurten, pp. 107–20. New York: Columbia University Press.

Gantt, D. G. (1986). Enamel thickness and ultrastructure in hominoids: with reference to form, function, and phylogeny. In *Systematics, Evolution, and Anatomy. Comparative Primate Biology*, vol. 1, ed. D. R. Swindler and J. Erwin, pp. 453–76. New York: Alan R. Liss, Inc.

Gantt, D. G., Strickland, C. P. and Rafter, J. A. (1998). Loss of lingual enamel in lower incisors of Papionini. *Dental Anthropology* **13**, 1–2.

Ganzhorn, J. U., Abraham, J. P. and Razanahoera-Rakotomalala, M. (1985). Some aspects of the natural history and food selection of *Avahi laniger*. *Primates* **25**, 452–63.

Garber, P. A. (1992). Vertical clinging, small body size, and the evolution of feeding adaptations in the Callitrichinae. *American Journal of Physical Anthropology* **88**, 469–82.

Garber, P. A. and Kinzey, W. G. (eds) (1992). Feeding adaptations in New World primates: An evolutionary perspective: Introduction. *American Journal of Physical Anthropology* **88**, 411–14.

Garn, S. M. and Lewis, A. B. (1957). Relationship between the sequence of calcification and the sequence of eruption of the mandibular molar and premolar teeth. *Journal of Dental Research* **36**, 992–105.

Garn, S. M., Kerewsky, R. S. and Swindler, D. R. (1966). Canine 'field' in sexual dimorphism of tooth size. *Nature* **212**, 1501–2.

Garn, S. M., Lewis, A. B. and Kerewsky, R. S. (1964). Sex differences in tooth size. *Journal of Dental Research* **43**, 306.

Garn, S. M., Lewis, A. B. and Kerewsky, R. S. (1965). Size interrelationships of the mesial and distal teeth. *Journal of Dental Research* **44**, 350–3.

Garn, S. M., Lewis, A. B. and Kerewsky, R. S. (1967). Shape similarities throughout the dentition. *Journal of Dental Research* **46**, 1481.

Gingerich, P. D. (1975). Dentition of *Adapis parisiensis* and the evolution of the lemuriform primates. In *Lemur Biology*, ed. I. Tattersall and R. W. Sussman, pp. 65–80. New York: Plenum Press.

Gingerich, P. D. (1977). Homologies of the anterior teeth in Indriidae and a functional basis for dental reduction in primates. *American Journal of Physical Anthropology* **47**, 387–94.

Glasstone, S. (1963). Regulative changes in tooth germs grown in tissue culture. *Journal of Dental Research* **42**, 1364–8.

Glasstone, S. (1966). Morphodifferentiation of teeth in embryonic mandibular segments in tissue culture. *Journal of Dental Research* **46**, 611–14.

Gleiser, I. and Hunt, E. E. (1955). The permanent mandibular first molar: its calcification, eruption and decay. *American Journal of Physical Anthropology*

13, 253–84.

Goodman, A. H. and Rose, J. C. (1990). Assessment of systematic physiological perturbations from dental enamel hypoplasias and associated histological structures. *Yearbook of Physical Anthropology* **33**, 59–110.

Goose, D. H. (1971). The inheritance of tooth size in British families. In *Dental Morphology and Evolution*, ed. A. A Dahlberg, pp. 263–70. Chicago: University of Chicago Press.

Gould, S. J. (1977). *Ever Since Darwin: Reflections in Natural History*. New York: W. W. Norton and Co.

Greene, D. L. (1973). Gorilla dental sexual dimorphism and early hominid taxonomy. In *Craniofacial Biology of Primates*, ed. M. R. Zingeser, vol. 3, pp. 82–100, Basel: Karger.

Greenfield, L. O. (1992). Relative canine size, behavior and diet in male ceboids. *Journal of Human Evolution* **23**, 469–80.

Greenfield, L. O. and Washburn, A. (1991). Polymorphic aspects of male anthropoid canines. *American Journal of Physical Anthropology* **84**, 17–34.

Greenfield, L. O. and Washburn, A. (1992). Polymorphic aspects of male anthropoid honing premolars. *American Journal of Physical Anthropology* **87**, 173–86.

Gregory, W. K. (1916). Studies on the evolution of the primates. *Bulletin of the American Museum of Natural History* **35**, 239–55.

Gregory, W. K. (1922). *The Origin and Evolution of the Human Dentition*. Baltimore: Williams and Wilkins Company.

Gregory, W. K. and Hellman, M. (1926). The dentition of *Dryopithecus*, and the origin of man. *Anthropological Papers of the American Museum of Natural History* **28**, 9–117.

Groves, C. P. (1993). Order Primates. In *Mammalian Species of the World: A Taxonomic and Geographic Reference*, 2nd edn, ed. D. E. Wilson and D. M. Reader, pp. 243–77. Washington, D.C.: Smithsonian Institution Press.

Guatelli-Steinberg, D. (2000). Linear enamel hypoplasia in Gibbons (*Hylobates lar carpenteri*). *American Journal of Physical Anthropology* **112**, 395–410.

Guatelli-Steinberg, D. (1998). Prevalence and etiology of linear enamel hypoplasia in non-human primates. Ph.D. dissertation, University of Oregon, Eugene, Oregon.

Guatelli-Steinberg, D. and Lukacs, J. R. (1999). Interpreting sex differences in enamel hypoplasia in human and non-human primates: Developmental, environmental, and cultural considerations. *Yearbook of Physical Anthropology* **42**, 73–126.

Guatelli-Steinberg, D. and Skinner, M. (2000). Prevalence and etiology of linear enamel hypoplasia in monkeys and apes from Asia and Africa. *Folia Primatologica* **71**, 115–32.

Gustafson, G. (1950). Age determination of teeth. *Journal of the American Dental Association* **41**, 45–62.

Haavikko, K. (1970). The formation and the alveolar and clinical eruption of the permanent teeth. *Proceedings of the Finnish Dental Society* **66**, 101–70.

Hall, B. K. (1992). *Evolutionary Developmental Biology*. London: Chapman and Hall.

Hanihara, T. and Natori, M. (1987). Preliminary analysis of numerical taxonomy of the genus *Saguinus* based on dental measurements. *Primates* **28**, 517–23.

Hanihara, T. and Natori, M. (1988). Numerical analysis of sexual dimorphism in *Saguinus* dentition. *Primates* **29**, 245–54.

Happel, R. (1988). Seed-eating by West African cercopithecines, with reference to the possible evolution of bilophodont molars. *American Journal of Physical Anthropology* **75**, 303–27.

Harvati, K. (2000). Dental eruption sequence among Colobine primates. *American Journal of Physical Anthropology* **112**, 69–86.

Henke, W. and Rothe, H. (1997). Zahnphylogenese der nichtmenschlichen Primaten, In *Die Evolution der Zähne: Phylogenie–Ontogenie–Variation*, ed. K. W. Alt and J. C. Turp, pp. 229–78. Berlin: Quintessence Verlag.

Hershkovitz, P. (1971). Basic crown patterns and cusp homologies of mammalian teeth. In *Dental Morphology and Evolution*, ed. A. A. Dahlberg, pp. 95–150. Chicago: The University of Chicago Press.

Hershkovitz, P. (1985). A preliminary taxonomic review of the South American bearded saki monkeys genus *Chiropotes* (Cebidae, Platyrrhini) with the description of a new subspecies. *Fieldiana* (NS) **27**, 1–46.

Hershkovitz, P. (1990). Titis: New World monkeys of the genus *Callicebus* (Cebidae, Platyrrhini): a preliminary taxonomic review. *Fieldiana* **55**, 1–109.

Hildebolt, C. F., Bate, G., McKnee, J. K. and Conroy, G. C. (1986). The microstructure of dentine in taxonomic and phylogenetic studies. *American Journal of Physical Anthropology* **70**, 39–46.

Hill, W. C. O. (1953). *Primates: Comparative Anatomy and Taxonomy*, vol. 1, *Strepsirhini*. New York: Interscience Publishers.

Hillson, S. (1996). *Dental Anthropology*. Cambridge University Press.

Hillson, S. and Bond, S. (1997). Relationship of enamel hypoplasia to the pattern of tooth crown growth: A discussion. *American Journal of Physical Anthropology* **104**, 89–104.

Hladik, C. M. (1979). Diet and ecology of prosimians. In *The Study of Prosimian Behavior*, ed. G. A. Doyle and R. D. Martin, pp. 307–57. New York: Academic Press.

Hladik, C. M., Charles-Dominique, P. and Peter, J. J. (1980). Feeding strategies of five nocturnal prosimians in the dry forest of the West Coast of Madagascar. In *Nocturnal Malagasy Primates*, ed. P. Charles-Dominique, H. M. Cooper, A. Hladik, C. M. Hladik, E. Pages, G. E. Pariente, A. Petter-Rousseaux and A. Schilling, pp. 41–74. New York: Academic Press.

Hooijer, D. A. (1948). Prehistoric teeth of man and the orang-utan from central Sumatra, with notes on the fossil orang-utan from Java and southern China. *Zoolog-Mededelingen* **29**, 173–301.

Hornbeck, P. V. and Swindler, D. R. (1967). Morphology of the lower fourth premolar of certain Cercopithecidae. *Journal of Dental Research* **46**, 979–83.

Hylander, W. L. (1975). Incisor size and diet in anthropoids with special reference to Cercopithecidae. *Science* **189**, 1095–8.

Jablonski, N. G. (1993). Evolution of the masticatory apparatus in *Theropithecus*.

In *Theropithecus: The Rise and Fall of a Primate Genus*, ed. N. G. Jablonski, pp. 299–330. Cambridge University Press.

Jablonski, N. G. (1994). Convergent evolution in the dentitions of grazing macropodine marsupials and the grass-eating primate *Theropithecus gelada*. *Journal of the Royal Society of Western Australia* **77**, 37–43.

Jablonski, N. G. and Crompton, R. H. (1994). Feeding behavior, mastication, and tooth wear in the Western Tarsier (*Tarsius bancanus*). *International Journal of Primatology* **15**, 29–59.

James, W. W. (1960). *The Jaws and Teeth of Primates*. London: Pitman Medical Publishing Co.

Janson, C. H. (1986). Capuchin counterpoint. *Natural History* **95**, 45–53.

Janson, C. H. and Boinski, S. (1992). Morphological and behavioral adaptations for foraging in generalist primates: the case of the cebines. *American Journal of Physical Anthropology* **88**, 483–98.

Jernvall, J. and Thesleff, I. (2000). Return of lost structure in the developmental control of tooth shape. In *Development, Function and Evolution of Teeth*, ed. M. F. Teaford, M. M. Smith and M. W. J. Ferguson, pp. 13–21. Cambridge University Press.

Jenkins, P. A. (1987). *Catalogue of Primates in British Museum (Natural History)*. Part IV. *Suborder Strepsirhini, including the subfossil Madagascan lemurs and Family Tarsiidae*. London: British Museum (Natural History).

Jenkins, P. A. (1990). *Catalogue of Primates in the British Museum (Natural History) and Elsewhere in the British Isles*. Part 5. London: Natural History Museums Publications.

Johanson, D. C. (1979). A consideration of the '*Dryopithecus* pattern.' *Ossa* **6**, 125–38.

Jolly, C. J. (1970). The seed-eaters: a new model of hominid differentiation based on a baboon analogy. *Man* **5**, 5–26.

Jolly, C. J. (1972). The classification and natural history of *Theropithecus* (*Simopithecus*) (Andrews, 1916), baboons of the African Plio-Pleistocene. *Bulletin of the British Museum (Natural History), Geology* **22**, 1–123.

Jorgensen, K. D. (1956). The *Dryopithecus* pattern in recent Danes and Dutchmen. *Journal of Dental Research* **34**, 195–208.

Kahumbu, P. and Eley, R. M. (1991). Teeth emergence in wild olive baboons in Kenya and formulation of a dental schedule for aging wild populations. *American Journal of Primatology* **23**, 1–9.

Kanazawa, E. and Rosenberger, A. L. (1988). Reduction index of the upper M2 in marmosets. *Primates* **29**, 525–33.

Kay, R. F. (1975). The functional adaptations of primate molar teeth. *American Journal of Physical Anthropology* **43**, 195–216.

Kay, R. F. (1977). The evolution of molar occlusion in the Cercopithecidae and early catarrhines. *American Journal of Physical Anthropology* **46**, 327–52.

Kay, R. F. (1978). Molar structure and diet in extant Cercopithecoidea. In *Development, Function and Evolution of Teeth*, ed. P. M. Butler and K. Joysey, pp. 309–39. London: Academic Press.

Kay, R. F. (1981). The nut-crackers — A new theory of the adaptations of the

Ramapithecinae. *American Journal of Physical Anthropology* **55**, 141–51.

Kay, R. F. and Hylander, W. L. (1978). The dental structure of mammalian folivores with special reference to primates and Phalangeroidea (Marsupialia). In *The Biology of Arboreal Folivores*, ed. G. Montgomery, pp. 173–91. Washington, D. C.: Smithsonian Institution Press.

Kay, R. F., Plavcan, J. M., Glander, K. E. and Wright, P. C. (1988). Sexual selection and canine dimorphism in New World monkeys. *American Journal of Physical Anthropology* **77**, 385–97.

Kay, R. F., Rasmussen, D. T. and Beard, K. C. (1984). Cementum annulus counts provide a means for age determination in *Macaca mulatta* (Primates, Anthropoidea). *Folia Primatologica* **42**, 85–95.

Kay, R. F., Ross, C. and Williams, B. A. (1997). Anthropoid origins. *Science* **275**, 797–804.

Kay, R. F., Sussman, R. W. and Tattersall, I. (1978). Dietary and dental variation in the genus *Lemur*, with comments concerning dietary-dental correlations among Malagasy Primates. *American Journal of Physical Anthropology* **49**, 119–28.

Kinzey, W. G. (1972). Canine teeth of the monkey, *Callicebus moloch*: Lack of sexual dimorphism. *Primates* **13**, 365–9.

Kinzey, W. G. (1973). Reduction of the cingulum in Ceboidea. In *Craniofacial Biology of Primates*, ed. M. R. Zingeser, pp. 101–27. Basel: Karger.

Kinzey, W. G. (1977). Diet and feeding behaviour of *Callicebus torquatus*. In *Primate Ecology: Studies of Feeding and Ranging Behaviour in Lemurs, Monkeys and Apes*, ed. T. H. Clutton-Brock, pp. 127–51. London: Academic Press.

Kinzey, W. G. (1992). Dietary and dental adaptations in the Pithecinae. *American Journal of Physical Anthropology* **88**, 499–514.

Kinzey, W. G. and Norconk, M. A. (1990). Hardness as a basis of fruit choice in two sympatric primates. *American Journal of Physical Anthropology* **81**, 5–16.

Kinzey, W. G. and Norconk, M. A. (1993). Physical and chemical properties of fruit and seeds eaten by *Pithecia* and *Chiropotes* in Surinam and Venezuela. *International Journal of Primatology* **14**, 207–28.

Kitahara-Frisch, J. (1973). Taxonomic and phylogenetic uses of the study of variability in the hylobatid dentition. In *Craniofacial Biology of Primates*, ed. M. R. Zingeser, vol. 3, pp. 128–47. Basel: Karger.

Korenhof, C. A. W. (1960). *Morphogenetical Aspects of the Human Upper Molar*. Utrecht: Druk Uitgeversmaatschappij Neerlandia.

Kraus, B. S. (1957). The genetics of the human dentition. *Journal of Forensic Sciences* **2**, 419–27.

Kraus, B. S. (1959a). Occurrence of the Carabelli trait in Southwest ethnic groups. *American Journal of Physical Anthropology* **17**, 117–23.

Kraus, B. S. (1959b). Differential calcification rates in the human primary dentition. *Archives of Oral Biology* **1**, 133–44.

Kraus, B. S. (1963). Morphogenesis of deciduous molar pattern in man. In *Dental Anthropology*, ed. D. R. Brothwell, pp. 87–104. New York: Pergamon Press.

Kraus, B. S. (1964). *The Basis of Human Evolution*. New York: Harper and Row.

Kraus, B. S. and Jordan, R. E. (1965). *The Human Dentition Before Birth.* Philadelphia: Lea and Febiger.

Kraus, B. S., Jordan, R. E. and Abrams, L. (1976). *A Study of the Masticatory System. Dental Anatomy and Occlusion.* Baltimore: The Williams and Wilkins Company.

Krause, D. W., Hartman, J. H. and Wells, N. A. (1997). Late Cretaceous vertebrates from Madagascar: Implications for biotic change in deep time. In *Natural Change and Human Impact in Madagascar*, ed. S. M. Goodman and B. D. Patterson, pp. 3–43. Washington, D.C.: Smithsonian Institution Press.

Krogman, W. M. (1930). Studies in growth changes in the skull and face of anthropoids: eruption of the teeth in anthropoids and Old World apes. *American Journal of Anatomy* **46**, 303–40.

Lasker, G. W. (1950). Genetic analysis of racial traits of the teeth. *Cold Spring Harbor Symposium on Quantitative Biology* **15**, 191–203.

Le Gros Clark, W. E. (1971). *The Antecedents of Man: An Introduction to the Evolution of the Primates*, 3rd edn. Chicago: Quadrangle Books.

Lowther, F. L. (1939). The feeding and grooming habits of the Galago. *Zoologica* **24**, 477–80.

Lucas, P. W. and Luke, D. A. (1984). Chewing it over – basic principles of food breakdown. In *Food Acquisition and Processing in Primates*, ed. D. I. Chivers, B. A. Wood and A. Bilsborough, pp. 283–302. New York: Plenum Press.

Lucas, P. W. and Teaford, M. (1994). Functional morphology of colobine teeth. In *Colobine Monkeys: Their Ecology, Behaviour and Evolution*, ed. A. G. Davies and J. F. Oates, pp. 173–203. Cambridge University Press.

Lucas, P. W., Corlett, R. T. and Luke, D. A. (1986a). Postcanine tooth size and diet in anthropoid primates. *Zeitschrift für Morphologie und Anthropologie* **76**, 253–76.

Lucas, P. W., Corlett, R. T. and Luke, D. A. (1986b). Sexual dimorphism of tooth size in anthropoids. *Human Evolution* **1**, 23–39.

Luo, Z. X., Cifelli, R. L. and Kielan-Jaworowska, Z. (2001). Dual origin of tribosphenic mammals. *Nature* **409**, 53–7.

Maas, M. C. (1993). Enamel microstructure and molar wear in the Greater Galago, *Otolemur crassicaudatus* (Mammalia, Primates). *American Journal of Physical Anthropology* **92**, 217–33.

Maas, M. C. (1994). Enamel microstructure in Lemuridae (Mammalia, Primates): Assessment of variability. *American Journal of Physical Anthropology* **95**, 221–41.

Mahler, P. E. (1980). Molar size sequence in the great apes: gorilla, orangutan, and chimpanzee. *Journal of Dental Research* **59**, 749–52.

Maier, W. (1977a). Die Evolution der bilophodonten Molaren der Cercopithecoidea. *Zeitschrift für Morphologie und Anthropologie* **1**, 24–56.

Maier, W. (1977b). Die bilophodonten Molaren der Indriidae (Primates) ein Evolutionsmorphologischer Modellfall. *Zeitschrift für Morphologie und Anthropologie* **3**, 307–44.

Maier, W. (1980). Konstruktionsmorphologische Untersuchungen am Gebiss der rezenten Prosimiae (Primates). *Abhandlungen der Senckenbergischen Natur-*

forschenden Gesellschaft **538**, 1–158.

Maier, W. (1984). Functional morphology of the dentition of the Tarsiidae. In *Biology of Tarsiers*, ed. C. Niemitz, pp. 45–58. Stuttgart: Gustav Fischer Verlag.

Martin, L. B. (1985). Significance of enamel thickness in hominoid evolution. *Nature* **314**, 260–3.

Martin, R. D. (1975). Ascent of the primates. *Natural History* **84**, 52–61.

Martin, R. D. (1990). *Primate Origins and Evolution: A Phylogenetic Reconstruction*. Princeton: Princeton University Press.

McCrossin, M. L. (1992). New species of bushbaby from the Middle Miocene of Maboko Island, Kenya. *American Journal of Physical Anthropology* **89**, 215–33.

McKusick, V. A. (1990). *Mendelian Inheritance in Man*, 9th edn. Baltimore: The Johns Hopkins University Press.

Meikle, W. E. (1977). Molar wear stages in *Theropithecus gelada*. *Kroeber Anthropological Society Papers* **50**, 21–5.

Midlo, C. and Cummins, H. (1942). *Palmar and Plantar Dermatoglyphics in Primates*. Philadelphia: Wistar Institute of Anatomy and Biology.

Miles, A. E. W. (1963). Dentition in the assessment of individual age in skeletal material. In *Dental Anthropology*, vol. 5, ed. D. R . Brothwell, pp. 191–210. New York: The Macmillan Company.

Miles, A. E. W. and Grigson, C. (1990). *Colyer's Variations and Diseases of the Teeth of Animals*, revised edn. Cambridge University Press.

Mills, J. R. E. (1963). Occlusion and malocclusion of the teeth of primates. In *Dental Anthropology*, ed. D. R. Brothwell, pp. 29–54. New York: The Macmillan Company.

Mittermeier, R. A., Tattersall, I., Konstant, W. R., Meyers, D. M. and Mast, R. B. (1994). *Lemurs of Madagascar*. Washington, D. C.: Conservation International.

Moore, G. J., McNeill, R. W. and D'Anna, J. A. (1972). The effects of digit sucking on facial growth. *Journal of the American Dental Association* **84**, 592–9.

Moorrees, C. F. A. (1957). *The Aleut Dentition: A Correlative Study of Dental Characteristics in an Eskimoid People*. Cambridge, MA: Harvard University Press.

Moorrees, C. F. A., Fanning, E. A. and Hunt, E. E. (1963). Age variation of formation stages for ten permanent teeth. *Journal of Dental Research* **42**, 1490–502.

Napier, J. R. and Walker, A. C. (1967). Vertical clinging and leaping: A newly recognized category of locomotor behaviour of primates. *Folia Primatologica* **21**, 250–76.

Nash, L. T. (1986). Dietary, behavorial, and morphological aspects of gummivory in Primates. *Yearbook of Physical Anthropology* **29**, 113–37.

Niemitz, C. (1984). Synecological relationships and feeding behaviour of the genus *Tarsius*. In *Biology of Tarsiers*, ed. C. Niemitz, pp. 59–75. Stuttgart: Gustav Fisher Verlag.

Niswander, J. D. and Chung, C. S. (1968). The effects of inbreeding on tooth size in Japanese children. *American Journal of Human Genetics* **17**, 390–8.

Noble, H. W. (1969). Comparative aspects of *amelogenesis imperfecta. Proceedings of the Royal Society of Medicine* **62**, 1295–7.

Nowak, R. M. (1999). *Walker's Primates of the World.* Baltimore: The Johns Hopkins University Press.

Oka, S. W. and Kraus, B. S. (1969). The circumnatal status of molar crown maturation among the Hominoidea. *Archives of Oral Biology* **14**, 639–55.

Orlosky, F. J. (1968). Comparative morphology and odontometrics of the deciduous dentition in the rhesus monkey (*Macaca mulatta*), olive baboon (*Papio anubis*) and king colobus (*Colobus polykomos*). Masters thesis, Michigan State University.

Orlosky, F. J. (1973). Comparative dental morphology of extant and extinct Cebidae. Ph.D. dissertation, University of Washington, Seattle.

Osborn, H. F. (1888). The evolution of the mammalian molars to and from the tritubercular type. *American Naturalist* **22**, 1067–79.

Osborn, H. F. (1907). *Evolution of Mammalian Molar Teeth from the Triangular Type.* New York: The Macmillan Company.

Osborn, J. H. (1978). Morphogenetic gradients: fields versus clones. In *Development, Function and Evolution of Teeth*, ed. P. M. Butler and K. A. Joysey, pp. 171–201. London: Academic Press.

Osborne, R. H., Horowitz, S. L. and De George, F. V. (1958). Genetic variation in tooth dimension. A twin study of the permanent anterior teeth. *American Journal of Human Genetics* **10**, 350–6.

Overdorff, D. J., Strait, S. G. and Telo, A. (1997). Seasonal variation in activity and diet in a small-bodied folivorous primate, *Hapalemur griseus*, in Southeastern Madagascar. *American Journal of Primatology* **43**, 211–23.

Oxnard, C. H., Crompton, R. H. and Lieberman, S. S. (1990). *Animal Lifestyles and Anatomies. The Case of the Prosimian Primates.* Seattle: University of Washington Press.

Pedersen, P. O. (1949). The East Greenland Eskimo dentition. *Meddelelser om Gronland* **142**, 1–244.

Petter, J. J. and Peyrieras, A. (1970). Observations eco-ethologiques sur les lemuriens Malgaches du genre *Hapalemur. Terre et Vie* **24**, 356–82.

Peyer, B. (1968). *Comparative Odontology.* (Translated and edited by R. Zangerl.) Chicago: The University of Chicago Press.

Phillips-Conroy, J. E. and Jolly, C. J. (1988). Dental eruption schedules of wild and captive baboons. *American Journal of Primatology* **15**, 17–29.

Plavcan, J. M. (1993). Canine size and shape in male anthropoid primates. *American Journal of Physical Anthropology* **92**, 201–16.

Plavcan, J. M. and Gomez, A. M. (1993). Dental scaling in the Callitrichinae, *International Journal of Primatology* **14**, 177–92.

Plavcan, J. M. and Van Schaik, C. P. (1994). Canine dimorphism. *Evolutionary Anthropology* **2**, 208–14.

Pocock, R. I. (1925). Additional notes on the external characters of some platyrrhine monkeys. *Proceedings of the Zoological Society of London* **15**, 27–47.

Reid, D. J. and Dean, M. C. (2000). The timing of linear hypoplasias on human anterior teeth. *American Journal of Physical Anthropology* **113**, 135–40.

Remane, A. (1951). Die Entstehung der Bilophodontie bei den Cercopithecidae. *Anatomischer Anzieger* **98**, 161–5.

Remane, A. (1960). Zähne und Gebiss. *Primatologia* 3, 637–846.

Richard, A. F. (1985). *Primates in Nature*. New York: W. H. Freeman and Company.

Roberts, D. (1941). The dental comb of lemurs. *Journal of Anatomy* **75**, 236–8.

Rose, K. D., Walker, A. and Jacobs, I. L. (1981). Function of the mandibular tooth comb in living and extinct mammals. *Nature* **289**, 583–5.

Rosenberger, A. L. (1977). Loss of incisor enamel in marmosets. *Journal of Mammology* **59**, 207–8.

Rosenberger, A. L. (1992). Evolution of feeding niches in New World monkeys. *American Journal of Physical Anthropology* **88**, 525–62.

Rosenberger, A. L. and Kinzey, W. G. (1976). Functional patterns of molar occlusion in platyrrhine primates. *American Journal of Physical Anthropology* **45**, 281–98.

Rosenberger, A. L. and Strasser, E. (1985). Toothcomb origins: Support for the grooming hypothesis. *Primates* **26**, 73–84.

Rosenberger, A. L. and Strier, K. B. (1989). Adaptive radiation of the ateline primates. *Journal of Human Evolution* **18**, 717–50.

Rowe, N. (1996). *The Pictorial Guide of the Living Primates*. New York: Pogonias Press.

Saheki, M. (1958). On the heredity of the tooth crown configuration studied in twins. *Acta Anatomica Nipponica* **33**, 456–70.

Saheki, M. (1966). Morphological studies of *Macaca fuscata*. IV. Dentition. *Primates* **7**, 407–22.

Sauther, M. L., Cuozzo, F. P. and Sussman, R. W. (2001). Analysis of dentition of a living wild population of ring-tailed lemurs (*Lemur catta*) from Beza Mahafaly, Madagascar. *American Journal of Physical Anthropology* **114**, 215–23.

Scheine, W. S. and Kay, R. (1982). A model for comparison of masticatory effectiveness in primates. *Journal of Morphology* **172**, 139–49.

Schour, I. and Hoffman, M. M. (1939). Studies in tooth development, II, The rate of apposition of enamel and dentine in man and other animals. *Journal of Dental Research* **18**, 161–75.

Schour, I. and Massler, M. (1940). Studies in tooth development: the growth pattern on human teeth. Part 1. *Journal of the American Dental Association* **27**, 1778–93.

Schultz, A. H. (1935). Eruption and decay of the permanent teeth in primates. *American Journal of Physical Anthropology* **19**, 489–581.

Schultz, A. H. (1958). Cranial and dental variability in *Colobus* monkeys. *Proceedings of the Zoological Society of London* **130**, 79–105.

Schultz, A. H. (1960). Age changes and variability in the skulls and teeth of the central American monkeys *Alouatta*, *Cebus* and *Ateles*. *Proceedings of the Zoological Society of London* **133**, 337–90.

Schuman, E. L. and Brace, C. L. (1955). Metric and morphologic variations in the

Liberian chimpanzee; comparisons with anthropoid and human dentitions. *Human Biology* **26**, 239–68.

Schwartz, G. T. and Dean, C. (2000). Interpreting the hominid dentition: ontogenetic and phylogenetic aspects. In *Development, Growth and Evolution*, ed. P. O'Higgins and M. Cohen, pp. 207–33. London: The Linnean Society of London.

Schwartz, J. H. (1974). Observations on the dentition of the Indriidae. *American Journal of Physical Anthropology* **41**, 107–14.

Schwartz, J. H. (1996). *Pseudopotto martini*: A new genus and species of extant lorisiform primate. *Anthropological Papers of the American Museum of Natural History* **78**, 2–14.

Schwartz, J. H. (1999). Homeobox genes, fossils, and the origin of species. *Anatomical Record* **257**, 15–31.

Schwartz, J. H. and Tattersall, I. (1985). Evolutionary relationships of the living lemurs and lorises and their potential affinities with the European Eocene Adapidae. *Anthropological Papers of the American Museum of Natural History* **60**, 1–100.

Schwarz, F. (1931). A revision of the genera and species of Madagascar Lemuridae. *Proceedings of the Zoological Society of London* **3**, 399–428.

Scott, G. R. and Turner, C. G. II. (1997). *Anthropology of Modern Human Teeth: Dental Morphology and its Variation in Recent Human Populations.* Cambridge University Press.

Seligsohn, D. (1977). Analysis of species-specific molar adaptations in strepsirhine primates. *Contributions to Primatology* **11**, 1–116. Basel: S. Karger.

Seligsohn, D. and Szalay, F. S. (1978). Relationship between natural selection and dental morphology: tooth function and diet in *Lepilemur* and *Hapalemur*. In *Development, Function and Evolution of Teeth*, ed. P. M. Butler and K. A. Joysey, pp. 289–307. New York: Academic Press.

Selmer-Olsen, R. (1949). *An Odontometrical Study on the Norwegian Lapps.* Oslo: I Kommisjon Hos Jacob Dybwad.

Serra, O. della (1951). Variacoes do articulado dos dentes incisivos nos macacos do genero *Alouatta* Lac. 1799. *Papeis avulsos do Derpmento Zoologia San Paulo* **10**, 139–46.

Serra, O. della (1952). O tuberculo intermedianio posterior (metaconulo) dos Platyrrhina do genero *Alouatta* Lac. *Anais Faculdade de Farmacia e Odontologia da Universita de San Paulo* **10**, 297–301.

Sharpe, P. T. (2000). Homeobox genes in initiation and shape of teeth during development in mammalian embryos. In *Development, Function and Evolution of Teeth*, ed. M. F. Teaford, M. M. Smith and M. W. J. Ferguson, pp. 3–12. Cambridge University Press.

Shellis, R. and Hiiemae, K. (1986). Distribution of enamel on the incisors of Old World monkeys. *American Journal of Physical Anthropology* **71**, 103–13.

Shellis, R. P., Beyon, A. D., Reid, D. J. and Hiiemae, K. M. (1998). Variations in molar enamel thickness among primates. *Journal of Human Evolution* **35**, 507–22.

Siebert, J. R. and Swindler, D. R . (1991). Perinatal dental development in the chimpanzee (*Pan troglodytes*). *American Journal of Physical Anthropology* **86**, 287–94.

Simons, E. L. (1997). Lemurs: old and new. In *Natural Change and Human Impact in Madagascar*, ed. S. M. Goodman and B. D. Patterson, pp. 142–68. Washington, D. C.: Smithsonian Institution Press.

Simons, E. L. and Bown, T. M. (1985). *Afrotarsius chatrathi*, first tarsiiform primate (? Tarsiidae) from Africa. *Nature* **313**, 475–7.

Simpson, G.G. (1936). Studies of the earliest mammalian dentition. *The Dental Cosmos* **78**, 791–953.

Simpson, G. G. (1937). The beginning of the age of mammals. *Biological Reviews* **12**, 1–47.

Simpson, G. G. (1955). The Phenacolemuridae, new family of early primates. *Bulletin of the American Museum of Natural History* **105**, 411–42.

Sirianni, J. E. (1974). Dental variability in African Cercopithecidae: A morphologic, metric and discriminant analysis. Ph.D. thesis, University of Washington, Seattle.

Sirianni, J. E. and Swindler, D. R. (1972). Inheritance of deciduous tooth size in *Macaca nemestrina*. *Journal of Dental Research* **52**, 179.

Sirianni, J. E. and Swindler, D. R. (1975). Tooth size inheritance in *Macaca nemestrina*. In *Contemporary Primatology*, ed. S. Kondo, M. Kawai and A. Ehara, pp. 12–19. Basel: S. Karger.

Sirianni, J. E. and Swindler, D. R. (1985). *Growth and Development of the Pigtailed Macaque*. Boca Raton: CRC Press.

Skaryd, S. M. (1971). Trends in the evolution of the pongid dentition. *American Journal of Physical Anthropology* **35**, 223–39.

Skinner, E. W. (1954). *The Science of Dental Materials*. Philadelphia: W.B. Saunders Co.

Skinner, M. F. (1986). Enamel hypoplasia in sympatric chimpanzee and gorilla. *Human Evolution* **1**, 289–312.

Smith, B. H. (1989). Dental development as a measure of life history in primates. *Evolution* **43**, 683–8.

Smith, B. H., Crummett, T. L. and Brandt, K. L. (1994). Ages of eruption of primate teeth: a compendium for aging individuals and comparing life histories. *Yearbook of Physical Anthropology* **37**, 177–232.

Smith, M. M. and Coates, M. I. (2000). Evolutionary origins of teeth and jaws: developmental modes and phylogenetic patterns. In *Development, Function and Evolution of Teeth*, ed. M. F. Teaford, M. M. Smith and M. W. J. Ferguson, pp. 133–51. Cambridge University Press.

Sofaer, J. A., Niswander, J. D., MacLean, C. J. and Workman, P. L. (1972). Population studies on Southwestern Indian tribes. V. Tooth morphology as an indicator of biological distance. *American Journal of Physical Anthropology* **37**, 357–66.

Sokal, R. R. and Rohlf, F. J. (1969). *Biometry: The Principles and Practice of Statistics in Biological Research*. San Francisco: W. H. Freeman and Co.

Strait, S. G. (1997). Tooth use and the physical properties of food. *Evolutionary Anthropology* **5**, 199–211.

Strasser, E. and Delson, E. (1987). Cladistic analysis of cercopithecid relations. *Journal of Human Evolution* **16**, 81–99.

Strier, K. B. (1992). Atelinae adaptations: Behavioral strategies and ecological constraints. *American Journal of Physical Anthropology* **88**, 515–24.

Swails, N. J. (1993). The Evolutionary Implications of Primate Tooth-Germ Development: Using Ontogenetic Data to Make Phylogenetic Inferences. Ph.D. thesis, University of Washington.

Swails, N. J. and Swindler, D. R. (2002). Development of the crista obliqua: implications of crista diversion and capture in primate dental evolution. *Dental Anthropology* (in press).

Swarts, J. D. (1988). Deciduous dentition: Implications for hominoid phylogeny. In *Orang-utan Biology*, ed. J. H. Schwartz, pp. 263–70. New York: Oxford University Press.

Swindler, D. R. (1961). Calcification of the permanent first mandibular molar in rhesus monkeys. *Science* **134**, 566.

Swindler, D. R. (1968). The maxillary incisors and evolution of Old World Monkeys. In *Taxonomy and Phylogeny of Old World Primates with References to the Origin of Man*, ed. B. Chiarelli, pp. 57–67. Torino: Rosenberg & Sellier.

Swindler, D. R. (1976). *Dentition of Living Primates*. London: Academic Press.

Swindler, D. R. (1979). The incidence of underbite occlusion in leaf-eating monkeys. *Ossa* **6**, 261–72.

Swindler, D. R. (1983). Variation and homology of the primate hypoconulid. *Folia Primatologia* **41**, 112–23.

Swindler, D. R. (1985). Nonhuman primate dental development and its relationship to human dental development. In *Nonhuman Primate Models for Human Growth and Development*, ed. E. S. Watts, pp. 67–94. New York: Alan R. Liss, Inc.

Swindler, D. R. (1995). Canine root development and morphology in *Macaca nemestrina*. *Primates*, **36**, 583–9.

Swindler, D. R. (1998). *Introduction to the Primates*. Seattle: University of Washington Press.

Swindler, D. R. and Beynon, A. D. (1993). The development and microstructure of the dentition of *Theropithecus*. In *Theropithecus: The Rise and Fall of a Primate Genus*, ed. N. G. Jablonski, pp. 351–81. Cambridge University Press.

Swindler, D. R., Emel, L. M. and Anemone, R. L. (1998). Dental variability of the Liberian Chimpanzee, *Pan troglodytes verus*. *Human Biology* **13**, 235–49.

Swindler, D. R., Gavan, J. A. and Turner, W. M. (1963). Molar tooth size variability in African monkeys. *Human Biology* **35**, 104–22.

Swindler, D. R. and McCoy, H. A. (1964). Calcification of deciduous teeth in rhesus monkeys. *Science* **144**, 1243–4.

Swindler, D. R. and Meekins, D. (1991). Dental development of the permanent mandibular teeth in the baboon, *Papio cynocephalus*. *American Journal of Human Biology* **3**, 571–80.

Swindler, D. R. and Olshan, A. F. (1982). Molar size sequence in Old World monkeys. *Folia Primatologica* **39**, 201–12.

Swindler, D. R. and Olshan, A. F. (1988). Comparative and evolutionary aspects of the permanent dentition. In *Orang-Utan Biology*, ed. J. H. Schwartz, pp. 271–82. New York: Oxford University Press.

Swindler, D. R. and Orlosky, F. J. (1974). Metric and morphological variability in the dentition of colobine monkeys. *Journal of Human Evolution* 3, 135–60.

Swindler, D. R. and Sassouni, V. (1962). Open bite and thumb sucking in rhesus monkeys. *The Angle Orthodontist* 32, 27–37.

Swindler, D. R. and Sirianni, J. E. (1968). Variability of maxillary incisors among primates. *Proceedings VIIIth International Congress of Anthropological and Ethnological Sciences, Anthropology*, vol. 1, pp. 308–11. Tokyo: Science Council of Japan.

Swindler, D. R. and Sirianni, J. E. (1975). Dental size and dietary habits of primates. *Yearbook of Physical Anthropology* 19, 166–82.

Swindler, D. R. and Ward, S. (1988). Evolutionary and morphological significance of the deflecting wrinkle in the lower molars of the Hominoidea. *American Journal of Physical Anthropology* 75, 405–11.

Szalay, F. S. (1969). Mixodectidae, Microsyopidae, and the insectivore-primate transition. *Bulletin of the American Museum of Natural History* 140, 193–330.

Szalay, F. S. and Seligsohn, D. (1977). Why did the strepsirhine toothcomb evolve? *Folia Primatologica* 27, 75–82.

Tarrant, L. H. and Swindler, D. R. (1972). The state of the deciduous dentition of a chimpanzee fetus (*Pan troglodytes*). *Journal of Dental Research* 51, 677.

Tarrant, L. H. and Swindler, D. R. (1973). Prenatal dental development in the black howler monkey (*Alouatta caraya*). *American Journal of Physical Anthropology* 38, 255–60.

Teaford, M. F. (1993). Dental microwear and diet in extant and extinct *Theropithecus*: preliminary analyses. In *Theropithecus: The Rise and Fall of a Primate Genus*, ed. N. G. Jablonski, pp. 331–49. Cambridge University Press.

Ten Cate, A. R. (1994). *Oral Histology, Development, Structure and Function*, 4th edn. St. Louis: C. V. Mosby Co.

Townsend, G. C., Richards, L. C., Brown, T. and Burgess, V. B. (1988). Twin zygosity determination on the basis of dental morphology. *Journal of Forensic Odonto-Stomatology* 6, 1–15.

Townsend, G. C., Richards, L. C., Brown, T., Burgess, V. B., Travan, G. R. and Rogers, J. R. (1992). Genetic studies of dental morphology in South Australian twins. In *Structure, Function and Evolution of Teeth*, eds. P. Smith and E. Tchernov, pp. 501–18. London: Freund Publishing House Ltd.

Tucker, A. S, Matthews, K. L and Sharpe, P. T. (1998). Transformation of tooth type induced by inhibition of BMP signaling. *Science* 282, 1136–8.

Turner, C. G. II and Hawkey, D. E. (1998). Whose teeth are these?: Carabelli's trait. In *Human Dental Development, Morphology, and Pathology*, ed. J. R. Lukacs, pp. 41–50, Eugene: The University of Oregon Anthropological Papers.

Turner, E. (1963). Crown development in human deciduous molar teeth. *Archives of Oral Biology* 8, 523–40.

Tuttle, R. H. (1969). Quantitative and functional studies on the hands of the Anthropoidea — I. The Hominoidea. *Journal of Morphology* 128, 309–64.

Unger, P. S. (1994). Patterns of ingestive behavior and anterior tooth use differences in sympatric anthropoid primates. *American Journal of Physical Anthropology* **95**, 197–219.

Unger, P. S. (1995). Fruit preference of four sympatric primate species at Ketambe. *International Journal of Primatology* **16**, 221–4.

Van Roosmalen, M. G. M., Mittermeier, R. A. and Fleagle, J. G. (1988). Diet of the northern bearded saki (*Chiropotes satanas chiropotes*): A neotropical seed predator. *American Journal of Primatology* **14**, 11–35.

Van Valen, L. (1966). Deltatheridia, a new order of mammals. *Bulletin of the American Museum of Natural History* **132**, 1–126.

Van Valen, L. M. (1982). Homology and causes. *Journal of Morphology* **173**, 305–12.

Vandebrock, G. (1961). The comparative anatomy of the teeth of lower and nonspecialized mammals. *Koninklijke Vlaamse Academie voor Watenschappen, Letteren en Scone Kunsten Van Belgie* **1**, 215–313.

Vitzthum, V. J. and Wikander, R. (1988). Incidence and correlates of enamel hyperplasia in nonhuman primates. *American Journal of Physical Anthropology* **75**, 284.

Wada, K., Ohtaishi, N. and Hachiya, N. (1978). Determination of age in the Japanese monkey from growth layers in the dental cementum. *Primates* **19**, 775–84.

Walker, A. (1984). Mechanisms of honing in the male baboon canine. *American Journal of Physical Anthropology* **65**, 47–60.

Walker, P. L. (1976). Wear striations on the incisors of cercopithecid monkeys as an index of diet and habitat preference. *American Journal of Physical Anthropology* **45**, 299–308.

Weidenreich, F. (1945). Giant early man from Java and China. *Anthropological Papers of the American Museum of Natural History* **40**, 5–134.

Weiss, K. M. (1990). Duplication, with variation: metameric logic in evolution from genes to morphology. *Yearbook of Physical Anthropology* **33**, 1–23.

Weiss, K. M. (1993). A tooth, a toe, and a vertebra: the genetic dimensions of complex morphological traits. *Evolutionary Anthropology* **2**, 121–34.

Winkler, L. A., Schwartz, J. H. and Swindler, D. R . (1996). Development of the orangutan permanent dentition: Assessing patterns and variation in tooth development. *American Journal of Physical Anthropology* **99**, 205–20.

Wolfheim, J. H. (1983). *Primates of the World*. Seattle: University of Washington Press.

Wood, B. A. and Abbott, S. A. (1983). Analysis of the dental morphology of Plio-Pleistocene hominids. I. Mandibular molars: crown area measurements and morphological traits. *Journal of Anatomy* **136**, 197–219.

Wright, P. C. and Martin, L. B. (1995). Predation, pollination and torpor in two nocturnal primates, *Cheirogaleus major* and *Microcebus rufus*, in the rain forest of Madagascar. In *Creatures of the Dark: The Nocturnal Prosimians*, ed. L. Alterman, G. A. Doyle and M. K. Izard, pp. 45–60. New York: Plenum Press.

Yamashita, N. (1998). Functional dental correlates of food properties in five

Malagasy lemur species. *American Journal of Physical Anthropology* **106**, 169–88.

Yoneda, M. (1982). Growth layers in dental cementum of *Sanguinus* monkeys in South America. *Primates* **23**, 460–4.

Yoshikawa, Y. and Deguchi, T. (1992). Sexual dimorphism of the canine roots in the *Macaca fuscata. Primates* **33**, 121–7.

Zhao, Z., Weiss, K. M. and Stock, D. W. (2000). Development and evolution of dentition patterns and their genetic basis. In *Development, Function and Evolution of Teeth*, ed. M. F. Teaford, M. M. Smith and M. W. J. Ferguson, pp. 152–72. Cambridge University Press.

Zingeser, M. R. (1968). Functional and phylogenetic significance of integrated growth and form in occluding monkey canine teeth. *American Journal of Physical Anthropology* **28**, 263–70.

Zingeser, M. R. (1973). Dentition of *Brachyteles arachnoides* with reference to Alouattine and Atelinine affinities. *Folia Primatologica* **20**, 351–90.

Taxonomic index